Statistical Quantitative Methods in Finance

Samit Ahlawat

Statistical Quantitative Methods in Finance

From Theory to Quantitative Portfolio Management

Apress®

Samit Ahlawat
Howley, NL, Canada

ISBN-13 (pbk): 979-8-8688-0961-3   ISBN-13 (electronic): 979-8-8688-0962-0
https://doi.org/10.1007/979-8-8688-0962-0

Managing Director, Apress Media LLC: Welmoed Spahr
Acquisitions Editor: Celestin Suresh John
Development Editor: Laura Berendson
Coordinating Editor: Kripa Joseph

Cover designed by eStudiocalamar

Cover image designed by Freepik

Distributed to the book trade worldwide by Apress Media, LLC, 1 New York Plaza, New York, NY 10004, U.S.A. Phone 1-800-SPRINGER, fax (201) 348-4505, e-mail orders-ny@springer-sbm.com, or visit www.springeronline.com. Apress Media, LLC is a California LLC and the sole member (owner) is Springer Science + Business Media Finance Inc (SSBM Finance Inc). SSBM Finance Inc is a **Delaware** corporation.

For information on translations, please e-mail booktranslations@springernature.com; for reprint, paperback, or audio rights, please e-mail bookpermissions@springernature.com.

Apress titles may be purchased in bulk for academic, corporate, or promotional use. eBook versions and licenses are also available for most titles. For more information, reference our Print and eBook Bulk Sales web page at http://www.apress.com/bulk-sales.

Any source code or other supplementary material referenced by the author in this book is available to readers on GitHub (https://github.com/Apress). For more detailed information, please visit https://www.apress.com/gp/services/source-code.

If disposing of this product, please recycle the paper

To my family, friends, and colleagues who encouraged and supported me in writing this book.

About the Author

Samit Ahlawat is a portfolio manager at QSpark Investment, specializing in US equity and derivative trading. He has extensive experience in quantitative asset management and market risk management, having previously worked at JP Morgan Chase and Bank of America. His research interests include artificial intelligence, risk management, and algorithmic trading strategies. Samit holds a master's degree in numerical computation from the University of Illinois Urbana-Champaign.

Samit has authored several research papers in artificial intelligence, finance, economics and numerical computation in addition to holding a patent for facial recognition technology. His research on using machine learning technologies to improve financial forecasting has enabled finance practitioners to leverage generative AI tools, such as variational auto-encoders (VAE), alongside statistical methodologies to model asset price distribution probabilities. Samit also mentors AI professionals at Kaggle and has delivered industry talks and presentations on artificial intelligence.

About the Technical Reviewer

Royana Anand is a distinguished leader in artificial intelligence and program management and a prominent figure in the tech industry, with an exceptional record of impactful contributions across top-tier organizations, such as Amazon Web Services, Google, Expedia Group, and Stanford University. She has made significant contributions to the fields of data science, AI, and finance. Her strategic leadership at AWS was instrumental in optimizing critical services like AWS Direct Connect, Diode, and AppConfig, which significantly enhanced global network performance, data transfer security, and application configuration management. Royana's innovative approach drove substantial operational improvements and set new benchmarks for efficiency and reliability in cloud services. At Google, Royana played a key role in advancing data science and AI initiatives, developing cutting-edge solutions that shaped industry standards. Her work at Expedia Group and Stanford University involved leading transformative projects that integrated advanced technologies into strategic frameworks, driving significant advancements in data analytics and operational excellence. As a technical reviewer for Springer Science and Business Media, Royana's meticulous evaluation of *Statistical Quantitative Methods in Finance* has elevated the quality of research publications, further establishing her as an authority in financial and technical research. Royana's contributions reflect her exceptional expertise and leadership in artificial intelligence, data science, finance, and technical program management, demonstrating her ability to drive innovation and achieve substantial impact. Her work has not only advanced technological solutions but also set new standards in the industry, establishing her as a prominent figure in her field.

Preface

I began writing this book with the encouragement and active support of my colleagues, including young professionals joining financial firms who frequently seek my recommendations on how to review all relevant statistical theory that is useful for financial modeling used at the workplace. There are reputed textbooks on individual topics such as regression, time series, and generalized linear models, but few covering the complete spectrum of statistical methods used in quantitative modeling in finance. This book was written to address that need and to provide a resource on practical quantitative modeling tasks that can be tackled using statistical methods. While most examples are drawn from the field of finance, students and practitioners from other disciplines may find it useful for understanding the concepts and identifying analogies to the modeling tasks they work on. This work strikes a balance between theory and practice, providing an overview of theory and illustrating it with practical examples.

This book assumes the reader is familiar with Python programming. Knowledge of libraries such as **statsmodels** and **sklearn** is not required. During the course of reading this book, the reader will acquire a synoptic understanding of frequently used APIs available for the model implementations supported by these libraries.

I welcome feedback from readers on what they found helpful in the book and where there is scope for improvement. A text focused on evolving practical concepts must necessarily evolve in future editions to cover new applications and models. This is particularly relevant with the increasing synergy between machine learning and data science. Statistical concepts such as entropy, cross-entropy, maximum likelihood, Gini impurity index, and Bayesian methods have become the vanguard of a new generation of artificial intelligence models, such as Glove (global vectors for word representation in NLP), double-Q learning in reinforcement learning (cross-entropy), and random forests (Gini impurity index), to name a few. On a parallel front, statistical models have been deployed as benchmarking tools during the calibration and model fitting phase of machine learning algorithms and as tools for ensuring ongoing quality controls during postproduction deployment.

Contents

Introduction

Statistical methods are the cornerstone of many quantitative models. Their success derives from clear and concise mathematical formulation that suggests implementation details, along with their ability to be implemented and executed on ubiquitous, readily available computational hardware. Mathematical formulation of such models endows them with a virtue of intelligibility – practitioners can readily explain their features and acquire an intuitive understanding of how the models will behave when used with different kinds of data. For example, statisticians can readily explain the importance of uncorrelated error terms in ordinary least squares and what kinds of data and model characteristics may exacerbate the problem of correlated residuals in ordinary least squares (e.g., missing explanatory variables). This intuitive understanding aids in the deployment of statistical models to appropriate use cases, serves as a valuable verification method to ensure correct implementation, and enables modelers to explain model choice to statisticians and non-statisticians alike. In addition, they also serve as indispensable benchmarking tools for artificial intelligence models and are frequently deployed as components of advanced machine learning models. A comprehensive understanding of statistical models is foundational for both data science and machine learning disciplines.

This book explains statistical modeling using a range of applications drawn from the field of finance. It covers a wide swath of statistical modeling, beginning from ordinary least squares (OLS) and culminating in generalized method of moments (GMM) models used in econometrics. While the exegetical approach to describing statistical modeling adopted in this book begins with an explanation of the model followed by mathematical formulation, it is not obscured by excessive mathematics and notation. It includes practical applications drawn from the field of finance, along with hands-on code accessible online to illustrate the salient features of the model. Implementing code leverages libraries such as **statsmodels**, **scipy**, and **sklearn**. It also includes pseudo-code to aid explanation of code and models. By adopting a practical, hands-on focus while leveraging widely used, open source model implementations, this book enables readers to become experts at understanding statistical models, judiciously decide when to use a particular model, and effectively implement it using numerical libraries. It also enables the readers to write their own model implementations, though in most circumstances, use of widely adopted numerical libraries is preferred.

This book also provides a foundation for machine learning students to appreciate the statistical foundations of some of the most widely used machine learning algorithms, such as the naive Bayes method and the expectation-maximization (EM) algorithm. Readers inclined toward machine learning applications will find a rich variety of synergistic content between statistical methods and machine learning algorithms. This is particularly true for models such as random forests that are inspired by machine learning fundamentals such as decision trees and bagging and also derive some of their most attractive properties, such as robustness to overfitting, from statistical concepts like entropy, Gini impurity reduction, and ensemble learning. By delineating the statistical properties of random forests, along with demonstrating illustrative practical examples, this book provides a vista into their versatility. It showcases a good example where machine learning and statistics are being leveraged hand in hand to tackle complex problems that had been heretofore regarded as intractable using only statistical methods.

The book concludes with a chapter showcasing how statistical models can be used as benchmarking tools for machine learning algorithms.

Overview

1

This book begins with foundational statistical methods, including ordinary least squares (OLS) and generalized linear models (GLM). A comprehensive discussion on these two topics is undertaken in the first two chapters that explain all aspects of theory, while drawing on relevant examples and coding exercises, to illustrate concepts related with model calibration, hypothesis testing, and predicting. A firm grasp of the concepts covered in the first two chapters is also critical for understanding the concepts discussed in subsequent chapters.

Following GLM, the book covers Markov dynamic regime switching models that are used in econometrics and in applied finance extensively. Sharing a number of features with hidden Markov models, they constitute a versatile quantitative methodology for fitting linear models for evolving environments. This chapter is followed by Tobit models that are used for fitting censored data using a regression-based model.

The following two chapters cover topics used more frequently by economists, but may also be helpful for applied quantitative finance professionals. This includes the generalized method of moments (GMM), including its applications in vector auto-regressions (VAR). This chapter is followed by dynamic stochastic general equilibrium models that have become the mainstay of econometric modeling used at most central banks.

Finally, the book concludes with a chapter on using statistical models as a benchmarking tool for machine learning models.

© Samit Ahlawat 2025
S. Ahlawat, *Statistical Quantitative Methods in Finance*,
https://doi.org/10.1007/979-8-8688-0962-0_1

Linear Regression

2

Linear regression is the workhorse of statistical model development due to its simplicity, intuitive appeal, and ubiquitous availability in numerical libraries and toolkits. The model defines a linear parametric relationship between an endogenous (dependent) variable and a group of exogenous (independent) variables, including an optional constant. Parameters defining this relationship are predicted by fitting the data using the linear model. In addition to predicting the parameters, the method also furnishes confidence intervals for those estimates. The model can be written as shown in Equation 2-1. ϵ represents the error between actual output, y, and predicted output, $\hat{y}$, as shown in Equation 2-2.

$$\mathbf{y} = \mathbf{X}\boldsymbol{\beta} + \boldsymbol{\epsilon}$$

$$\mathbf{y} \in \mathbb{R}^N, \mathbf{X} \in \mathbb{R}^{N \times (P+1)} \text{ and } \boldsymbol{\beta} \in \mathbb{R}^{P+1}$$

$$\epsilon \sim \text{IID random variable} \tag{2-1}$$

$$E[\epsilon] = 0 \text{ and } \text{var}(\epsilon) = \sigma^2$$

$$\hat{\mathbf{y}} = \mathbf{X}\boldsymbol{\beta} \tag{2-2}$$

In Equation 2-1, y denotes the endogenous or dependent variable because the model prescribes its definition in terms of exogenous or dependent variables. $\mathbf{X}$ denotes the matrix of exogenous or independent variables, with the ith row containing the values from the ith observation. This matrix is also called design matrix and is typically augmented by adding a column of 1 values to account for the constant term in the model. Let us assume there are P regressors or independent variables and N equations. Including a constant, $\mathbf{X}$ can be written as shown in Equation 2-3. $\mathbf{X}$ is a matrix of dimension $(N, P + 1)$ when the constant is included and of dimension (N, P) when it is not included.

Let us consider a linear model with a constant. When the number of observations, N, is equal to the number of parameters, P+1, there is a unique solution for the

© Samit Ahlawat 2025
S. Ahlawat, *Statistical Quantitative Methods in Finance*,
https://doi.org/10.1007/979-8-8688-0962-0_2

parameters β. This problem is known as interpolation because it fits the data points exactly and $\epsilon = 0$. If N is less than P+1, there are infinitely many solutions to fit the data points exactly with $\epsilon = 0$ for each of those. In practice, the number of observations, N, is typically greater than P+1, and the linear model must predict parameters β to produce the best fit to the data. Solving this problem is referred to as linear regression.

$$\mathbf{X} = \begin{bmatrix} 1 & X_{0,0} & X_{0,1} & \cdots & X_{0,P-1} \\ 1 & X_{1,0} & X_{1,1} & \cdots & X_{1,P-1} \\ & & \cdots & & \\ 1 & X_{N-1,0} & X_{N-1,1} & \cdots & X_{N-1,P-1} \end{bmatrix} \tag{2-3}$$

β represents the parameter vector and can be written as a column vector, as shown in Equation 2-4. Including the constant, there are P+1 parameters in the β column vector.

$$\beta = \begin{bmatrix} \alpha \\ \beta_0 \\ \cdots \\ \beta_{P-1} \end{bmatrix} \tag{2-4}$$

ϵ in Equation 2-1 is a random variable and represents the error. It has the following three properties, known as Gauss-Markov assumptions. The significance of the assumptions in solving the linear model using ordinary least squares (OLS) is described later.

1. $E[\epsilon|\mathbf{X}] = 0$. This condition states that the errors ϵ are independent of regressors $\mathbf{X}$ and have mean 0.
2. $\text{var}(\epsilon_i) = \sigma^2 < \infty$. This condition states that the errors are homoskedastic (constant variance).
3. $\text{cov}(\epsilon_i, \epsilon_j) = E[\epsilon_i \epsilon_j|\mathbf{X}] = 0$. This assumption states that the errors are uncorrelated. The last two assumptions jointly imply that the variance-covariance matrix of the errors, ϵ, is $\sigma^2 \mathbf{I}$ where $\mathbf{I}$ is the identity matrix.

2.1   Solving OLS

Ordinary least squares can be solved using the method of normal equations or QR factorization. Normal equations proceed from the mean-square objective function for minimizing the difference between observed and actual model outputs, as shown in Equations 2-5 and 2-6. Normal equations are commonly used for solving ordinary least squares due to their simplicity and intuitive geometric interpretation. Owing to the greater numerical robustness of QR factorization compared with normal equations, most numerical libraries use QR factorization.

2.1.1 Loss Function

In order to solve an ordinary least squares model, it is necessary to define an objective function. This is required because unlike interpolation, there is no solution for β that fits the data $\mathbf{y}$ and $\mathbf{X}$ exactly using the linear relationship $\mathbf{y} = \mathbf{X}\beta$. Let us try to minimize mean square errors of fitting the linear model to data, as shown in Equation 2-5.

$$MSE = \frac{\sum_{i=1}^{N} \epsilon_i^2}{N} \tag{2-5}$$

Model parameters, β, are obtained by minimizing this objective function with respect to the parameters. Let us rewrite the objective function in terms of the model parameters, β, as shown in Equation 2-6. SSR denotes the sum of square residuals, ϵ_i. The derivation uses the fact that $\mathbf{y}'\mathbf{X}\beta = \beta'\mathbf{X}'\mathbf{y}$ because both quantities are scalar and one side can be obtained by taking the transpose of the other.

$$\beta = \underset{\beta}{\text{argmin}} \frac{\sum_{i=1}^{N} \epsilon_i^2}{N}$$

$$= \underset{\beta}{\text{argmin}} \sum_{i=1}^{N} \epsilon_i^2$$

$$SSR = \sum_{i=1}^{N} \epsilon_i^2 \tag{2-6}$$

$$= (\mathbf{y} - \mathbf{X}\beta)' (\mathbf{y} - \mathbf{X}\beta)$$

$$= \mathbf{y}'\mathbf{y} - \mathbf{y}'\mathbf{X}\beta - \beta'\mathbf{X}'\mathbf{y} + \beta'\mathbf{X}'\mathbf{X}\beta$$

$$= \mathbf{y}'\mathbf{y} - 2\beta'\mathbf{X}'\mathbf{y} + \beta'\mathbf{X}'\mathbf{X}\beta$$

Let us minimize the objective function by setting its derivative with respect to model parameters, β, to zero as shown in Equation 2-7. It is customary to denote the computed value of β as $\hat{\beta}$ to underscore the fact that the original system of equations has no unique solution.

$$\frac{\partial SSR}{\partial \beta} = 0$$

$$-2\mathbf{X}'\mathbf{y} + 2\mathbf{X}'\mathbf{X}\hat{\beta} = 0 \tag{2-7}$$

$$\implies \mathbf{X}'\mathbf{y} = \mathbf{X}'\mathbf{X}\hat{\beta}$$

$$\implies \hat{\beta} = (\mathbf{X}'\mathbf{X})^{-1} \mathbf{X}'\mathbf{y}$$

Taking the second derivative of the objective function with respect to β shows that the optimum point is a minimum, as shown in Equation 2-8. $\mathbf{X'X}$ is a positive definite square matrix with an inverse.

$$\frac{\partial^2 SSR}{\partial \beta^2} = 2\mathbf{X'X} \geq 0 \tag{2-8}$$

2.1.2   Variance of OLS Estimator

OLS fits the data using an objective function; therefore, it is natural to ask the question: What is the variance of the estimated coefficients? This section derives an analytical formula for the variance. Under certain assumptions, the variance of the OLS estimator can be shown to be the least among a family of linear models. The assumptions are known as Gauss-Markov assumptions, and the OLS solution to a linear model satisfying those assumptions is said to be BLUE – **B**est **L**inear **U**nbiased **E**stimator. Before describing these properties in more detail, let us first derive an expression for the variance of parameters, β. The derivation is illustrated in Equation 2-9. $\hat{\beta}$ denotes the predicted values of model parameters from Equation 2-6. Due to the symmetry of the variance estimator in Equation 2-9, it is often called the sandwich estimator.

$$\mathbf{y} = \mathbf{X}\beta + \epsilon$$

$$\mathbf{X'y} = \mathbf{X'X}\hat{\beta} \text{ from Equation 2-6}$$

$$\therefore \mathbf{X'}(\mathbf{X}\beta + \epsilon) = \mathbf{X'X}\hat{\beta}$$

$$\hat{\beta} = \beta + (\mathbf{X'X})^{-1}\mathbf{X'}\epsilon$$

$$\therefore \hat{\beta} - \beta = (\mathbf{X'X})^{-1}\mathbf{X'}\epsilon$$

$$\implies \text{var}(\hat{\beta}) = E\left[\left(\hat{\beta} - \beta\right)\left(\hat{\beta} - \beta\right)'\right] \tag{2-9}$$

$$= E\left[(\mathbf{X'X})^{-1}\mathbf{X'}\epsilon\left((\mathbf{X'X})^{-1}\mathbf{X'}\epsilon\right)'\right]$$

$$= E\left[(\mathbf{X'X})^{-1}\mathbf{X'}\epsilon\epsilon'\mathbf{X}(\mathbf{X'X})^{-1}\right]$$

$$= (\mathbf{X'X})^{-1}\mathbf{X'}E\left[\epsilon\epsilon'\right]\mathbf{X}(\mathbf{X'X})^{-1}$$

In order to get the last equation in Equation 2-9, we assume that the errors, ϵ, are uncorrelated with exogenous variables, $\mathbf{X}$. Further, let us assume that the errors are homoskedastic, i.e., $E\left[\epsilon\epsilon'\right] = \sigma^2\mathbf{I}$, where $\mathbf{I}$ is the identity matrix. With these two assumptions, the variance-covariance matrix of error terms can be simplified to the expression shown in Equation 2-10.

$$\text{var}(\hat{\beta}) = (\mathbf{X}'\mathbf{X})^{-1} \mathbf{X}' \sigma^2 \mathbf{I} \mathbf{X} (\mathbf{X}'\mathbf{X})^{-1}$$
$$= \sigma^2 (\mathbf{X}'\mathbf{X})^{-1} \mathbf{X}'\mathbf{X} (\mathbf{X}'\mathbf{X})^{-1} \mathbf{I}$$
$$= (\mathbf{X}'\mathbf{X})^{-1} \sigma^2 \qquad (2\text{-}10)$$
$$\text{where } \sigma^2 = \frac{\sum_{i=1}^{N-P-1} \epsilon_i^2}{N}$$

$P + 1 =$ number of free parameters in the linear model

2.1.3 Gauss-Markov Assumptions

Gauss-Markov assumptions are a set of assumptions required to establish that the OLS estimator is the best linear unbiased predictor of a linear model's parameters. The assumptions are listed below:

1. Errors ϵ are independent of exogenous variables $\mathbf{X}$ and $E[\epsilon|\mathbf{X}] = \mathbf{0}$. This condition is required to establish that the OLS predictor $\hat{\beta}$ is an unbiased predictor of β. This can be seen from Equation 2-11.

$$\hat{\beta} - \beta = (\mathbf{X}'\mathbf{X})^{-1} \mathbf{X}'\epsilon$$
$$\implies E\left[\hat{\beta} - \beta\right] = E\left[(\mathbf{X}'\mathbf{X})^{-1} \mathbf{X}'\epsilon\right]$$
$$= (\mathbf{X}'\mathbf{X})^{-1} \mathbf{X}'E[\epsilon|\mathbf{X}] = 0 \qquad (2\text{-}11)$$
$$\implies E\left[\hat{\beta}\right] = E[\beta]$$

2. Errors ϵ have finite and constant variance, σ^2.
3. Errors ϵ have zero correlation. The above two assumptions imply that the variance-covariance matrix of errors is $\sigma^2 \mathbf{I}$.

In addition, if errors ϵ have a normal distribution, the OLS estimator is the best unbiased estimator. Normal errors, however, are not a part of Gauss-Markov assumptions, and even in its absence, OLS is the best linear unbiased estimator as discussed in the next subsection. However, without normality of errors, there could be better non-linear estimators of β.

2.1.4   BLUE: Best Linear Unbiased Estimator

When Gauss-Markov conditions are satisfied by a linear regression model, OLS is the best linear unbiased estimator of the model's parameters. The unbiased nature of parameter estimates was shown in Section 2.1.3. Parameter estimates are linear as can be seen from Equation 2-7.

In order to prove that OLS is the best linear estimator, one must show that it has the least variance in the family of linear estimators. The OLS predictor is given by Equation 2-12.

$$\hat{\beta} = \left(\mathbf{X'X}\right)^{-1}\mathbf{Xy} \tag{2-12}$$

Let us consider a general class of unbiased linear estimators, $\tilde{\beta}$, as shown in Equation 2-13.

$$\tilde{\beta} = \left(\mathbf{D} + \left(\mathbf{X'X}\right)^{-1}\mathbf{X'}\right)\mathbf{y} \tag{2-13}$$

with condition $\mathbf{DX} = 0$

Condition $\mathbf{DX} = 0$ in Equation 2-13 is required for the unbiased estimate. This can be seen by replacing y with $\mathbf{X}\beta + \epsilon$, as shown in Equation 2-14.

$$
\begin{aligned}
\tilde{\beta} &= \left(\mathbf{D} + \left(\mathbf{X'X}\right)^{-1}\mathbf{X'}\right)(\mathbf{X}\beta + \epsilon) \\
&= \mathbf{DX}\beta + \left(\mathbf{X'X}\right)^{-1}\mathbf{X'X}\beta + \left(\mathbf{D} + \left(\mathbf{X'X}\right)^{-1}\mathbf{X'}\right)\epsilon \\
&= \beta + \left(\mathbf{D} + \left(\mathbf{X'X}\right)^{-1}\mathbf{X'}\right)\epsilon \\
\therefore E\left[\tilde{\beta}\right] &= \beta
\end{aligned}
\tag{2-14}
$$

Now let us show that the variance of the OLS estimator is the minimum in the class of linear estimators. Using Equation 2-14, we can write the variance of the general linear estimator as shown in Equation 2-15. From Equation 2-10, $\left(\mathbf{X'X}\right)^{-1}\sigma^2$ is the variance of the OLS estimator. This establishes the assertion that the OLS estimator has the least variance in the class of linear estimators of β.

$$
\begin{aligned}
\mathrm{var}(\tilde{\beta}) &= \left(\tilde{\beta} - \beta\right)\left(\tilde{\beta} - \beta\right)' \\
&= \left(\mathbf{D} + \left(\mathbf{X'X}\right)^{-1}\mathbf{X'}\right)\epsilon\left(\left(\mathbf{D} + \left(\mathbf{X'X}\right)^{-1}\mathbf{X'}\right)\epsilon\right)' \\
&= \left(\mathbf{D} + \left(\mathbf{X'X}\right)^{-1}\mathbf{X'}\right)\epsilon\epsilon'\left(\mathbf{D'} + \mathbf{X}\left(\mathbf{X'X}\right)^{-1}\right)
\end{aligned}
$$

$$= \sigma^2 \left(\mathbf{D} + \left(X'X \right)^{-1} X' \right) \left(\mathbf{D}' + \mathbf{X} \left(X'X \right)^{-1} \right) \qquad (2\text{-}15)$$

$$= \sigma^2 \left(\mathbf{DD}' + \left(\mathbf{X}'\mathbf{X} \right)^{-1} \mathbf{X}'\mathbf{D}' + \mathbf{DX} \left(\mathbf{X}'\mathbf{X} \right)^{-1} + \left(\mathbf{X}'\mathbf{X} \right)^{-1} \right)$$

$$= \sigma^2 \left(\mathbf{DD}' + \left(\mathbf{X}'\mathbf{X} \right)^{-1} \right)$$

$$\because \mathbf{DX} = \mathbf{0} \text{ and } \mathbf{X}'\mathbf{D}' = \mathbf{0}$$

$$\geq \left(\mathbf{X}'\mathbf{X} \right)^{-1} \sigma^2$$

2.1.5   Residuals: Standardized and Studentized

In the linear model $\mathbf{y} = \mathbf{X}\boldsymbol{\beta} + \boldsymbol{\epsilon}$, $\boldsymbol{\epsilon}$ is the true residual, computed using the true but unknown parameter values, $\boldsymbol{\beta}$. This residual satisfies the Gauss-Markov assumptions. Using the computed value of parameters, $\hat{\boldsymbol{\beta}}$, we can calculate estimated residuals, $\hat{\boldsymbol{\epsilon}}$. The relationship between calculated residuals and actual residuals is shown in Equation 2-16. **Standardized residuals** are obtained by dividing calculated residuals with their variance. However, we only know that the variance of actual residuals is $\sigma^2 \mathbf{I}$, by Gauss-Markov assumption. The expression for the variance of calculated residuals is shown in Equation 2-17. The **hat matrix, H**, is defined as $\mathbf{X} \left(\mathbf{X}'\mathbf{X} \right)^{-1} \mathbf{X}'$. Standardized residuals follow Gauss-Markov assumptions, i.e., are independent and identically distributed and have mean 0 and a finite variance, σ^2. Variance is calculated using Equation 2-10, using calculated residuals $\hat{\epsilon}_i$.

$$\hat{\mathbf{y}} = \mathbf{X}\hat{\boldsymbol{\beta}}$$

$$\mathbf{y} = \mathbf{X}\boldsymbol{\beta} + \boldsymbol{\epsilon} \text{ required in Gauss-Markov assumptions}$$

$$\mathbf{y} = \mathbf{X}\hat{\boldsymbol{\beta}} + \hat{\boldsymbol{\epsilon}} \text{ known from OLS}$$

$$\therefore \hat{\boldsymbol{\epsilon}} = \mathbf{y} - \mathbf{X}\hat{\boldsymbol{\beta}}$$

$$= \left(\mathbf{I} - \mathbf{X} \left(\mathbf{X}'\mathbf{X} \right)^{-1} \mathbf{X}' \right) \mathbf{y} \text{ from Equation 2-12}$$

$$= (\mathbf{I} - \mathbf{H}) \mathbf{y} \qquad (2\text{-}16)$$

$$= (\mathbf{I} - \mathbf{H}) (\mathbf{X}\boldsymbol{\beta} + \boldsymbol{\epsilon})$$

$$= (\mathbf{I} - \mathbf{H}) \mathbf{X}\boldsymbol{\beta} + (\mathbf{I} - \mathbf{H}) \boldsymbol{\epsilon}$$

$$= \left(\mathbf{X} - \mathbf{X} \left(\mathbf{X}'\mathbf{X} \right)^{-1} \mathbf{X}'\mathbf{X} \right) \boldsymbol{\beta} + (\mathbf{I} - \mathbf{H}) \boldsymbol{\epsilon}$$

$$= \left(\mathbf{I} - \mathbf{X} \left(\mathbf{X}'\mathbf{X} \right)^{-1} \mathbf{X}' \right) \boldsymbol{\epsilon}$$

$$\therefore \hat{\boldsymbol{\epsilon}} = (\mathbf{I} - \mathbf{H}) \boldsymbol{\epsilon}$$

$$\begin{aligned}
\text{var}(\hat{\epsilon}_i) &= E\left[(\mathbf{I} - \mathbf{H})\,\epsilon\epsilon'\,(\mathbf{I} - \mathbf{H})'\right] \\
&= (\mathbf{I} - \mathbf{H})\,E\left[\epsilon\epsilon'\right](\mathbf{I} - \mathbf{H})' \\
&= \sigma^2\,(\mathbf{I} - \mathbf{H})\,(\mathbf{I} - \mathbf{H})' \\
&= \sigma^2\left(\mathbf{I} - 2\mathbf{H} + \mathbf{H}\mathbf{H}'\right) \\
&= \sigma^2\left(\mathbf{I} - 2\mathbf{H} + \mathbf{X}(\mathbf{X}'\mathbf{X})^{-1}\mathbf{X}'\mathbf{X}(\mathbf{X}'\mathbf{X})^{-1}\mathbf{X}'\right) \\
&= \sigma^2\,(\mathbf{I} - 2\mathbf{H} + \mathbf{H}) \\
&= \sigma^2\,(\mathbf{I} - \mathbf{H})
\end{aligned} \tag{2-17}$$

Standardized residuals, $\epsilon_{i,\text{std}}$, are calculated by dividing computed residuals with their variance, as shown in Equation 2-18.

$$\begin{aligned}
\epsilon_{i,\text{std}} &= \frac{\hat{\epsilon}_i}{\sqrt{\text{var}(\hat{\epsilon}_i)}} \\
&= \frac{\hat{\epsilon}_i}{\sqrt{\sigma^2\,(\mathbf{I} - \mathbf{H})}} \\
&= \frac{\hat{\epsilon}_i}{\sigma\sqrt{1 - h_{ii}}}
\end{aligned} \tag{2-18}$$

$$\text{where } \sigma = \sqrt{\frac{\sum_{i=1}^{N}\left(\hat{\epsilon}_i\right)^2}{N - P - 1}}$$

A high or low value of standardized residual indicates that the point is an outlier. It is customary to consider a point with standardized residual greater than three in magnitude an outlier. Standardized residuals, though effective if ferreting out outliers, cannot spot outliers that are also highly influential points. For that, we need studentized residuals.

Studentized residuals, t, are used to assess the presence of points that influence a linear model's computed parameters to such an extent that they can no longer be deemed as outliers using standardized residuals. Studentized residuals are computed by excluding a point from regression, followed by computation of residual as $y_i - \hat{y}_i$, where $\hat{y}_i$ denotes the predicted response value using the linear model that was fitted after excluding the ith point. Standardizing the residuals computed in this fashion gives studentized residuals. Calculating studentized residuals by performing N regressions, excluding each data point one at a time is computationally expensive. Instead, the expression shown in Equation 2-19 can be used for this purpose that circumvents the calculation of auxiliary regressions. $\epsilon_{i,\text{std}}$ denotes the standardized residual.

Asymptotically, studentized residuals follow the **Student's t-distribution** with zero mean, one standard deviation, and $N - P - 1$ degrees of freedom. For large N, this distribution is close to a standard normal distribution.

$$t_i = \epsilon_{i,\text{std}} \sqrt{\frac{N - P - 2}{N - P - 1 - \epsilon_{i,\text{std}}^2}} \tag{2-19}$$

2.1.6 Influential Points

Influential points are those that have an outsized impact on fitted parameter values of a model. While fitting a linear model, two kinds of influential points can be identified: those with extreme values of response variable, y, and those with extreme values of input variable, $\mathbf{X}$.

1. **Outlier**: These are points with an extreme value of response variable, y. As a rule of thumb, points falling outside three standard deviations of y around its mean, i.e., $y \notin \mu \pm 3\sigma$, may be regarded as outliers. For a more rigorous definition, one could use standardized residuals and classify points with $\epsilon_{i,\text{std}}$ greater than three as an outlier.
2. **Points with high leverage**: The leverage of a point is defined as the corresponding diagonal value of the hat matrix, $\mathbf{H}$, and denotes the extent to which independent values of an observation differ from that of other observations. Points with high leverage are not necessarily influential. In order to understand leverage, two noteworthy properties of the hat matrix must be mentioned:

- All diagonal elements of the hat matrix, $h_{i,i}$, obey the relation $0 \le h_{i,i} \le 1$. This is established in Equation 2-20.

$$\mathbf{H}^2 = \left(\mathbf{X}(\mathbf{X}'\mathbf{X})^{-1}\mathbf{X}'\mathbf{X}(\mathbf{X}'\mathbf{X})^{-1}\mathbf{X}' \right)$$

$$= \mathbf{H}$$

$$\therefore h_{i,i} = \sum_{k=1}^{p+1} h_{i,k} h_{k,i}$$

$$= \sum_{k=1}^{p+1} h_{i,k} h_{i,k} \tag{2-20}$$

$$= h_{i,i}^2 + \sum_{k=1, k \ne i}^{p+1} h_{i,k}^2 \quad \text{because } \mathbf{H} \text{ is symmetric}$$

$$\ge h_{i,i}^2$$

$$\therefore 0 \le h_{i,i} \le 1$$

- The average of leverage values is equal to $\frac{P+1}{N}$. This is established in Equation 2-21.

$$\sum_{i=1}^{P+1} h_{i,i} = \text{trace}(\mathbf{X}(\mathbf{X}'\mathbf{X})^{-1}\mathbf{X}')$$

$$= \text{trace}((\mathbf{X}'\mathbf{X})^{-1}\mathbf{X}\mathbf{X}') \qquad (2\text{-}21)$$

$$= \text{trace}(\mathbf{I}) = P + 1$$

$$\therefore \text{average}(h_{i,i}) = \frac{P+1}{N}$$

Using these properties, if the leverage value of a point exceeds thrice its average value, $3\frac{P+1}{N}$, it can be deemed to be an influential point.

It can be shown that leverage values, $h_{i,i}$, are given by Equation 2-22.

$$h_{i,i} = \frac{1}{P+1} + \frac{(x_i - \bar{x})^2}{\sum_{j=1}^{P+1}(x_j - \bar{x})^2} \qquad (2\text{-}22)$$

2.2   Normal Equations

Normal equations are a method of solving OLS by inverting the matrix obtained in Equation 2-7, $\mathbf{X}'\mathbf{X}$. This method derives its name from the geometric interpretation of equation obtained after minimizing the mean square error.

As was observed earlier, a least squares problem involves an overdetermined system of linear equations. If the dimension of design matrix $\mathbf{X}$ is $(M, N + 1)$ – where M denotes the number of observations or rows, N denotes the number of free parameters, and the model includes a constant – M must be larger than $N + 1$, and the rank of $\mathbf{X}$ is $N + 1$. This condition ensures that the system has a unique least squares solution. From Equation 2-7, the solution to the linear model using OLS was derived to be $(\mathbf{X}'\mathbf{X})\,\hat{\boldsymbol{\beta}} = \mathbf{X}'\mathbf{y}$. The dimension of $\mathbf{X}'\mathbf{X}$ is $(N + 1, N + 1)$ with a rank of $N + 1$. Being a square matrix with full column and row rank, it has a unique inverse and the OLS solution for β can be written as $(\mathbf{X}'\mathbf{X})^{-1}\mathbf{X}'\mathbf{y}$. If the rank of $\mathbf{X}$ is less than $N + 1$, there are infinitely many solutions to the linear model that minimize the sum of residuals.

Geometrically, $\mathbf{X}\boldsymbol{\beta}$ denotes the space of vectors spanned by matrix $\mathbf{X}$. For different values of $\boldsymbol{\beta}$, $\mathbf{X}\boldsymbol{\beta}$ is a vector of dimension M but rank $N + 1$. To understand this fact, observe that $\mathbf{X}$ has a column span of $N+1$, i.e., $N+1$ linearly independent columns. A linear combination of $N + 1$ columns can only span a subspace of dimension $N + 1$. $\mathbf{y}$, on the other hand, has a full rank of M, which, according to our assumption, is greater than $N + 1$. Depicting the subspace spanned by $\mathbf{X}\boldsymbol{\beta}$

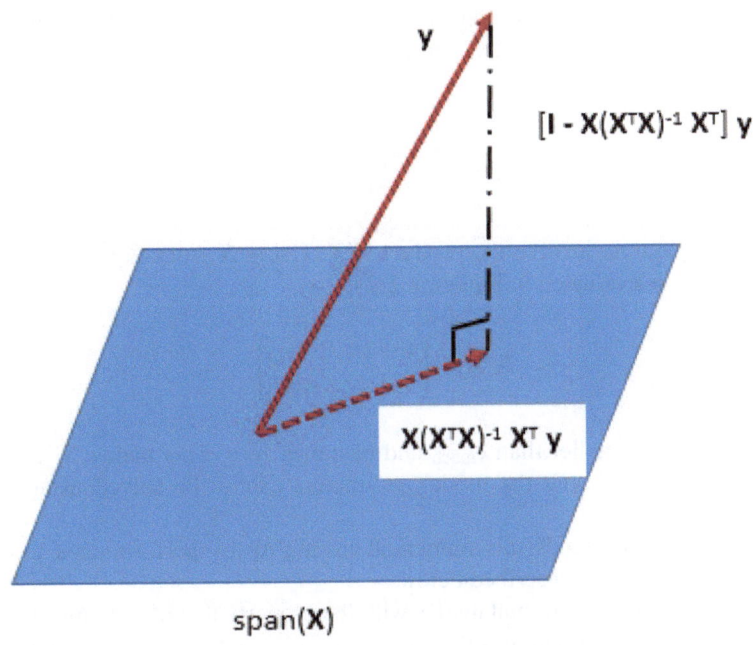

$$[I - X(X^TX)^{-1} X^T] y$$

$$X(X^TX)^{-1} X^T y$$

span(X)

Fig. 2-1. Geometric Depiction of OLS Solution Using Normal Equations

as a hyperplane of dimension $N + 1$, vector **y** lies outside that plane, as shown in Figure 2-1.

Similarly, $X'y$ is a vector of rank $N+1$, lying in the subspace spanned by **X**. This vector is shown in Figure 2-1. A solution to OLS is given by $\hat{\beta} = \left(X'X\right)^{-1} X'y$, as was shown in Equation 2-7. Premultiplying β by **X** gives $X\left(X'X\right)^{-1} X'y$ which is a vector in the subspace spanned by **X** that minimizes the residual between this vector and **y**. Intuitively, this should be the vector obtained by orthogonally projecting vector **y** onto the subspace spanned by **X**. This shows that the $X\left(X'X\right)^{-1} X'$ matrix is an orthogonal projector for **X**, projecting an out-of-plane vector **y** orthogonally onto its subspace as depicted in Figure 2-1. Normal equations derive their name from this geometric property.

2.3   QR Factorization

QR factorization is numerically more stable than OLS for solving a linear regression model. To see why this is true, consider a matrix **X** as shown in Equation 2-23. ϵ_{mach} represents the smallest floating-point number that can be represented in machine precision. For a 32-bit floating-point number, this is 2^{-32}, and for a 64-bit floating-point number, it is 2^{-64}. If a number is smaller than ϵ_{mach}, it is represented as 0. $\sqrt{\epsilon_{mach}}$ is bigger than ϵ_{mach}. Let δ represent a number smaller than $\sqrt{\epsilon_{mach}}$.

$$\mathbf{X} = \begin{bmatrix} 1 & 1 \\ \delta & 0 \\ 0 & \delta \end{bmatrix} \tag{2-23}$$

Normal equations involve the product $\mathbf{X'X}$. Using $\mathbf{X}$ from Equation 2-23, this matrix product is evaluated in Equation 2-24.

$$\mathbf{X'X} = \begin{bmatrix} 1 + \delta^2 & 1 \\ 1 & 1 + \delta^2 \end{bmatrix} \tag{2-24}$$

δ^2 is a number smaller than ϵ_{mach} and becomes zero. This makes $\mathbf{X'X}$ singular to machine precision, and the model parameters cannot be solved using normal equations.

QR factorization avoids this numerical instability by performing a triangular decomposition on $\mathbf{X}$, as described below.

Let $\mathbf{Q}$ denote an orthonormal matrix with the property that its column vectors are orthogonal to each other and have a norm of 1. This property implies that $\mathbf{Q'Q} = \mathbf{I}$. $\mathbf{R}$ is an upper triangular matrix. As before, denote the dimensions of $\mathbf{X}$ by (M, N+1) with M > N+1. Adding 1 to N accounts for the additional row due to inclusion of constant in $\boldsymbol{\beta}$. $\mathbf{Q}$ is of dimension (M, M), while $\mathbf{R}$ is of dimension (M, N+1). Only the upper triangular entries in $\mathbf{R}$ have non-zero values. Writing $\mathbf{Q}$ as $[\mathbf{Q_1}, \mathbf{Q_2}]$ where $\mathbf{Q_1}$ is a matrix comprising first N+1 columns of $\mathbf{Q}$ and $\mathbf{Q_2}$ comprises the remaining columns, we get Equation 2-25. $\mathbf{R_1}$ is a reduced form of upper triangular matrix $\mathbf{R}$. Dimensions of $\mathbf{R_1}$ are (N+1, N+1). Being the upper triangular matrix, equation $\mathbf{Q'_1 y}$ = $\mathbf{R_1}\boldsymbol{\beta}$ can be solved using back-substitution.

$$\begin{aligned} \mathbf{y} &= \mathbf{X}\boldsymbol{\beta} \\ &= \mathbf{QR}\boldsymbol{\beta} \\ &= [\mathbf{Q_1 Q_2}] \begin{bmatrix} R_1 \\ 0 \end{bmatrix} \boldsymbol{\beta} \\ \therefore \begin{bmatrix} R_1 \\ 0 \end{bmatrix} \boldsymbol{\beta} &= [\mathbf{Q'_1 Q'_2}] \mathbf{y} \\ \begin{bmatrix} R_1 \\ 0 \end{bmatrix} \boldsymbol{\beta} &= \begin{pmatrix} \mathbf{Q'_1 y} \\ \mathbf{Q'_2 y} \end{pmatrix} \end{aligned} \tag{2-25}$$

The OLS solution is obtained by minimizing the norm of the residual vector. The residual vector of Equation 2-25 is shown in Equation 2-26, and the minimum norm solution corresponds to solving the linear system $\mathbf{Q}_1'\mathbf{y} = \mathbf{R}_1\boldsymbol{\beta}$.

$$
\begin{aligned}
&\min_{\boldsymbol{\beta}} \|\mathbf{y} - \mathbf{X}\boldsymbol{\beta}\|^2 \\
&= \min_{\boldsymbol{\beta}} \left\|\mathbf{R}_1\boldsymbol{\beta} - \mathbf{Q}_1'\mathbf{y}\right\|^2 + \left\|\mathbf{Q}_2'\mathbf{y}\right\|^2 \\
&= \min_{\boldsymbol{\beta}} \left\|\mathbf{R}_1\boldsymbol{\beta} - \mathbf{Q}_1'\mathbf{y}\right\|^2 \\
&\therefore \text{solve } \mathbf{R}_1\boldsymbol{\beta} = \mathbf{Q}_1'\mathbf{y}
\end{aligned}
\tag{2-26}
$$

There are three different methods of performing QR decomposition on $\mathbf{X}$: Householder transformation, Givens rotation, and Gram-Schmidt orthogonalization. These are discussed below.

2.3.1   Householder Transformation

Householder transformation constructs the orthogonal matrix $\mathbf{Q}$ by eliminating all entries shown with $*$ and modifying the element marked by x in Equation 2-27.

$$
\begin{bmatrix}
x & \cdot & \cdot & \cdot \\
* & x & \cdot & \cdot \\
* & * & x & \cdot \\
* & * & * & x \\
* & * & * & * \\
* & * & * & *
\end{bmatrix}
\tag{2-27}
$$

Let $\mathbf{a}$ denote the column vector at or below position x, as was shown in Equation 2-27. Denoting the unit vector that has one in position x and zero elsewhere as $\mathbf{e}_1$, we want to derive an expression for a matrix $\mathbf{H}$ that rotates $\mathbf{a}$ to a vector along $\mathbf{e}_1$. Because the transformation must preserve vector length, we can write $\mathbf{H}$ as shown in Equation 2-28. The geometric interpretation of the operation of matrix $\mathbf{H}$ on $\mathbf{a}$ is shown in Figure 2-2.

$$
\begin{aligned}
\mathbf{Ha} &= \|\mathbf{a}\|^2 \mathbf{e}_1 \\
&= -\mathbf{v} + \mathbf{a} \\
\therefore \mathbf{v} &= \mathbf{a} - \|\mathbf{a}\|^2 \mathbf{e}_1
\end{aligned}
\tag{2-28}
$$

Fig. 2-2. Geometric
Interpretation of Householder
Transformation of Vector a

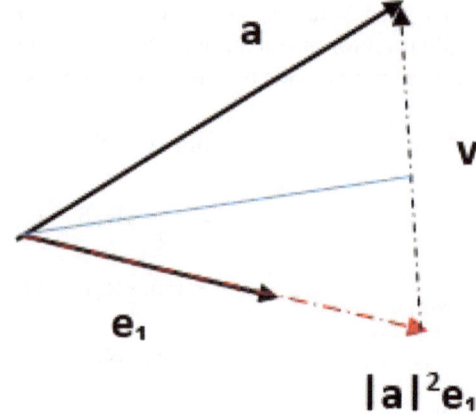

Using geometric properties, vector $\mathbf{v}$ can be written as twice the length of the projection of vector $\mathbf{a}$ on vector $\mathbf{v}$, as shown in Equation 2-29.

$$\mathbf{v} = 2\frac{\mathbf{v}'\mathbf{a}}{\mathbf{v}'\mathbf{v}}\mathbf{v} \tag{2-29}$$

Substituting the expression for $\mathbf{v}$ from Equation 2-29 into Equation 2-28, we get an expression for Householder transformation matrix $\mathbf{H}$ in terms of $\mathbf{v}$, as shown in Equation 2-30.

$$\mathbf{Ha} = \mathbf{a} - 2\frac{\mathbf{v}'\mathbf{a}}{\mathbf{v}'\mathbf{v}}\mathbf{v}$$

$$= \mathbf{a} - 2\frac{\mathbf{v}\mathbf{v}'}{\mathbf{v}'\mathbf{v}}\mathbf{a}$$

$$= \left(\mathbf{I} - 2\frac{\mathbf{v}\mathbf{v}'}{\mathbf{v}'\mathbf{v}}\right)\mathbf{a} \tag{2-30}$$

$$\therefore \mathbf{H} = \left(\mathbf{I} - 2\frac{\mathbf{v}\mathbf{v}'}{\mathbf{v}'\mathbf{v}}\right)$$

$$\mathbf{v} = \mathbf{a} \pm \|\mathbf{a}\|^2 \mathbf{e}_1$$

The sign in the expression for vector $\mathbf{v}$ in Equation 2-30 is chosen to avoid cancellation with the component of $\mathbf{a}$ along the direction of $\mathbf{e}_1$.

Using Householder transformations for column vectors of $\mathbf{X}$, we transform it to upper triangular matrix $\mathbf{R}$. Inverting the transformations, we can get a representation of $\mathbf{Q}$ and $\mathbf{R}$ matrices, as shown in Equation 2-31.

$$\mathbf{H}_{N+1} \cdots \mathbf{H}_1 \mathbf{X} = \mathbf{R}$$

$$\mathbf{X} = (\mathbf{H}_{N+1} \cdots \mathbf{H}_1 \mathbf{X})' \, \mathbf{R}$$

$$\implies \mathbf{Q} = (\mathbf{H}_{N+1} \cdots \mathbf{H}_1 \mathbf{X})' \tag{2-31}$$

Listing 2-1. Applying Householder Tranformations to Obtain QR Decomposition of a Matrix

```python
import numpy as np


class HouseHolder(object):
    def __init__(self, dtype=np.int32):
        self.dtype = dtype

    def houseHolderVecToQ(self, V):
        X0 = np.eye(V.shape[0], dtype=self.dtype)
        factor = -2 / np.dot(V[:, 0], V[:, 0])
        X0 += factor * np.einsum("i,j->ij", V[:, 0], V[:, 0])
        for j in range(1, V.shape[1]):
            factor = -2 / np.dot(V[:, j], V[:, j])
            X1 = np.eye(V.shape[0], dtype=self.dtype)
            X1 += factor * np.einsum("i,j->ij", V[:, 0], V[:, 0])
            X0 = np.einsum("ij,jk->ik", X1, X0)
        return X0

    def qrDecompose(self, matrix):
        nrow, ncol = matrix.shape
        V = matrix.copy()
        R = np.zeros((ncol, ncol), dtype=self.dtype)
        for j in range(ncol):
            R[0:j+1, j] = V[0:j+1, j]
            V[0:j, j] = 0
            v = V[:, j]
            xlen = np.dot(v, v)
            mult = 1 if v[j] >= 0 else -1
            v[j] += mult * xlen
            # v is now the vector in householder transformation
            vlen = np.dot(v, v)
            for i in range(j+1, ncol):
                V[:, i] += -2 * (np.dot(v, V[:, i]) / vlen) * v

        Q = self.houseHolderVecToQ(V)
        return Q, R
```

The code for applying Householder transformations to compute QR decomposition of a matrix is shown in Listing 2-1.

Code Explanation

Method **qrDecompose** of class **HouseHolder** performs QR decomposition in Listing 2-1. This method accepts a two-dimensional matrix passed as a numpy array. It returns Q and R matrices and leaves the input matrix unmodified. The first for loop inside the method iterates over each row of the matrix and applies transformation shown in Equation 2-28 to the elements on and below the diagonal in column **j**. Elements above diagonal will be reduced to zero by Householder transformation called later. In order to avoid cancellation of significant digits, it uses a multiplier that is set to -1 if the vector element on diagonal is negative. Iterating over all the columns, we obtain the transformed V matrix with all elements above diagonal set to zero. This matrix contains the HouseHolder vectors as columns.

After this, method **houseHolderVecToQ** is called. This method applies the transformations shown in Equation 2-31 to obtain matrices Q and R.

2.3.2 Givens Rotation

Givens rotation involves selecting θ_i to transform the ith column vector, $\mathbf{a_i}$, into a vector along an orthogonal direction by eliminating all entries below a diagonal element, as shown in Equation 2-32. Due to the property $\cos^2 \theta_i + \sin^2 \theta_i = 1$, the transformation preserves vector norm and yields an upper triangular matrix **R** in the end. However, it is computationally more expensive than Householder transformation or Gram-Schmidt orthogonalization.

$$
\begin{pmatrix}
0 \cdots & 0 & \cdots & 0 & \cdots & 0 \\
\cdot \cdots & \cdot & \cdots & \cdot & \cdots & \\
0 \cdots & \cos\theta_i & \cdots & \sin\theta_i & \cdots & 0 \\
\cdot \cdots & \cdot & \cdots & \cdot & \cdots & \\
0 \cdots & -\sin\theta_i & \cdots & \cos\theta_i & \cdots & 0 \\
\cdot \cdots & \cdot & \cdots & \cdot & \cdots &
\end{pmatrix}
\begin{pmatrix}
a_{1i} \\
\cdots \\
a_{ii} \\
\cdots \\
a_{ji} \\
\cdots
\end{pmatrix}
=
\begin{pmatrix}
a_{1i} \\
\cdots \\
\tilde{a_i i} \\
\cdots \\
0 \\
\cdots
\end{pmatrix}
\tag{2-32}
$$

2.3.3 Gram-Schmidt Orthogonalization

Gram-Schmidt orthogonalization obtains QR decomposition of a matrix **X** by making the columns X_j orthogonal to column X_i for all columns $j > i$. There are two flavors of this process: classical Gram-Schmidt and modified Gram-Schmidt. Classical Gram-Schmidt is illustrated in pseudo-code 1.

Algorithm 1 Classical Gram-Schmidt Orthogonalization

Require: Matrix **X** of dimension (M, N), with column vectors $= (\mathbf{x_1}, \cdots, \mathbf{x_N})$.
1: **for** $j = 1, 2, \cdots, N$ **do**
2:      $\mathbf{v} = \mathbf{x_j}$
3:      **for** $i = 1, 2, \cdots, j\text{-}1$ **do**
4:          $R_{ij} = \mathbf{q'_i x_j}$
5:          $\mathbf{v} = \mathbf{v} - R_{ij}\mathbf{q_i}$
6:      **end for**
7:      $R_{jj} = \|\mathbf{v}\|$
8:      $\mathbf{q_j} = \frac{\mathbf{v}}{R_{jj}}$
9: **end for**

Classical Gram-Schmidt is numerically unstable because it suffers from loss of orthogonality of vectors $\mathbf{q_i}$. Also, if two column vectors of $\mathbf{X}$, $\mathbf{x_i}$ and $\mathbf{x_j}$, are nearly parallel, $\mathbf{v}$ may become close to zero due to cancellation.

Modified Gram-Schmidt is numerically more stable. It involves using the existing vector $\mathbf{v}$ that has components along previous orthogonal vectors $\mathbf{q_j}$ removed in place of $\mathbf{a_j}$. The algorithm is sketched in pseudo-code 2.

Algorithm 2 Modified Gram-Schmidt Orthogonalization

Require: Matrix $\mathbf{X}$ of dimension (M, N), with column vectors = $(\mathbf{x_1}, \cdots, \mathbf{x_N})$.

1: **for** j = 1, 2, $\cdots$, N **do**
2:     $\mathbf{v} = \mathbf{x_j}$
3:     **for** i = 1, 2, $\cdots$, j-1 **do**
4:         $R_{ij} = \mathbf{q_i'v}$
5:         $\mathbf{v} = \mathbf{v} - R_{ij}\mathbf{q_i}$
6:     **end for**
7:     $R_{jj} = \|\mathbf{v}\|$
8:     $\mathbf{q_j} = \frac{\mathbf{v}}{R_{jj}}$
9: **end for**

The implementation of classical and modified Gram-Schmidt orthogonalization code is shown in Listing 2-2.

Listing 2-2. Classical and Modified Gram-Schmidt Orthogonalization Method to Obtain QR Decomposition of a Matrix

---

```
import numpy as np


class GramSchmidt(object):
    def __init__(self, dtype=np.int32):
        self.dtype = dtype

    def qrDecomSimple(self, matrix):
        nrow, ncol = matrix.shape
        Q1 = matrix.copy()
        R = np.zeros((ncol, ncol), dtype=self.dtype)
        for j in range(ncol):
            for i in range(j):
                R[i, j] = np.dot(Q1[:, i], matrix[:, j])
                Q1[:, j] -= R[i, j] * Q1[:, i]
            R[j, j] = np.dot(Q1[:, j], Q1[:, j])
            Q1[:, j] /= R[j, j]
        return Q1, R

    def qrDecompose(self, matrix):
        ''' Modified (stable) version of Gram Schmidt decomposition '''
        nrow, ncol = matrix.shape
        Q1 = matrix.copy()
        R = np.zeros((ncol, ncol), dtype=self.dtype)
        for j in range(ncol):
```

```
26        for i in range(j):
27            R[i, j] = np.dot(Q1[:, i], Q1[:, j])
28            Q1[:, j] −= R[i, j] * Q1[:, i]
29        R[j, j] = np.dot(Q1[:, j], Q1[:, j])
30        Q1[:, j] /= R[j, j]
31    return Q1, R
```

Code Explanation

Inside class **GramSchmidt**, method **qrDecomSimple** applies the classical Gram-Schmidt orthogonalization sketched in pseudo-code 1. This method accepts a two-dimensional numpy array as input and returns Q and R matrices, leaving the input matrix unmodified. The outer j for loop iterates over all columns of the matrix, while the inner i for loop makes the columns from 1 to $j - 1$ orthogonal to column j. After each inner i loop completes an iteration, it calculates q_j. In this way, it keeps a set of orthogonal vectors q_j that are mutually orthogonal among all vectors from 1 to j. Finally, when the outer loop finishes, it has the final set of orthogonal vectors.

Modified Gram-Schmidt orthogonalization is implemented inside method **qrDe-compose** according to pseudo-code 2. As explained earlier, it uses the computed orthogonal vectors v_i in place of x_i inside the inner for loop.

2.4   Singular Value Decomposition

Singular value decomposition, or SVD, is another method of solving OLS that is more stable than QR factorization. If matrix X is rank deficient, i.e., if two or more of its column vectors are linearly dependent, Gram-Schmidt orthogonalization may give division-by-zero errors. Using SVD, we can identify singular values that lead to rank deficiency.

SVD of a matrix X is shown in Equation 2-33. If X has dimensions (M, N); U, Σ, and V are matrices of dimensions (M, M), (M, N), and (N, N), respectively. Further, U and V are orthogonal matrices, while Σ is a diagonal matrix with diagonal entries $\sigma_{ii} \geq 0$ ordered by descending value. Diagonal entries, σ_{ii}, are also called singular values.

$$X = U\Sigma V'$$

$$\Sigma = \begin{pmatrix} \sigma_{11} & 0 & \cdots & 0 \\ 0 & \sigma_{22} & \cdots & 0 \\ . & . & \cdots & . \\ 0 & 0 & 0 & \sigma_{NN} \\ 0 & 0 & 0 & 0 \end{pmatrix} \tag{2-33}$$

Using SVD, the solution to $\mathbf{X}\boldsymbol{\beta} = \mathbf{y}$ can be written as shown in Equation 2-34.

$$\boldsymbol{\beta} = \boldsymbol{\Sigma}^{-1}\mathbf{U}'\mathbf{y}\mathbf{V}$$

$$= \sum_{i=1}^{N} \frac{\mathbf{u_i'}\mathbf{y}\mathbf{v_i}}{\sigma_{ii}} \text{ where } \sigma_{ii} \neq 0 \tag{2-34}$$

SVD can be obtained using the eigenvalue decomposition of $\mathbf{X}'\mathbf{X}$. This can be observed by multiplying the SVD of $\mathbf{X}$ and observing that $\mathbf{U}'\mathbf{U} = \mathbf{I}$, as shown in Equation 2-35. Similarly, the eigenvalue decomposition of $\mathbf{X}\mathbf{X}'$ yields the square of singular vectors, $\boldsymbol{\Sigma}^2$ and $\mathbf{U}$.

$$\mathbf{X}'\mathbf{X} = \mathbf{V}\boldsymbol{\Sigma}'\mathbf{U}'\mathbf{U}\boldsymbol{\Sigma}\mathbf{V}'$$

$$= \mathbf{V}\boldsymbol{\Sigma}^2\mathbf{V}'$$

$$\therefore (\mathbf{X}'\mathbf{X})\,\mathbf{V} = \boldsymbol{\Sigma}^2\mathbf{V}$$

$$\mathbf{X}\mathbf{X}' = \mathbf{U}\boldsymbol{\Sigma}\mathbf{V}'\mathbf{V}\boldsymbol{\Sigma}'\mathbf{U}' \tag{2-35}$$

$$= \mathbf{U}\boldsymbol{\Sigma}^2\mathbf{U}'$$

$$\therefore (\mathbf{X}'\mathbf{X})\,\mathbf{U} = \boldsymbol{\Sigma}^2\mathbf{U}$$

A more numerically stable algorithm is the Golub-Kahan method that involves using Householder transformations on $\mathbf{X}$ to obtain a bidiagonal matrix, followed by an iterative algorithm to obtain a diagonal matrix.

2.5   Maximum Likelihood

The maximum likelihood method is based on maximizing the probability of observing the data, assuming the residuals belong to a family of parametric probability distribution. It is common to use the Gaussian distribution parameterized by its mean and variance as the probability distribution of residuals. Under this assumption, the maximum likelihood method applied to the linear model is equivalent to OLS.

In order to appreciate this, consider the linear model with residuals, ϵ, following a normal distribution. Assuming the model is unbiased, the mean of the normal distribution is zero. Assuming the residuals are homoskedastic, the variance of ϵ can be denoted by σ^2. The likelihood of observing the data is shown in Equation 2-36.

$$\mathbf{y} = \mathbf{X}'\boldsymbol{\beta} + \epsilon \text{ where } \epsilon \sim N(0, \sigma^2)$$

$$\Pr(\mathbf{X}|\boldsymbol{\beta}, \sigma^2) = \prod_{i=1}^{M} \frac{1}{\sqrt{2\pi\sigma^2}} \exp\left(\frac{1}{2}\left(\frac{y_i - x_i\beta}{\sigma}\right)^2\right) \tag{2-36}$$

For the exponential class of probability functions, it is more convenient to work with log-likelihood which is the natural logarithm of probability derived in Equation 2-36. The logarithm being a monotonic increasing function, the maximum of log-likelihood corresponds to the maximum of probability. The expression for log-likelihood is shown in Equation 2-37.

$$L(\mathbf{X}|\boldsymbol{\beta}, \sigma^2) = \log(\Pr(\mathbf{X}|\boldsymbol{\beta}, \sigma^2))$$

$$= -\frac{M \log(2\pi\sigma^2)}{2} - \frac{1}{2}\sum_{i=1}^{M}\left(\frac{y_i - x_i\beta}{\sigma}\right)^2 \tag{2-37}$$

Maximizing the log-likelihood with respect to σ^2 and $\boldsymbol{\beta}$ gives Equation 2-38. σ^2 is the residual variance as observed from the data, and $\boldsymbol{\beta}$ corresponds to the system of normal equations, establishing the equivalence of the maximum likelihood method and OLS under the assumption of normal, homoskedastic residuals. $\boldsymbol{\beta}$ being a vector of parameters, we must maximize L with respect to each component of $\boldsymbol{\beta}$, as denoted by β_j, in Equation 2-38.

$$\frac{\partial L}{\partial \sigma^2} = 0$$

$$\implies -\frac{M}{2} + \frac{1}{2}\sum_{i=1}^{M}\left(\frac{y_i - x_i\beta}{\sigma}\right)^2 = 0$$

$$\implies M = \frac{\sum_{i=1}^{M}(y_i - x_i\beta)^2}{M} \tag{2-38}$$

$$\frac{\partial L}{\partial \beta_j} = 0$$

$$\implies x_j\sum_{i=1}^{M}(y_i - x_i\beta) = 0$$

$$\implies \mathbf{X'y} = \mathbf{X'X}\boldsymbol{\beta}$$

2.6   Confidence Intervals

Assuming a normal distribution for residuals, one can derive confidence intervals for parameter estimates obtained using OLS. However, once Gauss-Markov assumptions of homoskedasticity and zero autocorrelation among residuals are relaxed, it becomes necessary to modify expressions for confidence intervals using heteroskedasticity and autocorrelation (HAC) consistent standard errors, as described in this section.

A confidence interval indicates the range of parameter values within which the actual but unknown parameter value is located with some level of probability. OLS uses an empirical batch of data $\mathbf{y}$ and $\mathbf{X}$ to predict β. It is natural to ask that if one used a different batch of data, how much could parameter estimates vary from the estimates obtained using the random sample. In order to answer this question, we must make an assumption that the empirical data is a random (unbiased) sample from the underlying true but unknown data distribution.

Under Gauss-Markov assumptions, it was shown in Equation 2-10 that the estimated parameters of the linear model, $\hat{\beta}$, are normally distributed with mean β and variance $(\mathbf{X}'\mathbf{X})^{-1}\sigma^2$. This implies that with confidence α, parameter estimate $\hat{\beta}$ lies in the interval shown in Equation 2-39, where Φ^{-1} denotes the inverse of CDF (cumulative density function) of the Gaussian distribution.

$$\hat{\beta} \sim N(\beta, (\mathbf{X}'\mathbf{X})^{-1}\mathbf{I}\sigma^2)$$

$$-\Phi^{-1}\left(\frac{1-\alpha}{2}\right) \leq \frac{\hat{\beta}-\beta}{\sigma\sqrt{(\mathbf{X}'\mathbf{X})^{-1}\mathbf{I}}} \leq \Phi^{-1}\left(\frac{1-\alpha}{2}\right) \tag{2-39}$$

with probability α

$$\left[\beta_{\text{low}}, \beta_{\text{high}}\right] = \hat{\beta} \pm \sigma\sqrt{(\mathbf{X}'\mathbf{X})^{-1}\mathbf{I}}\,\Phi^{-1}\left(\frac{1-\alpha}{2}\right)$$

2.6.1   Heteroskedasticity Consistent Standard Errors

When residuals ϵ no longer have same variance, the second Gauss-Markov assumption is violated. The OLS solution is still consistent because the first Gauss-Markov assumption of independence between ϵ and exogenous regressors $\mathbf{X}$ holds (refer to Equation 2-11). However, the solution is no longer best, i.e., it does not have minimum variance. To see this fact, notice that $E\left[\epsilon\epsilon'\right]$ in Equation 2-9 is a diagonal matrix with entries ϵ_i^2. These entries represent statistical estimates of variance for each residual. The expression can no longer be simplified and is greater than the minimum variance of a linear estimator, shown in Equation 2-15. Consequently, confidence intervals obtained using variance $(\mathbf{X}'\mathbf{X})^{-1}\mathbf{I}\sigma^2$ in Equation 2-39 are overly optimistic (i.e., small) because the actual variance of estimated parameters shown in Equation 2-40 is larger. White [1] derived this heteroskedasticity consistent variance estimator, also known as the HC0 estimator.

$$\text{Variance}\left(\hat{\beta}\right) = (\mathbf{X}'\mathbf{X})^{-1}\mathbf{X}'\text{diag}\left(\epsilon_i^2\right)\mathbf{X}\left(\mathbf{X}'\mathbf{X}\right)^{-1} \tag{2-40}$$

In practice, owing to finite sample size, White's heteroskedasticity consistent variance estimator gives biased estimates of variance. To fix this problem, variants of this estimator have been proposed.

The HC1 estimator accounts for the finite sample bias in HC0 by multiplying it with $\frac{M}{M-N-1}$ as shown in Equation 2-41, where M is the number of observations or rows and $N+1$ is the number of parameters, including the constant.

$$\text{HC1} = \frac{M}{M-N-1} \left(\mathbf{X}'\mathbf{X}\right)^{-1} \mathbf{X}'\text{diag}\left(\epsilon_i^2\right) \mathbf{X} \left(\mathbf{X}'\mathbf{X}\right)^{-1}$$

(2-41)

where $\mathbf{X}$ has dimensions $M \times (N+1)$

The HC2 estimator uses a different weighing for each diagonal element ϵ_i^2. Unlike HC1 which uses a constant weight, HC2 uses $\frac{1}{1-h_{ii}}$ as the weight of the ith diagonal element ϵ_i^2. h_{ii} is defined as the ith diagonal element of matrix $\mathbf{X}\left(\mathbf{X}'\mathbf{X}\right)^{-1}\mathbf{X}'$, or the hat matrix. This is shown in Equation 2-42.

$$\text{HC2} = \left(\mathbf{X}'\mathbf{X}\right)^{-1}\mathbf{X}'\text{diag}\left(\frac{\epsilon_i^2}{1-h_{ii}}\right)\mathbf{X}\left(\mathbf{X}'\mathbf{X}\right)^{-1}$$

(2-42)

where h_{ii} is the ith diagonal element of $\mathbf{X}\left(\mathbf{X}'\mathbf{X}\right)^{-1}\mathbf{X}'$

HC3 modifies HC2 by using $\frac{1}{(1-h_{ii})^2}$ as the weight for each diagonal element ϵ_i^2. Davidson and MacKinnon [2] showed that this variance estimator has less bias than HC2. The HC4 estimator – proposed by Cribari-Neto [3] – uses $\frac{1}{(1-h_{ii})^{\delta_i}}$ as the weight, where $\delta_i = \min(4, \frac{M}{N+1}h_{ii})$.

2.6.2   HAC Consistent Standard Errors

If residuals from OLS have autocorrelation in addition to heteroskedasticity, the parameter estimates are still unbiased and consistent because the first Gauss-Markov condition holds. However, because the last two Gauss-Markov assumptions are violated, the OLS estimator no longer has the least variance and confidence intervals obtained using Equation 2-10 are biased. Using the correction for heteroskedasticity introduced in Equation 2-41 or 2-42 is inadequate to cure the bias due to the presence of autocorrelation in residuals.

Newey and West [4] proposed a correction to variance estimation in order to account for heteroskedasticity and autocorrelation. Their estimator, known as the Newey-West estimator, depends on a parameter L specifying the number of auto-correlation lags to consider. The Newey-West estimator is shown in Equation 2-43, where $\mathbf{x_i}$ denotes the ith row of design matrix $\mathbf{X}$ written as a column vector.

The expression in Equation 2-43 gives a matrix with a diagonal element and L subdiagonal entries. Setting $L = 0$ gives the heteroskedasticity consistent estimator, HC1.

$$\text{HAC} = \left(X'X\right)^{-1} X' \epsilon\epsilon' X \left(X'X\right)^{-1}$$

$$X' \epsilon\epsilon' X = \text{diag}\left(\frac{1}{M} \sum_{i=1}^{M} \epsilon_i^2 x_i x_i'\right) + \frac{1}{M} \sum_{l=1}^{L} \sum_{i=l+1}^{M} w_l \epsilon_i \epsilon_{i-l} \left(x_i x_{i-1}' + x_{i-1} x_i'\right)$$

$$w_l = 1 - \frac{l}{L+1}$$

$$(2\text{-}43)$$

Equation 2-43 uses Bartlett kernel w_l. There are other choices of kernels w_l, giving different estimates of the covariance matrix that may yield lower bias for small sample sizes. For example, Gallant [5] uses Parzen kernel.

2.7 Regularization

By increasing the number of parameters in a linear model, one can boost regression R^2 by improving the model fit over the training (or calibration) dataset. However, this makes the model overfit the data, fitting the data outliers as though they were aspects of the underlying data distribution. Because they are artifacts of measurement errors, however, they are often not replicated in the test dataset. This leads to a significant deterioration in the model fit on the test dataset. In order to guard against overfitting in the training dataset, it is customary to use regularization.

Regularization involves adding a penalty term to the objective function that penalizes models with higher number of parameters or models with higher value of parameters. Adding the sum of model parameters' absolute values, $\|\beta_i\|$, gives **L1** or **Lasso** regularization. The objective function used in Lasso regression is shown in Equation 2-44. Similarly, **L2** or **ridge** regression involves using $\|\beta_i\|^2$ as the penalty function (Equation 2-45) and imposes greater penalty of higher parameter estimates as compared with L1 regression.

$$\min_{\beta} \frac{1}{N} \sum_{i=1}^{N} \left(y_i - \sum_{j=0}^{P} \beta_j x_{j,i}\right)^2 + \sum_{j=0}^{P} \|\beta_j\| \qquad (2\text{-}44)$$

$$\min_{\beta} \frac{1}{N} \sum_{i=1}^{N} \left(y_i - \sum_{j=0}^{P} \beta_j x_{j,i}\right)^2 + \sum_{j=0}^{P} \|\beta_j\|^2 \qquad (2\text{-}45)$$

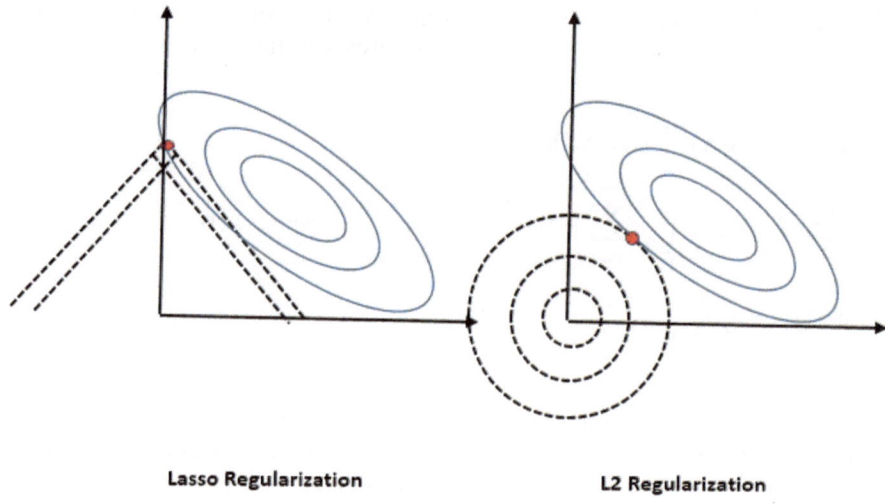

Lasso Regularization   **L2 Regularization**

Fig. 2-3. Lasso (L1) Regularization Setting Parameters to Zero

Equation 2-44 is able to set the parameters β_j of insignificant exogenous variables x_j to zero, thereby obtaining a **parsimonious** model. This is in contrast to L2 regularization that can only reduce the estimated parameter magnitude of insignificant exogenous variables. To understand this property of Lasso regression, let us look at the contours of the penalty term and OLS term (elliptical due to quadratic objective), as shown in Figure 2-3. Because the L2 penalty term is a circle (or sphere in higher dimensions), it cannot set parameter estimates to zero. In Lasso (L1) regularization, however, the contours of the penalty term are lines, and the optimum value of the combined objective function will lie on a corner that involves setting some parameters to zero.

Elastic net or L1–L2 regularization involves a weighted combination of L1 and L2 penalty terms.

2.8   Goodness of Fit

Goodness of fit metrics furnish a quantitative measure of how well a model fits the data. There are a variety of goodness of fit metrics in use, some focused solely on quantifying the model's ability to produce output that matches the known output, while others take the number of model parameters into account, in addition to prediction accuracy. Furthermore, models fitted using maximum likelihood have a different set of goodness of fit metrics. These are discussed below.

2.8.1 Coefficient of Determination, R^2

The most commonly used goodness of fit metric for a linear model is R^2, or the **coefficient of determination**. It is defined as shown in Equation 2-46. Its value lies between zero and one, with values closer to one indicating better model fit to the data. SS_{res} can be regarded as the unexplained variance, and when it is zero, the model fits the data perfectly, giving an R^2 of one.

$$R^2 = 1 - \frac{SS_{res}}{SS_{tot}}$$

$$SS_{tot} = \sum_{i=1}^{M} (y_i - \bar{y})^2$$

$$SS_{res} = \sum_{i=1}^{M} \epsilon_i^2 \tag{2-46}$$

$$\bar{y} = \frac{1}{M} \sum_{i=1}^{M} y_i$$

2.8.2 Adjusted R^2

Increasing the number of parameters will generally reduce the residual variance which is proportional to SS_{res}. This may lead to overfitting, giving higher bias in testing data. In order to account for the number of model parameters, adjusted R^2 is calculated as shown in Equation 2-47, penalizing the models that achieve low SS_{res} by increasing the number of parameters.

$$R^2_{adj} = 1 - \frac{M-1}{M-N-1} \frac{SS_{res}}{SS_{tot}} = 1 - \left(1 - R^2\right) \frac{M-1}{M-N-1} \tag{2-47}$$

2.8.3 Pseudo R^2

For models fitted using the maximum likelihood method, the corresponding measure is **pseudo R^2**. It is defined using the likelihood of an existing model, $L(\beta)$, and the likelihood of a hypothetical model that uses only an intercept, $L(\beta_0)$, using Equation 2-48.

$$\text{pseudo } R^2 = 1 - \left(\frac{L(\beta_0)}{L(\beta)}\right)^{\frac{2}{M}} \tag{2-48}$$

2.8.4   Information Criterion

For models that use regularization, adjusted R^2 does not adequately account the benefit of adding penalty terms that reduce variance of forecast at the cost of increasing bias. Also, R^2 and adjusted R^2 are not applicable to models fit using log-likelihood. The **information criterion** handles both of these shortcomings by accounting for the likelihood of observing data and penalizing models with higher number of parameters. There are several flavors of information criterion in use:

1. **Akaike Information Criterion**: AIC is defined as shown in Equation 2-49, where k denotes the number of free parameters and $L(\beta)$ is the likelihood. Smaller AIC values are better. In an OLS model, k is equal to $P + 1$, according to the notation used earlier. β denotes the model's free parameters.

$$\text{AIC} = 2k - 2\ln(L(\beta)) \tag{2-49}$$

2. **Bayesian Information Criterion**: BIC is similar to AIC except with regard to penalty terms, as shown in Equation 2-50. BIC uses $k\ln(M)$ where k is the number of model parameters and N is the number of observations. This criterion is also known as the **Schwarz information criterion**.

$$\text{BIC} = k\ln(N) - 2\ln(L(\beta)) \tag{2-50}$$

2.8.5   Wald Test

The Wald test is used to assess the statistical significance of estimated parameter values. The statistic is defined in Equation 2-51. $\hat{\beta}$ denotes the estimated parameter value, and β_0 denotes the hypothesized parameter value. The variance of the estimated parameter, $\text{var}(\hat{\beta})$, is computed by taking the second derivative of the likelihood function.

$$W = \frac{\hat{\beta} - \beta_0}{\text{var}\left(\hat{\beta}\right)} \tag{2-51}$$

Under the null hypothesis, the estimated parameter value is statistically identical to β_0. Asymptotically, the Wald statistic follows a χ^2 distribution with one degree of freedom. Using a confidence threshold, one can compare the Wald statistic's value to the critical value assuming a $\chi^2(1)$ distribution to accept or reject the null hypothesis.

2.8.6   LM Test

The Lagrange multiplier, or LM, test is also known as the score test. Score is defined as the derivative of the log-likelihood function, $l(y, X, \theta)$, with respect to

Fig. 2-4. Log-Likelihood
Function and Its Maximum

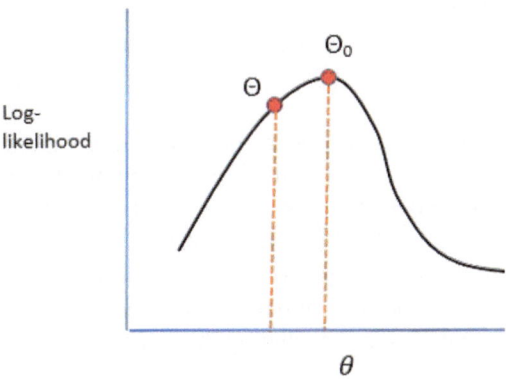

Log-
likelihood

θ

the model's parameters, θ, as denoted by Equation 2-52. In general, this is a vector
because θ is a vector of model parameters.

$$s(\theta) = \frac{\partial l(y, X, \theta)}{\partial \theta} \qquad (2\text{-}52)$$

A typical log-likelihood function may look as shown in Figure 2-4. Maximum
likelihood attempts to find θ_0 at which $l(y, X, \theta)$ attains maximum. This is obtained
by setting the first derivative to zero, as shown in Equation 2-53.

$$\frac{\partial l(y, X, \theta)}{\partial \theta} = 0 \qquad (2\text{-}53)$$

Expanding the log-likelihood function in the neighborhood of maxima, θ_0, using
Taylor expansion, one obtains Equation 2-54. $(\theta - \theta_0)^2$ is the variance of parameter
estimate θ_0.

$$l(y, X, \theta) = l(y, X, \theta_0) + \frac{\partial l(y, X, \theta)}{\partial \theta_0} (\theta - \theta_0) + \frac{1}{2} \frac{\partial^2 l(y, X, \theta)}{\partial \theta_0 \partial \theta_0} (\theta - \theta_0)^2 +$$

$$O((\theta - \theta_0)^3)$$

$$= l(y, X, \theta_0) + \frac{1}{2} \frac{\partial^2 l(y, X, \theta)}{\partial \theta_0 \partial \theta_0} \text{var}(\theta_0)$$

$$\therefore \text{var}(\theta_0) = 2 \left(\frac{\partial^2 l(y, X, \theta)}{\partial \theta_0 \partial \theta_0} \right)^{-1} (l(y, X, \theta) - l(y, X, \theta_0))$$

$$(2\text{-}54)$$

Intuitively, the high variance of θ_0 implies a lower value of the second derivative,
i.e., a rounded and diffuse log-likelihood function near its maximum. Conversely, a
higher value of the second derivative of the log-likelihood function translates to a
lower variance and a sharply pointed log-likelihood curve near θ_0.

Lagrange multiplier (LM) computes the metric shown in Equation 2-55 and
compares it against a χ^2 distribution, where $l(y, \theta)$ denotes the log-likelihood

function.

$$LM = \left(\frac{\partial l(y, X, \theta)}{\partial \theta}\right)' \left(\frac{\partial^2 l(y, \theta)}{\partial \theta \partial \theta}\right) \left(\frac{\partial l(y, \theta)}{\partial \theta}\right) \sim \chi^2(N) \qquad (2\text{-}55)$$

The LM test can be used to validate the statistical significance of a randomly chosen θ vector to be the maximum of the log-likelihood function. Under the null hypothesis, the choice of a random θ gives a statistically similar value of log-likelihood as the true maxima, θ_0. Therefore, under the null hypothesis, the expression in Equation 2-55 is equivalent to $\frac{s(\theta)^2}{\text{var}(\theta)}$ which follows a $\chi^2(N)$ distribution with N degrees of freedom, where $P + 1$ denotes the number of free parameters in θ, including the constant (intercept).

2.9   Diagnostic Measures

1. **Durbin-Watson metric**: This metric tests if the residuals are correlated. It only tests for one-lag autocorrelation. If the model exhibits autocorrelation at two or higher lags, the Durbin-Watson test will not detect it. For that purpose, the **Ljung-Box** or **Box-Pierce** test should be used. The Durbin-Watson test calculates the metric shown in Equation 2-56. The metric d is approximately equal to $2(1 - \rho)$. Assuming the null hypothesis is true, $\rho = 0$, and the metric should be close to two. The test statistic is compared against a table of values, and the null hypothesis is accepted or rejected subject to a confidence level.

$$d = \frac{\sum_{i=2}^{M} \left(\hat{\epsilon}_i - \hat{\epsilon}_{i-1}\right)^2}{\sum_{i=1}^{M} \hat{\epsilon}_i^2} \qquad (2\text{-}56)$$

$$\hat{\epsilon}_i = y_i - \hat{y}_i$$

The asymptotic limit of the Durbin-Watson statistic, $2(1 - \rho)$, is derived in Equation 2-57. In deriving this limit, the number of observations, M, goes to infinity.

$$\begin{aligned}
d &= \frac{\sum_{i=2}^{M} \left(\hat{\epsilon}_i - \hat{\epsilon}_{i-1}\right)^2}{\sum_{i=1}^{M} \hat{\epsilon}_i^2} \\
&= \frac{\sum_{i=2}^{M} \hat{\epsilon}_i^2 + \hat{\epsilon}_{i-1}^2 - 2\hat{\epsilon}_i \hat{\epsilon}_{i-1}}{\sum_{i=1}^{M} \hat{\epsilon}_i^2} \\
&= 1 + \frac{\sum_{i=2}^{M} \hat{\epsilon}_{i-1}^2}{\sum_{i=1}^{M} \hat{\epsilon}_i^2} - 2\rho \frac{\sum_{i=2}^{M} \hat{\epsilon}_{i-1}^2}{\sum_{i=1}^{M} \hat{\epsilon}_i^2} \\
&\approx 2(1 - \rho)
\end{aligned} \qquad (2\text{-}57)$$

2. **Breusch-Pagan test**: This is an LM test and involves a two-step regression to compute the metric NR^2, where N denotes the number of observations or equations and R^2 is the coefficient of determination of the second OLS. This metric is compared against critical values of a χ^2 distribution, as discussed below:

   (a) Perform the primary regression $y = X\beta + \epsilon$ and calculate the standardized residuals, ϵ.

   (b) Compute the variance of residuals and divide the residuals, ϵ, by the variance. Let us denote the processed residuals having unit variance by v.

   (c) Perform the auxiliary regression $v = X\alpha + \eta$, as shown in Equation 2-58. X typically includes all the regressors of the original OLS. Calculate R^2 of the auxiliary regression.

   $$v_i = \alpha_{0,i} + \alpha_{1,i}x_{1,i} + \cdots + \alpha_{N,i}x_{N,i} + \eta_i \qquad (2\text{-}58)$$

   (d) The metric is defined as NR^2, where N denotes the number of equations. Under the null hypothesis of no autocorrelation, computed parameters $\alpha = 0$, except perhaps α_0. The metric is compared against the critical value of a $\chi^2(N)$ distribution with N degrees of freedom using the specified confidence level. If the metric is below the critical value, the null hypothesis is rejected, showing the presence of autocorrelation.

3. **White's heteroskedasticity test**: White's test is used to detect the presence of heteroskedasticity that can cause the OLS estimator to lose its efficiency, though it retains its consistency. This is an LM test that calculates a metric and compares it against a χ^2 distribution. The test begins by calculating residuals, $\hat{\epsilon}_i$, from OLS. These residuals are then plugged into an auxiliary linear model shown in Equation 2-59. The auxiliary regression includes a constant, all regressors used in the original OLS, squares of regressors used in the original OLS, and their cross products. The R^2 metric is computed using residuals from the auxiliary regression, and the test metric NR^2 is calculated, as shown in Equation 2-60, where N denotes the number of equations or observations. Under the null hypothesis of homoskedasticity, $\hat{\epsilon}_i$ should be nearly independent of the regressors and their coefficients should be close to zero. Therefore, NR^2 should behave like a sum of $P - 1$ normally distributed independent random variables with constant variance, i.e., a χ^2 distribution with $P - 1$ degrees of freedom where P denotes the number of free parameters in the auxiliary regression Equation 2-59.

$$y_i = \beta_{0,i} + \beta_{1,i}x_{1,i} + \cdots + \beta_{P,i}x_{P,i} + \hat{\epsilon}_i \text{ Original OLS}$$

$$\hat{\epsilon}_i = \delta_{0,i} + \delta_{1,i}x_{1,i} + \cdots + \delta_{P,i}x_{P,i} + \delta_{P+1,i}x_{1,i}^2 + \cdots +$$

$$\delta_{2P,i}x_{P,i}^2 + \delta_{2P+1,i}x_{1,i}x_{2,i} + \cdots + \qquad (2\text{-}59)$$

$$\delta_{2P+P(P-1)/2,i}x_{P-1,i}x_{P,i} + \hat{v}_i \text{ Auxiliary OLS}$$

$$\text{Metric} = NR^2 \sim \chi^2(2P + \frac{P(P-1)}{2} + 1)$$

(2-60)

R^2 is calculated using auxiliary OLS

4. **Box-Pierce test for correlation**: Box-Pierce is a portmanteu test, which means that it can detect the joint presence or absence of correlations up to a certain lag. This is in contrast to the Durbin-Watson test that can only detect first-order correlation. The test metric is shown in Equation 2-61 and uses h autocorrelation lags. It computes statistical estimates of autocorrelations, $\hat{\rho}_k$, ranging from 1 to h lags, using Equation 2-62. Under the null hypothesis of no autocorrelation, the metric is distributed as a $\chi^2(h)$ distribution with h degrees of freedom. This provides a statistical test to accept or reject the null hypothesis based on the metric value.

$$Q_{BP} = N \sum_{i=1}^{h} \hat{\rho}_k^2$$

(2-61)

$$\hat{\rho}_k = \frac{\sum_{i=k+1}^{N} \hat{\epsilon}_i \hat{\epsilon}_{i-k}}{N-k}$$

(2-62)

$$\hat{\epsilon}_i = y_i - \hat{y}_i$$

5. **Ljung-Box test**: The Ljung-Box test is functionally similar to the Box-Pierce test but gives better small-sample and large-sample performance than the Box-Pierce test. Due to this advantage, it's almost universally preferred over the Box-Pierce test for determining serial correlation in residuals up to a lag h. The test metric is shown in Equation 2-63. Asymptotically, it is distributed as a $\chi^2(h)$ distribution with h degrees of freedom, where h denotes the maximum autocorrelation lag to be tested.

$$Q_{LB} = N(N+2) \sum_{k=1}^{h} \frac{\hat{\rho}_k^2}{N-k} \sim \chi^2(h)$$

(2-63)

6. **F-statistic**: The F-test is used to jointly test the statistical significance of all parameters in an OLS regression. The null hypothesis is that all parameters are insignificant (i.e., 0). Rejection of the null hypothesis signifies that at least some parameters of the OLD model are significant, and therefore, the OLS furnishes a statistically significant relationship between $\mathbf{y}$ and $\mathbf{X}$. The test metric is shown in Equation 2-64. Under the null hypothesis of insignificant OLS parameters,

the metric is distributed as an F-distribution with P and $N - P - 1$ degrees of freedom. The F-distribution is the ratio of χ^2 distributions.

$$F = \frac{\sum_{i=1}^{N} (y_i - \bar{y})^2 - \sum_{i=1}^{N} \left(y_i - \hat{y}_i\right)^2}{\sum_{i=1}^{N} \left(y_i - \hat{y}_i\right)^2} \sim F(P, N - P - 1) \qquad (2\text{-}64)$$

7. **Skew and kurtosis**: Skew and kurtosis measure the degree of closeness of the residual distribution with the normal distribution. While OLS estimation does not assume a normal distribution of errors, in the absence of normality, there could be non-linear parameter estimators that are more efficient than OLS, notwithstanding the fact that OLS is BLUE. Assuming a normal distribution, skew should be close to zero, and kurtosis should be close to three. Skew and kurtosis are defined as the second and third moments of the residual distribution, respectively, as shown in Equation 2-65.

$$S = \frac{1}{N\sigma^3} \sum_{i=1}^{N} (\epsilon_i - \bar{\epsilon})^3$$

$$K = \frac{1}{N\sigma^4} \sum_{i=1}^{N} (\epsilon_i - \bar{\epsilon})^4 \qquad (2\text{-}65)$$

$$\text{where } \sigma^2 = \frac{\sum_{i=1}^{N} (\epsilon_i - \bar{\epsilon})^2}{N}$$

8. **Jarque-Bera**: The Jarque-Bera test is used to check the hypothesis that residuals follow a normal distribution. It uses skewness and kurtosis of residuals to compute the metric shown in Equation 2-66. Under the null hypothesis, the Jarque-Bera statistic, JB, has a $\chi^2(2)$ distribution with two degrees of freedom. S and K refer to skewness and kurtosis, respectively, as defined in Equation 2-65.

$$JB = \frac{N}{6} \left(S^2 + \frac{1}{4}(K - 3)^2\right) \qquad (2\text{-}66)$$

9. **Condition number**: This metric indicates how sensitive estimated parameters are to small changes in matrix $\mathbf{X'X}$. The condition number is estimated as the ratio of the largest absolute eigenvalue of the matrix to the smallest absolute eigenvalue. A large condition number indicates a near matrix collinearity, i.e., a column-rank deficiency in matrix $\mathbf{X'X}$. This is a forewarning of conditioning issues and possible linear dependence of some exogenous (independent) variables, $\mathbf{X}$.

10. **Non-parametric tests of normality**: Non-parametric tests avoid specification of parameter values of a hypothetical normal distribution and offer greater

flexibility in testing whether two distributions are identical. Two non-parametric tests are popular: Kolmogorov-Smirnov test and quantile-quantile plots.

Kolmogorov-Smirnov test: Abbreviated as KS test, it compares the CDF of two distributions and constructs a metric, shown in Equation 2-67, to discriminate between the two distributions. Under the null hypothesis of identical distributions, the metric follows a KS distribution.

$$D = \sup_x (CDF_1(x) - CDF_2(x)) \qquad (2\text{-}67)$$

Quantile-quantile plot: The quantile-quantile plot is a graphical representation of CDF of a distribution on the X-axis, plotted against the CDF of a hypothetical normal distribution on the Y-axis. Under the null hypothesis that the distribution is normal, the plot should fall on a line at a 45° angle.

11. **Variance inflation factor**: This metric quantifies the level of multicollinearity present in explanatory variables. OLS assumes that exogenous or independent variables, $\mathbf{X}$, are linearly independent. Without linear independence, $\mathbf{X'X}$ will be singular and cannot be inverted. To compute the variance inflation factor for x_j, it is regressed against other independent variables using OLS shown in Equation 2-68.

$$x_j = \beta_0 + \beta_1 x_1 + \cdots + \beta_{j-1} x_{j-1} + \beta_{j+1} x_{j+1} + \cdots + \beta_p x_p + \epsilon \qquad (2\text{-}68)$$

The variance inflation factor is then defined using Equation 2-69, where R_j^2 is computed from OLS in Equation 2-68. Values greater than ten indicate the presence of multicollinearity.

$$\text{VIF}_j = \frac{1}{1 - R_j^2} \qquad (2\text{-}69)$$

12. **Cook's distance**: This metric is used to quantify the importance of a data point in OLS regression. Functionally, it is similar to the studentized residual obtained after deleting a point in OLS, as described in a previous section. It is calculated by excluding a point, i, from regression and obtaining the fitted response values for each point, $\hat{y}_{j,(i)}$. This value is subtracted from the fitted response value obtained using all data points in an OLS, $\hat{y}_j$. Finally, the metric is computed using Equation 2-70.

$$D_i = \frac{\sum_{j=1}^{N} \left(\hat{y}_j - \hat{y}_{j,(i)} \right)^2}{P\sigma^2}$$

$$\sigma^2 = \frac{1}{N - P - 1} \sum_{k=1}^{N} \left(y_k - \hat{y}_k \right)^2 \qquad (2\text{-}70)$$

In practice, one does not run the auxiliary regression with i point excluded. Instead, one uses the analytical formula shown in Equation 2-71 that uses leverage values, $h_{i,i}$, computed using the hat matrix.

$$D_i = \frac{(y_i - \hat{y}_i)^2}{P\sigma^2} \frac{h_{i,i}}{(1 - h_{i,i})^2} \tag{2-71}$$

Values greater than one are usually regarded as influential points.

2.10   Cointegration and Error Correction Model

Generally, OLS requires the exogenous or independent variables to be I(0), i.e., weakly stationary series. This is because of the first two Gauss-Markov assumptions that imply that residuals are weakly stationary with zero mean and a constant, finite variance. This requirement generally means that exogenous variables must typically be weakly stationary, or I(0), as well. I(0) refers to integrated series of order zero, which means that the series has constant mean and variance. In contrast, an I(1) process does not have a constant mean because it has a trend. Differencing it, $x_t - x_{t-1}$, yields an I(0) or weakly stationary process.

2.10.1 Spurious Regression

If we regress I(1) variables using OLS, the regression will yield parameter estimates with high t-statistics and low p-values, indicating a highly significant relationship. R^2 will be moderate to high, adding to the false sense of significant relationship. The Durbin-Watson statistic, however, will indicate the presence of serial autocorrelation. In spite of t-statistics and F-statistic showing the presence of statistically significant relationship, the regression is spurious.

To illustrate this phenomenon, let us consider two I(1) variables, y_t and x_t, as defined in Equation 2-72.

$$y_t = y_{t-1} + \epsilon_{1,t}$$
$$x_t = x_{t-1} + \epsilon_{2,t} \tag{2-72}$$
$$\epsilon_{j,t} \sim N(0, 1) \text{ for } j = 1, 2$$

The variables are plotted in Figure 2-5. OLS yields a high R^2 of 0.998 (Listing 2-3), with estimated coefficients appearing highly significant. The Durbin-Watson statistic, however, is 0.043, indicating the presence of serial autocorrelation. This also reveals that the overly optimistic t-statistics and confidence intervals for parameter estimates may be biased.

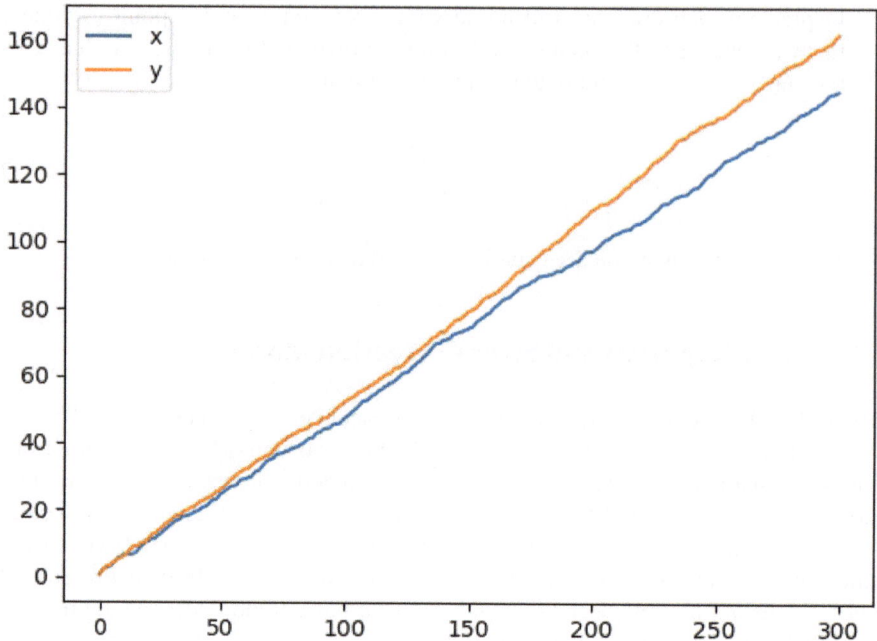

Fig. 2-5. Spurious Regression Between Two I(1) Variables

Listing 2-3. OLS Results for Spurious Regression

```
 1
 2  OLS Regression Results
 3  ====================================================
 4  Dep. Variable :                        y
 5  R−squared:                            0.998
 6  Adj. R−squared:                       0.998
 7  F− statistic :                        1.457e+05
 8  Prob (F− statistic ):                 0.00
 9  Log−Likelihood:                      −654.40
10  No. Observations :                    300
11  AIC:                                  1313.
12  Df Residuals :                        298
13  BIC:                                  1320.
14  Df Model:                             1
15  Covariance Type:               nonrobust
16  ====================================================
17        coef    std err     t      P>|t|   [0.025   0.975]
18  − − − − − − − − − − − − − − − − − − − − − − − − − − −
19  const  −2.0394 0.250    −8.151   0.000  −2.532  −1.547
20  x1      1.1296 0.003   381.725   0.000   1.124   1.135
21  ====================================================
22  Omnibus:                             16.174
23  Durbin−Watson:                        0.043
```

24	Prob(Omnibus):			0.000	
25	Jarque−Bera (JB):			15.534	
26	Skew:			−0.504	
27	Prob(JB):			0.000424	
28	Kurtosis :			2.526	
29	Cond. No.			170.	

30 ==

31

32 Notes:

33 [1] Standard Errors assume that the covariance matrix of the errors is correctly specified .

Regressing the first difference of y against the first difference of x reveals the true relationship, with the estimated parameters becoming insignificant. This should have been expected because after taking the first difference, we are regressing two stationary I(0) time series that are independent and, therefore, have no relationship to one another. Results in Listing 2-4 illustrate this fact. R^2 is close to zero, the F-statistic fails to reject the null hypothesis if no significant relationship between the dependent and independent variables, and the p-value for parameter estimate is insignificant.

Listing 2-4. OLS Results After Taking First Difference

1 OLS Regression Results

2 ==

3	R−squared:		0.000	
4	Adj. R−squared:		−0.003	
5	F− statistic :		0.1397	
6	Prob (F− statistic):		0.709	
7	Log−Likelihood:		−42.085	
8	No. Observations :		299	
9	AIC:		88.17	
10	Df Residuals :		297	
11	BIC:		95.57	
12	Df Model:		1	
13	Covariance Type:		nonrobust	

14 ==

	coef	std err	t	P>|t|	[0.025	0.975]
15						
16						
17	const	0.5493	0.030	18.109	0.000	0.490	0.609
18	x1	−0.0199	0.053	−0.374	0.709	−0.125	0.085

19 ==

20	Omnibus:		96.288	
21	Durbin−Watson:		1.959	
22	Prob(Omnibus):		0.000	
23	Jarque−Bera (JB):		16.435	
24	Skew:		−0.151	
25	Prob(JB):		0.000270	
26	Kurtosis :		1.892	
27	Cond. No.		4.12	

28 ==

29

Notes:
[1] Standard Errors assume that the covariance matrix of the errors is correctly
     specified .

2.10.2 ADF Test

The last section on spurious regression highlights the importance of ensuring the variables used in regression are I(0), i.e., weakly stationary. An augmented Dickey-Fuller test can be used to establish that a series is I(0) or, equivalently, is not I(1) and does not have a unit root. The test regresses the first difference, Δy_t on a constant, α, a trend term, βt, lagged value of the variable, y_{t-1}, and lagged first differences up to lag p, as shown in Equation 2-73.

$$\Delta y_t = \alpha + \beta t + \gamma y_{t-1} + \delta_1 \Delta y_{t-1} + \delta_2 \Delta y_{t-2} + \cdots + \delta_p \Delta y_{t-p} + \epsilon_t \qquad (2\text{-}73)$$

Under the null hypothesis of unit root presence, $\gamma \approx 0$. The ADF test calculates the metric $\frac{\gamma}{\sigma_\gamma}$, where σ_γ denotes the standard error of parameter γ. OLS estimation yields estimates of both γ and σ_γ. Under the null hypothesis, the computed metric follows a Dicker-Fuller distribution. Rejection of the null hypothesis is regarded as proof that series y_t is I(0), i.e., does not have a unit root.

2.10.3 Cointegration

The only exception to the requirement for variables in OLS to be I(0) variables occurs when there is a cointegrating relationship present among the variables. This is defined as the existence of a vector β that makes $y - \mathbf{X}\beta \sim I(0)$ become a weakly stationary series with no unit root, as shown in Equation 2-74. The vector $[1, \beta]$ is called the cointegration vector. When cointegration is present, OLS estimates are super-consistent and all parameter estimates and confidence intervals are valid. Cointegrating vector is unique only up to a multiplicative constant.

$$y - \mathbf{X}\beta \sim I(0)$$
$$y, \mathbf{X} \sim I(1) \qquad (2\text{-}74)$$

2.10.4 Error Correction Model

Every cointegration relationship implies the existence of an error correction model. Viewing cointegration relation as a long-run equation, deviations from the long-run equation must give rise to correction terms that drive the variables in a direction toward the restoration of a long-run relationship. This is known as the error correction model.

Let us assume y_t and x_t are I(1) variables with $y_t - \beta x_t \sim I(0)$ as the cointegration equation. If $y_t - \beta x_t$ begins increasing at a time instant, the following time instant y_{t+1} should be pulled toward βx_{t+1} for the cointegration equation to hold over the long term. This leads to the formulation of an error correction model, as shown in Equation 2-75.

$$\Delta y_t = \alpha_1 + \beta_1 (y_{t-1} - \beta x_{t-1}) + \sum_{k=1}^{p} \gamma_{1,k} \Delta y_{t-k} + \sum_{k=1}^{p} \gamma_{2,k} \Delta x_{t-k} + \epsilon_{1,t}$$

$$\Delta x_t = \alpha_2 + \beta_2 (y_{t-1} - \beta x_{t-1}) + \sum_{k=1}^{p} \gamma_{3,k} \Delta y_{t-k} + \sum_{k=1}^{p} \gamma_{4,k} \Delta x_{t-k} + \epsilon_{2,t}$$

$$(2\text{-}75)$$

In order to estimate the error correction model, one must first estimate the cointegration vector, β. This is done by running an OLS on $y_t = \alpha + x\beta + \epsilon$. With the estimate of β in hand, equations comprising the error correction model in Equation 2-75 are estimated, once again using OLS.

2.11 Instrumental Variables

First Gauss-Markov assumption assumes that errors are uncorrelated with exogenous (independent) variables. If this assumption is violated, OLS estimates will be biased and inconsistent. Instrumental variables help remediate the problem when error terms are correlated with one or more independent variables, x. To see why the presence of correlation between error terms and independent variables causes OLS to be biased and inconsistent, let us consider a two-variable OLS, $y = \alpha + \beta x + \epsilon$. OLS minimizes $(y - \alpha - \beta x)^2$ with respect to parameters α and β. Differentiating with respect to β yields $x(y - \alpha - \beta x) = 0$, i.e., $x\epsilon = 0$. This shows that OLS tries to select residuals that are orthogonal to independent variable x. The computed value of parameter β, denoted as $\hat{\beta}$, is computed as shown in Equation 2-76. As seen from Equation 2-76, computed parameter estimate $\hat{\beta}$ does not converge to the true parameter value β, causing OLS to be biased and inconsistent because $\text{cov}(x, \epsilon) \neq 0$.

$$\hat{\beta} = \frac{\text{cov}(y, x)}{\text{var}(x)}$$

$$= \frac{\text{cov}(\alpha + \beta x + \epsilon, x)}{\text{var}(x)}$$

$$= \frac{\text{cov}(\alpha + \beta x, x)}{\text{var}(x)} + \frac{\text{cov}(\epsilon, x)}{\text{var}(x)} \tag{2-76}$$

$$= \beta + \frac{\text{cov}(\epsilon, x)}{\text{var}(x)}$$

In practice, correlation between independent variables and error terms could arise because of a missing variable in the OLS equation that is correlated with an exogenous variable, x. A missing variable causes the residual to absorb the values of that variable, creating correlation between the exogenous variable and residual.

Instrumental variables, or IVs, involve selecting another input variable, z, that is correlated with the input variable x but is uncorrelated with the error term, ϵ. This is followed by a two-step regression procedure, as described below:

1. Regress the instrumental variable, z, on the original regressors, x, as shown in Equation 2-77.

$$\mathbf{X} = \mathbf{Z}\delta + u$$

Z is correlated with X and uncorrelated with u

$$\therefore \delta = \left(\mathbf{Z'Z}\right)^{-1} \mathbf{Z'X} \tag{2-77}$$

$$\therefore \hat{\mathbf{X}} = \mathbf{Z}\left(\mathbf{Z'Z}\right)^{-1} \mathbf{Z'X}$$

$$= \mathbf{P_z X}$$

In Equation 2-77, $\mathbf{P_z}$ is the orthogonal projector for Z. We obtain a projection of exogenous variables, $\hat{\mathbf{X}}$, on subspace spanned by Z.

2. Regress the original dependent variable, y, on $\hat{\mathbf{X}}$ obtained from step 1, as shown in Equation 2-78, to obtain unbiased and consistent parameter estimates of $\boldsymbol{\beta}$.

$$y = \hat{\mathbf{X}}\boldsymbol{\beta} + \epsilon$$

$$\boldsymbol{\beta} = \left(\hat{\mathbf{X}}'\hat{\mathbf{X}}\right)^{-1} \hat{\mathbf{X}}'y$$

$$= \left(\mathbf{X'P_z'P_z X}\right)^{-1} \mathbf{X'P_z y} \tag{2-78}$$

$$= \left(\mathbf{X'P_z X}\right)^{-1} \mathbf{X'P_z y}$$

because P_z is idempotent

Generalized Linear Model

3

A generalized linear model, or GLM, extends the linear model to cover certain non-linear modeling tasks. The extension of the linear model is necessitated by the following functional requirements:

1. Model output could be range bound. For example, a model that predicts the classification of a data point into a class is going to produce a label belonging to a discrete set of classes. Similarly, a model that predicts the probability of an event for an observation will produce an output that lies between zero and one. A linear model, on the other hand, has a default range from $-\infty$ to ∞.
2. There could be a non-linear relationship between model output and exogenous (independent) variables. A linear model formulates the model output as a dot product of exogenous variables with a parameter vector. One could retain the linear product between the parameter vector and exogenous variables, but the output could be a non-linear function of the dot product.
3. Error distribution of residuals is assumed to be independent and identically distributed (IID) and weakly stationary. In many applications, the error distribution is assumed to be normal. Could this error distribution belong to a parametric family of distributions? Certain modeling tasks seem naturally compatible with certain distributions. For example, a model that predicts an event's probability of occurrence is likely to follow a binomial distribution, with the event occurring or not occurring in a training dataset.

© Samit Ahlawat 2025
S. Ahlawat, *Statistical Quantitative Methods in Finance*,
https://doi.org/10.1007/979-8-8688-0962-0_3

A generalized linear model addresses the foregoing shortcomings of a linear model by introducing the following enhancements:

1. A non-linear function, g, maps the output, y, to the linear part of the predictor, denoted by $\eta = \mathbf{X}\boldsymbol{\beta}$. This is shown in Equation 3-1.

$$\eta = \mathbf{X}\boldsymbol{\beta} = \beta_0 + \sum_{i=1}^{p} \beta_i x_i$$

$$= g(E[y])$$

(3-1)

To compare and contrast, in a linear model, g was a unit function with $E[y] = \mathbf{X}\boldsymbol{\beta}$ or $y = \mathbf{X}\boldsymbol{\beta} + \epsilon$. g is called the link function because it links the output to the linear predictor, η.

2. Function g has a range spanning $(-\infty, \infty)$ and domain specific to the problem. For example, if the modeling task involves predicting a probability, the domain of g would be [0, 1]. Similar to the linear model, parameters $\boldsymbol{\beta}$ only occur as a linear combination with exogenous variables, $\mathbf{X}$.

3. The probability distribution of output follows an exponential class of probability distributions, as shown in Equation 3-2. Equation 3-2 gives the **general form** of the exponential family of distributions.

$$f(y; \theta) = h(y) \exp\left(\eta(\theta)T(y) - A(\theta)\right)$$

(3-2)

In Equation 3-2, $T(y)$ is called **sufficient statistic** because it represents how data impacts the distribution through model parameters, θ. $\eta(\theta)$ is called the **natural parameter**. $A(\theta)$ is referred to as the **log-partition function** because it represents a normalization constant that ensures that the probability density function adds to 1, i.e., $\int f(y; \theta)dy = 1$. $h(y)$ is known as the **tilting parameter**. The **natural form** of the exponential family of distributions is obtained from the general form by setting $\eta(\theta) = \theta$ and $T(y) = y$, as illustrated in Equation 3-3. An additional **dispersion parameter** is introduced in natural form.

$$f(y; \theta, \phi) = \exp\left(\frac{y\theta - b(\theta)}{a(\phi)} + c(y, \phi)\right)$$

(3-3)

The exponential class of distributions is fairly generic. It includes Gaussian, Poisson, binomial, gamma, and inverse gamma distributions, as shown in a subsequent section. The product of model parameters and exogenous (independent) variables is represented by θ. $a(\phi)$ is a scale function that determines the variance. $c(y, \phi)$ is a normalizing constant that ensures the integral of the probability distribution function over θ sums to 1. Hence, it is a function of y and ϕ.

3.1   Score Equations

Score equations are applicable to a log-likelihood function and furnish a generic method for calculating the mean and variance of parameter estimates. There are two score equations, as described below. Let $l(\theta, \phi; y)$ denote the log-likelihood function corresponding to the likelihood function $f(y; \theta, \phi)$ as shown in Equation 3-4. θ denotes a vector of model parameters related to mean, while ϕ denotes model parameters related to scale or variance. In a GLM, only model parameters denoted by θ are calibrated to fit the data. ϕ is specified by the specific probability distribution function.

$$l(\theta, \phi; y) = \log f(y; \theta, \phi) \tag{3-4}$$

1. Maximum likelihood involves finding a maxima of the log-likelihood function with respect to model parameters, θ, as shown in Equation 3-5.

$$E\left[\frac{\partial l(\theta, \phi; y)}{\partial \theta}\right] = 0 \tag{3-5}$$

For the exponential family of probability distributions, we have the log-likelihood function as $\frac{y\theta - b(\theta)}{a(\phi)} + c(y, \phi)$. Applying Equation 3-5 gives Equation 3-6.

$$E[y] = \frac{db(\theta)}{d\theta} = b'(\theta) \tag{3-6}$$

2. The variance of the log-likelihood function can be related to the second derivative. Because we are finding a maxima for the log-likelihood function, its second derivative should be negative. This gives Equation 3-7.

$$E\left[\frac{\partial^2 l(\theta, \phi; y)}{\partial \theta^2}\right] + E\left[\left(\frac{\partial l(\theta, \phi; y)}{\partial \theta}\right)^2\right] = 0 \tag{3-7}$$

Applying Equation 3-7 to the exponential family of probability distributions, we get Equation 3-8. Equation 3-8 uses the result from Equation 3-6 to derive an expression for the variance of y.

$$E\left[\left(\frac{y - b'(\theta)}{a(\phi)}\right)^2\right] + E\left[-\frac{b''(\theta)}{a(\phi)}\right] = 0$$

$$E\left[(y - E[y])^2\right] = b''(\theta)a(\phi) \tag{3-8}$$

$$\therefore \operatorname{var}(y) = b''(\theta)a(\phi)$$

The expression for the mean and variance of y shown in Equation 3-8 can be derived using the probability density function. Recall that the probability density function integrates to 1 over y as shown in Equation 3-9.

$$\int_{-\infty}^{\infty} \exp\left(\frac{y\theta - b(\theta)}{a(\phi)} + c(y, \phi)\right) dy = 1$$

$$\therefore \exp\frac{b(\theta)}{a(\phi)} = \int_{-\infty}^{\infty} \exp\left(\frac{y\theta}{a(\phi)} + c(y, \phi)\right) dy$$

(3-9)

Differentiating Equation 3-9 with respect to θ, we get the expression for the mean of y, as shown in Equation 3-10.

$$\exp\left(\frac{b(\theta)}{a(\phi)}\right)\frac{b'(\theta)}{a(\phi)} = \int_{-\infty}^{\infty} \frac{y}{a(\phi)} \exp\left(\frac{y\theta}{a(\phi)} + c(y, \phi)\right) dy$$

$$\therefore \frac{b'(\theta)}{a(\phi)} = \int_{-\infty}^{\infty} \frac{y}{a(\phi)} \exp\left(\frac{y\theta - b(\theta)}{a(\phi)} + c(y, \phi)\right) dy$$

$$\therefore b'(\theta) = E[y]$$

(3-10)

Differentiating Equation 3-9 a second time with respect to θ, we get the expression for the variance of y, as shown in Equation 3-11. We have used the fact that $b'(\theta) = E[y]$ in deriving Equation 3-11.

$$\exp\left(\frac{b(\theta)}{a(\phi)}\right)\left(\left(\frac{b'(\theta)}{a(\phi)}\right)^2 + \frac{b''(\theta)}{a(\phi)}\right) = \int_{-\infty}^{\infty} \frac{y^2}{a(\phi)^2} \exp\left(\frac{y\theta}{a(\phi)} + c(y, \phi)\right) dy$$

$$\therefore \frac{b''(\theta)}{a(\phi)} = \int_{-\infty}^{\infty} \frac{y^2 - (E[y])^2}{a(\phi)^2} \exp\left(\frac{y\theta - b(\theta)}{a(\phi)} + c(y, \phi)\right) dy$$

$$\therefore b''(\theta)a(\phi) = E\left[y^2 - (E[y])^2\right] = \text{var}(y)$$

(3-11)

3.2   Exponential Family of Probability Distributions

The exponential family of probability distributions constitutes a parametric family defined by Equation 3-3. Normal, Poisson, binomial, gamma, and inverse gamma probability distributions belong to this parametric family. It is instructive to write the probability distribution function of each of these distributions and compare it with the canonical form shown in Equation 3-3. Doing so allows us to immediately apply Equations 3-6 and 3-8 to derive expressions for the expected value and variance of model output y.

The canonical link function is obtained by setting $\eta = \mathbf{X}\boldsymbol{\beta} = b'(\theta)$.

Table 3-1. Commonly Used GLM Models

Data Distribution	Range of Output	Canonical Link Name	Canonical Link $g(\mu) = \mathbf{X}\beta$
Normal	$\mathcal{R} \in (-\infty, \infty)$	identity	$\mu = \mathbf{X}\beta$
Poisson	Integers $\in (0, 1, \cdots, \infty)$	log	$\log \mu = \mathbf{X}\beta$
Gamma	$\mathcal{R} \in (0, \cdots, \infty)$	negative inverse	$-\frac{1}{\mu} = \mathbf{X}\beta$
Inverse gamma	$\mathcal{R} \in (0, \cdots, \infty)$	inverse square	$\frac{1}{\mu^2} = \mathbf{X}\beta$
Binomial	Integers $\in (0, \cdots, N)$	logit	$-\frac{\mu}{N-\mu} = \mathbf{X}\beta$
Bernoulli	Integer $\in (0, 1)$	logit	$-\frac{\mu}{1-\mu} = \mathbf{X}\beta$
Multinomial	Integer tuple $(p_1, \cdots, p_K)$ with $\sum_{i=1}^{K} p_i = 1$ Integer p_i 0 or 1	logit	$-\frac{\mu_i}{1-\mu_i} = \mathbf{X}\beta_i$. $\sum_{i=1}^{K} \mu_i = 1$
Normal CDF (Probit)	Integer $\in (0, 1)$	normal CDF inverse	$\Phi^{-1}(\mu) = \mathbf{X}\beta$ $\Phi(z) =$ $\int_{-\infty}^{z} \frac{1}{\sqrt{2\pi}} \exp\left(-\frac{x^2}{2}\right) dx$

A concise table showing different kinds of generalized linear models with their data distributions and canonical link functions is shown in Table 3-1. $\mathcal{R}$ denotes the space of real numbers. log denotes natural logarithm.

3.2.1 Normal Distribution

A normal distribution's probability density function can be written in the form of an exponential distribution as shown in Equation 3-12. θ denotes the mean of the distribution and σ^2 is its variance. Figure 3-1 shows the effect of changing σ for $\theta = 0$. Changing θ centers the distribution around a different mean.

$$f(y_i; \theta_i, \phi) = \frac{1}{\sqrt{2\pi\sigma^2}} \exp\left(-\frac{(y_i - \theta_i)^2}{2\sigma^2}\right) \text{ where } \theta_i = x_i \beta_i$$

$$l(\theta_i, \phi; y_i) = -\frac{1}{2} \log 2\pi\sigma^2 - \frac{(y_i - \theta_i)^2}{2\sigma^2}$$

$$= -\frac{1}{2} \log 2\pi\sigma^2 - \frac{y_i^2 + \theta_i^2 - 2y_i\theta_i}{2\sigma^2} \tag{3-12}$$

$$= \frac{y_i\theta_i - \frac{\theta_i^2}{2}}{\sigma^2} - \frac{y_i^2}{2\sigma^2} - \frac{1}{2} \log 2\pi\sigma^2$$

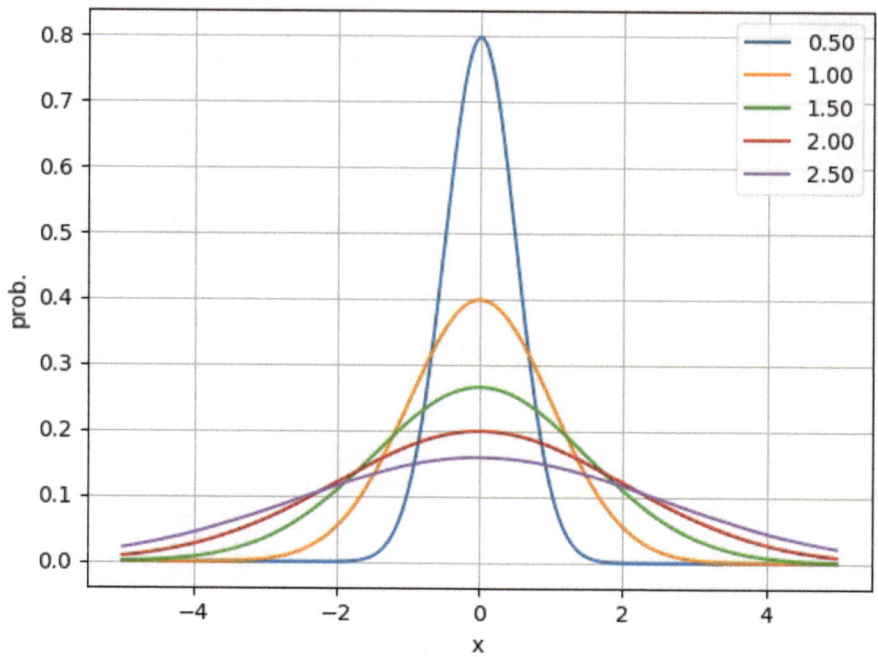

Fig. 3-1. Normal Distribution Using Mean $= 0$

$$\therefore b(\theta_i) = \frac{\theta_i^2}{2}$$

$$\text{and } a(\phi) = \sigma^2$$

Applying Equations 3-6 and 3-8 to the normal distribution gives the expression for the expected value and variance of y, as shown in Equation 3-13.

$$E[y] = b'(\theta) = \theta$$

$$\text{var}(y) = \sigma^2 \tag{3-13}$$

The canonical link function for a normal distribution is the identity function. Setting $\eta = \mathbf{X}\boldsymbol{\beta}$ equal to $\theta = E[y]$, we see that $E[y] = \eta$, i.e., the identity link function.

Application

In this example, let us predict the daily change in the yield of 10-year constant maturity treasury bonds, Δy_{10Yr}. Let us use the daily change in the yield of 1-month treasury bills (Δy_{3Mo}), the daily change in the yield of 30-year treasury bills (Δy_{30Yr}), and the daily return on S&P 500 (r_m) as predictor variables (exogenous variables). The model is formulated in Equation 3-14.

$$\Delta y_{10Yr}(t) = \alpha + \beta_1 \Delta y_{1Mo}(t) + \beta_2 \Delta y_{30Yr}(t) + \beta_3 r_m(t) + \epsilon_t$$

$$r_m = \frac{P_{SP500}(t) - P_{SP500}(t-1)}{P_{SP500}(t-1)} \tag{3-14}$$

Data from 2014 to 2024 is used for the purpose, with the first 90% of observations used for training (i.e., fitting) the model and the remaining 10% used for testing. Statistics related with model fitting are shown in Listing 3-1. As seen from the statistics, the R^2 of the model is 0.99. However, only the change in 30-year bond yield is statistically significant with 95% confidence. The market return has a negative correlation with the change in 10-year bond yield, as can be seen from the negative coefficient, –0.54. This conforms with the general observation that when equity markets rise, yields on long maturity bonds fall and vice versa.

Plots of residuals from the training dataset show that the error distribution is close to normal, as seen in Figure 3-2.

In the test dataset, the error distribution of residuals is close to the normal distribution as well, as can be seen from Figure 3-3. We can test the assumption of normality of residuals more formally using the Kolmogorov-Smirnov test. As shown in Listing 3-1, the KS test for test data yields a p-value of 0.1875. Therefore, the null hypothesis of normality cannot be rejected at 95% confidence interval because the p-value is greater than 0.05.

The code for fitting and testing the model is shown in Listing 3-2.

Listing 3-1. OLS Results for Spurious Regression

```
Generalized Linear Model Regression Results
===============================================================
Dep. Variable :     y   No. Observations :          2346
Model:            GLM Df Residuals :                 2342
Model Family:   Gaussian   Df Model:                   3
Link Function:  identity   Scale:              0.00033633
Method:              IRLS  Log–Likelihood:        6054.2
Date:    Sat, 10 Feb 2024  Deviance:             0.78768
Time:              11:00:43  Pearson chi2:          0.788
No. Iterations :       3   Pseudo R–squ. (CS): 0.9976
Covariance Type: nonrobust
===============================================================
          coef   std err     z      P>|z|   [0.025   0.975]
---------------------------------------------------------------
const   0.0004   0.000    1.041     0.298   -0.000   0.001
x1      0.0088   0.011    0.784     0.433   -0.013   0.031
x2      0.9473   0.008  115.201     0.000    0.931   0.963
x3     -0.0540   0.035   -1.549     0.121   -0.122   0.014
===============================================================
```

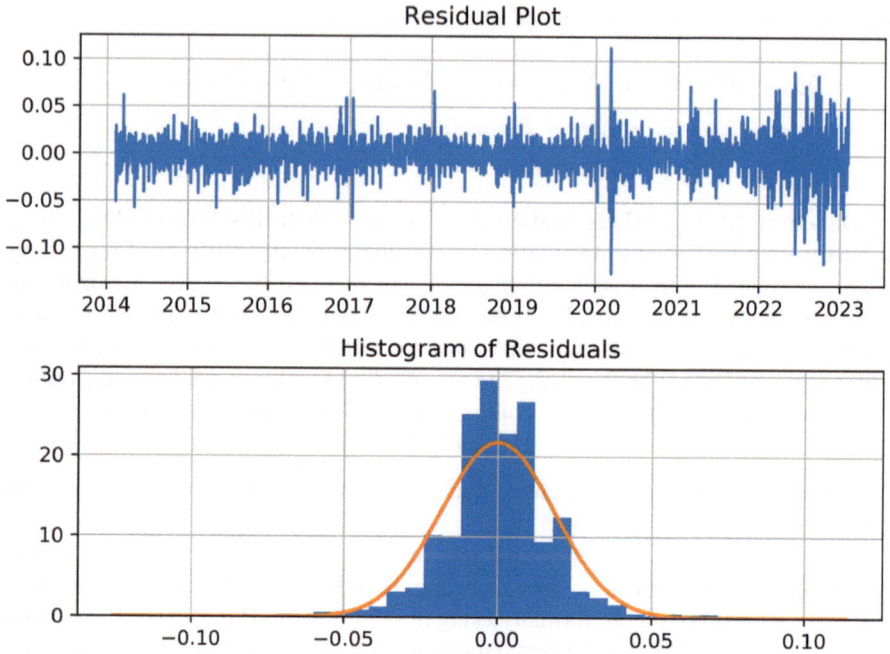

Fig. 3-2. Residual Plots of GLM with Normal Distribution in Training Dataset

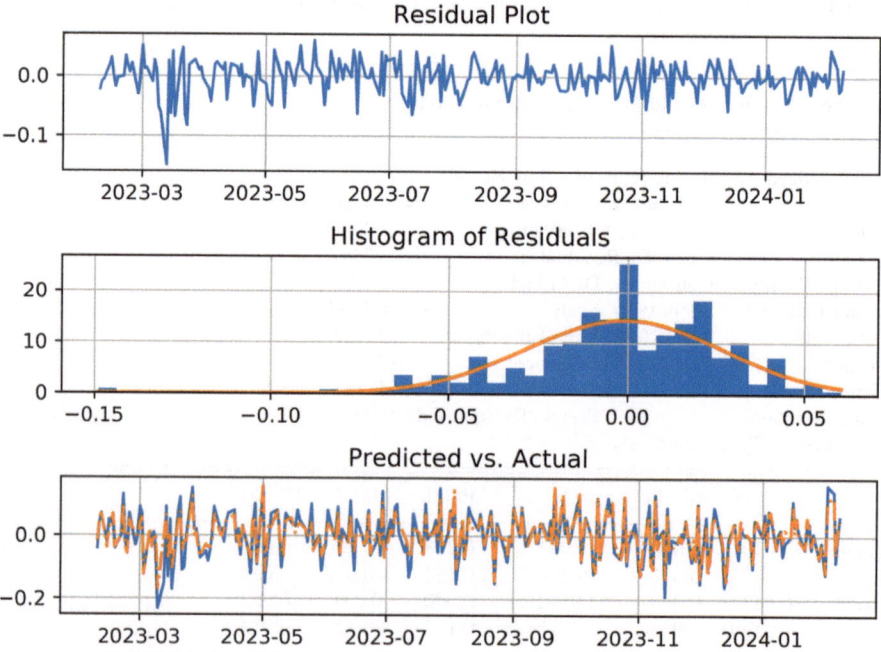

Fig. 3-3. Residual Plots of GLM with Normal Distribution in Test Dataset

Listing 3-2. Normal Generalized Linear Model

```
 1   import numpy as np
 2   import pandas as pd
 3   import statsmodels.api as sm
 4   import matplotlib.pyplot as plt
 5   import os
 6   import logging
 7   import scipy.stats as ss
 8
 9
10   DATADIR = r"C:\prog\cygwin\home\samit_000\latex\book_stats\code\data"
11   PLOTDIR = r"C:\prog\cygwin\home\samit_000\latex\book_stats\code\plots"
12   logging.basicConfig(level=logging.DEBUG)
13
14
15   class NormalGLM(object):
16       def __init__(self, endogName, exogNames, trainPerc=0.9):
17           filename = os.path.join(DATADIR, endogName + ".csv")
18           y = pd.read_csv(filename, parse_dates=["DATE"])
19           for xi in exogNames:
20               filename = os.path.join(DATADIR, xi + ".csv")
21               x = pd.read_csv(filename, parse_dates=["DATE"])
22               y = pd.merge(y, x, on=["DATE"], how="inner")
23
24           y.replace(".", np.nan, inplace=True)
25           for col in y.columns:
26               if col != "DATE":
27                   y.loc[:, col] = y.loc[:, col].astype(np.float64)
28           y.ffill(inplace=True)
29           self.endog = endogName
30           self.exog = exogNames
31           y = self.calculatePercChange(y)
32           self.df = y
33
34           self.testdata = int(trainPerc * self.df.shape[0]) - 1
35           self.logger = logging.getLogger(self.__class__.__name__)
36           self.model = None
37
38       def calculatePercChange(self, y):
39           yval = y.loc[:, self.endog].values
40           ygrowth = yval[1:] - yval[0:-1]
41           xgrowth = []
42           for x in self.exog:
43               xval = y.loc[:, x].values
44               if x == "SP500":
45                   xgrowthi = xval[1:]/xval[0:-1] - 1
46               else:
47                   xgrowthi = xval[1:] - xval[0:-1]
48               xgrowth.append(xgrowthi)
49           datadict = {"DATE": y.DATE[1:], self.endog: ygrowth}
50           for i, x in enumerate(self.exog):
51               datadict[x] = xgrowth[i]
52           return pd.DataFrame(datadict)
```

```
53
54    def  fit ( self ):
55        y = self .df. loc [: self . testdata ,  self .endog]. values
56        X = self .df. loc [: self . testdata ,  self .exog]. values
57        X = sm.add_constant(X, has_constant="add")
58        glm = sm.GLM(y, X, family=sm.families.Gaussian( link=sm. families . links . identity
              ( )))
59        glm = glm. fit ()
60        self . logger . info (glm.summary())
61        summaryfile = os. path . join (PLOTDIR, self.__class__.__name__ + ".txt")
62        with  open(summaryfile,  'w')  as  fh:
63            fh . write (glm.summary(). as_text ())
64        self .model = glm
65
66    def  plotResid ( self ):
67        fig , axs = plt . subplots (nrows=2, ncols=1)
68        resid  = self .model. resid_response
69        meanval = resid .mean()
70        sd = resid . std ()
71        resid_std  = ( resid  − meanval)/sd
72        res = ss . kstest ( resid_std ,  ss .norm.cdf)
73        self . logger . info ( res )
74
75        xv = np. linspace ( resid .min(),  resid .max(), 100)
76        yv = ss .norm.pdf(xv, meanval, sd)
77        dates  = self .df. loc [0: self . testdata ,  "DATE"].values
78        axs [0]. plot ( dates ,  resid )
79        axs [0]. grid ()
80        axs [0].  set_title ("Residual  Plot")
81        axs [1]. hist ( resid ,  bins=40, density =True)
82        axs [1]. plot (xv, yv,  lw=2)
83        axs [1]. grid ()
84        axs [1].  set_title ("Histogram of  Residuals")
85        plt . tight_layout ()
86        plt . savefig (os. path . join (PLOTDIR, "trainResidNormal.jpeg"),
87                      dpi=500)
88
89    def  plotTestResults ( self , y, ypred):
90        fig , axs = plt . subplots (nrows=3, ncols=1)
91        resid  = y − ypred
92        meanval = resid .mean()
93        sd = resid . std ()
94        resid_std  = ( resid  − meanval)/sd
95        ksres = ss . kstest ( resid_std ,  ss .norm.cdf)
96        self . logger . info ( ksres )
97        xv = np. linspace ( resid .min(),  resid .max(), num=100)
98        yv = ss .norm.pdf(xv, loc=meanval, scale =sd)
99        dates  = self .df. loc [ self . testdata +1:,  "DATE"].values
100       axs [0]. plot ( dates ,  resid )
101       axs [0]. plot ()
102       axs [0]. grid ()
103       axs [0].  set_title ("Residual  Plot")
104       axs [1]. hist ( resid ,  bins=40, density =True)
```

```
105        axs [1]. plot (xv, yv, lw=2)
106        axs [1]. grid ()
107        axs [1]. set_title ("Histogram of Residuals")
108        axs [2]. plot (dates, y, label="y")
109        axs [2]. plot (dates, ypred, "-.", label="ypred")
110        axs [2]. grid ()
111        axs [2]. set_title ("Predicted vs. Actual")
112        plt . tight_layout ()
113        plt . savefig (os. path . join (PLOTDIR, "testResidNormal.jpeg"),
114                    dpi=500)
115
116    def  test ( self ):
117        testdata  =  self . testdata  + 1
118        y = self . df. loc [ testdata  :,  self . endog]. values
119        X = self . df. loc [ testdata  :,  self . exog]. values
120        X = sm.add_constant(X, has_constant="add")
121        ypred  =  self . model. predict (X)
122        self . plotTestResults (y,  ypred)
123
124
125 if __name__ == "__main__":
126    normal = NormalGLM("DGS10", ["DGS1MO", "DGS30", "SP500"])
127    normal. fit ()
128    normal. plotResid ()
129    normal. test ()
```

Code Explanation

Salient features of the code are explained below. Let us step through the code in the order of execution:

1. Inside the main section, the code creates an instance of class **NormalGLM**, providing an endogenous variable and a list of exogenous variables as arguments.
2. The endogenous variable is **DGS10**. This variable name represents a dataset containing the daily closing yields on US government nominal bonds with ten-year maturity. The data is obtained from the FRED database made available by the Federal Reserve Bank of St. Louis at [6]. This is raw data; it is processed to create the final endogenous variable.
3. Data for constructing exogenous variables is a list containing the following datasets: **DGS1MO**, **DGS30**, and **SP500**. **DGS1MO** is the dataset containing the daily closing yields on US government nominal bills with 1-month maturity. **DGS30** is the dataset containing the daily closing yields on US government nominal bonds with 30-year maturity, and **SP500** contains daily closing prices for the S&P 500 index.
4. Inside the constructor of class **NormalGLM**, the code reads the comma-separated files using the pandas library and joins the dataframes on date.
5. After some data cleanup such as performing a forward fill for unavailable values represented as ".", it calculates the daily growth rate of S&P 500.
6. For the endogenous variable, it calculates the daily change in yield.

7. For rate-based exogenous variables (all exogenous variables except S&P 500 growth rate), it calculates the daily change in yield and uses the difference as exogenous variables, as explained earlier.
8. It fits the GLM model using the **GLM** class from the **statsmodels** library and calling the **fit** method on it. It uses the **Gaussian** family of link functions in order to fit a normal GLM model. This family takes an argument specifying the link function to use. The default link function for the **Gaussian** family is the **identity** function. The code uses identity as the link function, provided as an argument to the **Gaussian** family's constructor. This argument is provided for clarity; it is not required due to the fact that this is the default link function for the **Gaussian** family.
9. It writes the summary statistics from the model fitting step.
10. Following this, it plots the residuals obtained on the training dataset and uses a Kolmogorov-Smirnov test to check for the normality of residuals. The API is available using the **kstest** method from the **scipy** library.
11. Following this, the code makes predictions on the test dataset, predicting the change in yield on ten-year constant maturity government bonds, and plots the results against observed values.

3.2.2   Poisson Distribution

The Poisson distribution is used to model the probability of observing a specified number of event occurrences when there is a known average number of occurrences per unit time. It is a discrete probability over the set of nonnegative integers which represents the number of occurrences, as shown for a few examples in Figure 3-4. For discrete probabilities, the equivalent of probability density is called probability mass function. The probability mass function of the Poisson distribution is shown in Equation 3-15, along with an illustration of why it belongs to the exponential family of probability distributions.

$$f(y; \theta, \phi) = \frac{e^{-\mu}\mu^y}{y!} = \exp\left(-\mu + y\log\mu - \log y!\right)$$

$$l(\theta, \phi; y) = y\log\mu - \mu - \log y!$$

$$\therefore \theta = \log\mu \text{ or } \exp(\theta) = \mu$$

$$b(\theta) = \exp(\theta) \tag{3-15}$$

$$a(\phi) = 1$$

$$E[y] = b'(\theta) = \exp(\theta) = \mu$$

$$\mathrm{var}(y) = b''(\theta)a(\phi) = \exp(\theta) = \mu$$

Equation 3-15 shows that the mean and variance of the Poisson distribution are both equal to μ. The canonical link function is obtained by setting $\eta = \theta = \log\mu =$

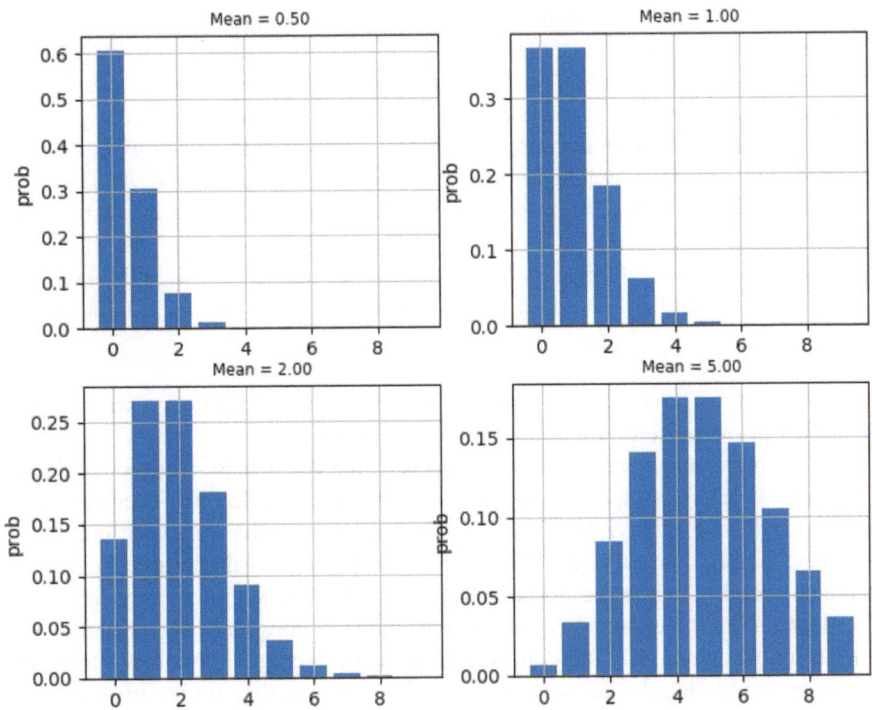

Fig. 3-4. Poisson Distribution

$\log(E[y])$, i.e., a log link function. The range of the Poisson distributed variable is $[0, \infty]$, and the logarithm function transforms the range to $[-\infty, \infty]$.

Application

Let us estimate the delinquency rate on credit card debt in the United States using the Poisson distribution. Credit card delinquency is a key econometric data tracked by economists to gauge the financial health of consumers. Rising credit card delinquencies point to weakening of the consumer's ability to service credit card debt, which could be a precursor to a decline in consumer spending and economic growth. Because the delinquency rate cannot be negative, let us use the log link. Furthermore, let us normalize the delinquency rate by weekly credit card debt and revolving debt. This normalization is necessary because greater credit card debt origination will increase the delinquency rate, everything else being the same.

First, let us select the explanatory variables. This choice is governed by data and intuitive understanding of the drivers behind the response variable. Let us use the following explanatory variables:

1. Quarterly GDP growth rate, $\frac{GDP(t)-GDP(t-1)}{GDP(t-1)}$.

2. Change in the interest rate charged by banks on credit card debt relative to 2-year (or 8-quarter) average, $\Delta \text{INT}(t)$.
3. Change in the PCE price index relative to prior 2-year average, $\Delta PCE(t)$. Personal consumption expenditure (PCE) is the Federal Reserve's preferred metric to track inflation because it excludes volatile food and energy prices.
4. Change in real disposable income relative to prior 2-year average, $\Delta \text{RDI}(t)$.
5. Change in the interest rate on 30-year fixed-rate mortgage relative to 2-year average, $\Delta 30\text{YrMort}(t)$.
6. Indicator variable to flag years before the Great Financial Crisis (GFC) of 2009, $I(t)$. This variable is required because prior to GFC, lending standards were relatively lax.

The model is formulated in Equation 3-16. All the explanatory variables must be known prior to the prediction of the response variable for a period in order to avoid in-sample bias. Credit card delinquency rate divided by weekly credit card debt issuance $\mu(t)$ is the response variable and is available once each quarter (3 months) with one quarter lag. The interest rate charged by banks on credit card debt is available quarterly. The PCE price index and real disposable income are available monthly, while the 30-year mortgage rate is available weekly. Therefore, in order to align the exogenous variables available at non-quarterly frequency with ones available at quarterly interval, we calculate the average of monthly and weekly variables within the relevant quarter. N_w and N_m denote the number of weekly and monthly observations in a quarter in Equation 3-16.

$$\log(\mu(t)) = \alpha + \beta_1 \frac{GDP(t) - GDP(t-1)}{GDP(t-1)} + \beta_2 \Delta\text{INT}(t) + \beta_3 \Delta\text{PCE}(t) +$$

$$\beta_4 \Delta\text{RDI}(t) + \beta_5 \Delta 30\text{YrMort}(t) + \beta_6 I(t) + \epsilon_t$$

$$\Delta\text{INT}(t) = \text{INT}(t) - \overline{INT}(t)$$

$$\overline{INT}(t) = \frac{\sum_{i=1}^{8} \text{INT}(t-i)}{8}$$

$$(3\text{-}16)$$

In order to verify that the endogenous variable follows the Poisson distribution, we calculate the mean and standard deviation of credit card delinquency rate normalized by debt issuance for the dataset. The mean is 0.018194 and the standard deviation is 0.013455, and the two values are close enough to justify the assumption of the Poisson distribution. The code for fitting the model and testing it on test data is shown in Listing 3-3.

Listing 3-3. Poisson Generalized Linear Model

```python
import numpy as np
import pandas as pd
import statsmodels.api as sm
import matplotlib.pyplot as plt
import os
import logging
import scipy.stats as ss


DATADIR = r"C:\prog\cygwin\home\samit_000\latex\book_stats\code\data"
PLOTDIR = r"C:\prog\cygwin\home\samit_000\latex\book_stats\code\plots"
logging.basicConfig(level=logging.DEBUG)


class PoissonGLM(object):
    def __init__(self, endogName, exogNames, trainPerc=0.9):
        filename = os.path.join(DATADIR, endogName + ".csv")
        y = pd.read_csv(filename, parse_dates=["DATE"])
        y.loc[:, "year"] = y.DATE.dt.year
        y.loc[:, "month"] = y.DATE.dt.month
        y.loc[:, "quarter"] = (y.month.values - 1) // 3
        self.convertColumnToFloat(y, endogName)
        for xi in exogNames:
            filename = os.path.join(DATADIR, xi + ".csv")
            x = pd.read_csv(filename, parse_dates=["DATE"])
            x.loc[:, "year"] = x.DATE.dt.year
            x.loc[:, "month"] = x.DATE.dt.month
            x.loc[:, "quarter"] = (x.month.values - 1) // 3
            self.convertColumnToFloat(x, xi)
            x = x[["year", "quarter", xi]].groupby(["year", "quarter"]).mean().
                reset_index(drop=False)
            y = pd.merge(y, x, on=["year", "quarter"], how="inner")

        y.replace(".", np.nan, inplace=True)
        floatcols = set(exogNames + [endogName])
        for col in y.columns:
            if col in floatcols:
                y.loc[:, col] = y.loc[:, col].astype(np.float64)
        y.ffill(inplace=True)
        self.endog = endogName
        self.exog = exogNames
        y = self.calculateTransformedVars(y)
        self.df = y

        self.testdata = int(trainPerc * self.df.shape[0]) - 1
        self.logger = logging.getLogger(self.__class__.__name__)
        self.model = None

    def convertColumnToFloat(self, df, col):
        df.loc[:, col] = df.loc[:, col].replace(".", np.nan).astype(np.float64).ffill()

    def calculateDiffOverAvg(self, df, col, lag, newcolname):
```

```
52          vals  = df.loc [:,  col ]. values
53          avg = np.zeros ( vals.shape [0],  dtype=np.float64 )
54          for  i  in  range(lag):
55              avg[i]  = vals [0: i+1].sum() /  (i+1)
56
57          sumv = vals [0: lag ]. sum()
58          for  i  in  range(lag ,  vals.shape [0]) :
59              sumv += vals[i]  − vals[i−lag]
60              avg[i] = sumv/lag
61          df.loc [:,  newcolname] = vals/avg − 1
62          return  df
63
64      def  calculateTransformedVars ( self , y):
65          # convert  credit  card  delinq  rate  to  decimal
66          y.loc [:,  self.endog] = y.loc [:,  self.endog] /  100.0
67
68          # calculate  GDP growth rate
69          y. sort_values (by=["year",  " quarter" ],  inplace=True)
70          gdp = y.loc [:,  "GDP"].values
71          growthRate = gdp[1:] /  gdp[0:−1] − 1
72          y.loc [:,  "GDPGrowthRate"] = 0.0
73          y.loc [1:,  "GDPGrowthRate"] = growthRate
74
75          # convert  TERMCBCCALLNS: CB interest rate on credit cards (monthly) to decimal
76          y.loc [:,  "TERMCBCCALLNS"] = y.TERMCBCCALLNS / 100.0
77
78          # calculate  int  rate − trailing 8 quarter (2 year) average
79          y = self.calculateDiffOverAvg(y,  "TERMCBCCALLNS", 8, "IntRateDiff")
80
81          # divide  CCLACBW027SBOG: Loan on credit card and other revolving plans (weekly
                ) by 200
82          y.loc [:,  "CCLACBW027SBOG"] = y.CCLACBW027SBOG / 200.0
83
84          # divide  PCEPI: PCE price index  (monthly) by 100
85          y.loc [:,  "PCEPI"] = y.PCEPI / 100.0
86          y = self.calculateDiffOverAvg(y,  "PCEPI", 8,  " InflDiff ")
87
88          # divide  DSPIC96: Real disposable  income (monthly) by 2000
89          y.loc [:,  "DSPIC96"] = y.DSPIC96 / 2000.0
90          y = self.calculateDiffOverAvg(y,  "DSPIC96", 8,  "RealDispIncDiff")
91
92          # convert  MORTGAGE30US: 30 year mortgage rate (weekly) to decimal
93          y.loc [:,  "MORTGAGE30US"] = y.MORTGAGE30US / 100.0
94          y = self.calculateDiffOverAvg(y,  "MORTGAGE30US", 8, "Mort30Diff")
95
96          y.loc [:,  "BeforeGFC"] = np.where(y.year < 2010 ,  1, 0)
97
98          # divide  y by normalized  credit  card  outstanding  loans
99          y.loc [:,  self.endog] = y.loc [:,  self.endog] / y.loc [:,  "CCLACBW027SBOG"]
100
101         self.exog = ["GDPGrowthRate", "IntRateDiff", " InflDiff ", "RealDispIncDiff", "
                Mort30Diff",
102                     "BeforeGFC"]
```

```
103          return  y
104
105      def  fit ( self ):
106          y = self . df . loc [8: self . testdata ,  self . endog]. values
107          X = self . df . loc [8: self . testdata ,  self . exog]. values
108          X = sm. add_constant (X, has_constant ="add")
109          glm = sm.GLM(y, X, family=sm. families . Poisson ( link =sm. families . links . log ( ) ) )
110          glm = glm. fit ()
111          self . logger . info (glm. summary ())
112          summaryfile = os. path . join (PLOTDIR, self.__class__.__name__ + ".txt")
113          with  open(summaryfile, 'w')  as  fh:
114              fh . write (glm. summary (). as_text ())
115          self . model = glm
116
117      def  plotResid ( self ):
118          fig ,  axs = plt . subplots (nrows=1, ncols=1,  figsize =(10, 10))
119          yendog = self . model. model. endog
120          yhatv = self . model. predict ( self . model. model. exog)
121          resid  = np. log (yendog
122                           /yhatv)
123          self . logger . info ("mu = %f, sd = %f", self . df . loc [:,  self . endog]. mean(), self . df
                           . loc [:,  self . endog]. std ())
124
125          dates = self . df . loc [8: self . testdata ,  "DATE"]. values
126
127          axs . plot (dates ,  resid )
128          axs . grid ()
129          axs . set_title ("Residual  Plot  (Training  Dataset)")
130          plt . tight_layout ()
131          plt . savefig (os. path . join (PLOTDIR, "trainResidPoisson. jpeg"),
132                           dpi=500)
133
134      def  plotTestResults ( self ,  y,  ypred):
135          fig ,  axs = plt . subplots (nrows=2, ncols=1,  figsize =(10, 10))
136          resid  = np. log (y /  ypred)
137
138          dates = self . df . loc [ self . testdata +1:,  "DATE"]. values
139          axs [0]. plot (dates ,  resid )
140          axs [0]. plot ()
141          axs [0]. grid ()
142          axs [0]. set_title ("Residual  Plot  (Test  Dataset)")
143          axs [1]. plot (dates ,  y,  label ="y")
144          axs [1]. plot (dates ,  ypred,  "−.",  label ="ypred")
145          axs [1]. grid ()
146          axs [1]. legend ()
147          axs [1]. set_title ("Predicted  vs.  Actual")
148          plt . tight_layout ()
149          plt . savefig (os. path . join (PLOTDIR, "testResidPoisson. jpeg"),
150                           dpi=500)
151
152      def  test ( self ):
153          testdata  = self . testdata  + 1
154          y = self . df . loc [ testdata :,  self . endog]. values
```

```
155        X = self.df.loc[ testdata :,  self.exog]. values
156        X = sm.add_constant(X, has_constant="add")
157        ypred =  self.model.predict (X)
158        self. plotTestResults (y,  ypred)
159
160
161   if __name__ == "__main__":
162        poisson = PoissonGLM("DRCCLACBS", ["TERMCBCCALLNS", "CCLACBW027SBOG
               ", "PCEPI", "GDP",
163                                            "DSPIC96", "MORTGAGE30US"])
164        # TERMCBCCALLNS: CB interest rate on credit cards (monthly)  100
165        # CCLACBW027SBOG: Loan on credit card and other revolving plans (weekly) 200
166        # PCEPI: PCE price index  (monthly) 100
167        # DSPIC96: Real disposable  income (monthly) 2000
168        # MORTGAGE30US: 30 year mortgage rate (weekly) 100
169
170        poisson. fit ()
171        poisson. plotResid ()
172        poisson. test ()
```

Code Explanation

A code walk-through is presented below following the execution sequence:

1. The code instantiates an object of class **PoissonGLM**. The constructor accepts arguments for an endogenous variable and a list of exogenous variables. These arguments point to datasets that are read and processed to create endogenous and exogenous variables, respectively.
2. Training-testing data split is set at 0.9, which means 90% of the data is used for training and the final 10% used for testing.
3. The endogenous variable is the delinquency rate on credit card loans for all commercial banks in the United States. The dataset can be downloaded as part of the FRED database using the symbol **DRCCLACBS** from [7].
4. Exogenous variables are constructed from data for the interest rate charged by commercial banks on credit cards **TERMCBCCALLNS**, total outstanding credit card debt **CCLACBW027SBOG**, PCE price index **PCEPI**, gross domestic product **GDP**, real disposable income **DSPIC96**, and rate on 30-year fixed-rate mortgage **MORTGAGE30US**. This data is available from the FRED database: [8–12], and [13].
5. After reading the datasets in comma-separated format, the constructor of class **PoissonGLM** converts the data to floating-point format.
6. After this, the datasets are joined. But before this can be done, year, month, and quarter corresponding to each date are extracted. This is needed because the data follow different reporting frequencies. Credit card delinquency rate is quarterly, interest rate on credit cards is monthly, total outstanding credit card debt is weekly, PCE index data is monthly, GDP is quarterly, real disposable income is monthly, and 30-year fixed-rate mortgage rate is weekly.

7. Data is converted to quarterly by appropriate transformations and joined in the constructor, **__init__**.

8. Transformations are applied inside the method **calculateTransformedVars** to get the final endogenous and exogenous variables.

9. Credit card delinquency rate is converted into a decimal from a percentage, GDP growth rate is computed, and interest rate on credit cards is converted to a decimal.

10. The interest rate charged on credit card debt is transformed inside the method **calculateDiffOverAvg**. This method computes a trailing 2-year (or 8-quarter) average and calculates the growth rate relative to this moving average as $\frac{v(t)}{\bar{v}(t)} - 1$.

11. Data on credit card debt is normalized by dividing with 200. This is because the series begins at around 200 value. This variable is not used by itself. It is used to normalize the endogenous variable – delinquency rate on credit card debt. As explained earlier, this normalization is required to take into account changing amount of credit card debt. This is because the calculation of delinquency rate does not take the total outstanding credit card debt into account.

12. The PCE price index is transformed using method **calculateDiffOverAvg** which calculates the growth rate of the PCE index relative to a trailing 2-year average.

13. The real disposable income and 30-year fixed-rate mortgage rate are also transformed using the method **calculateDiffOverAvg**.

14. Indicator variable **BeforeGFC** is computed indicating if the date occurred before the Great Financial Crisis.

15. Finally, the data is fitted using the **GLM** class of the **statsmodels** library. The constructor is passed endogenous and exogenous variables, in addition to the family of models to use: **Poisson**. It is provided a log link. Before passing the exogenous variables, a constant is added to include the intercept term in regression.

16. Statistics for fitting the model are reported.

17. A residual plot for training data is produced.

18. The model is used to predict credit card default rates on the test dataset, and the results are plotted.

The plot of residuals observed on the training dataset is shown in Figure 3-5. It can be seen from the plot that there is a spike in residual during the year 2010. This is because the indicator variable $I(t)$ changes value in January 2010. Aside from dates around 2010, training residual is around 0.

The residual plot observed on the test dataset is shown in Figure 3-6. It also plots predicted vs. actual delinquencies. One can observe that prediction accuracy improves markedly after October 2023. Before that, the model overpredicts delinquency rate. This is perhaps due to the fact that most economists and market participants viewed a mild recession as a likely event in the midst of the Federal

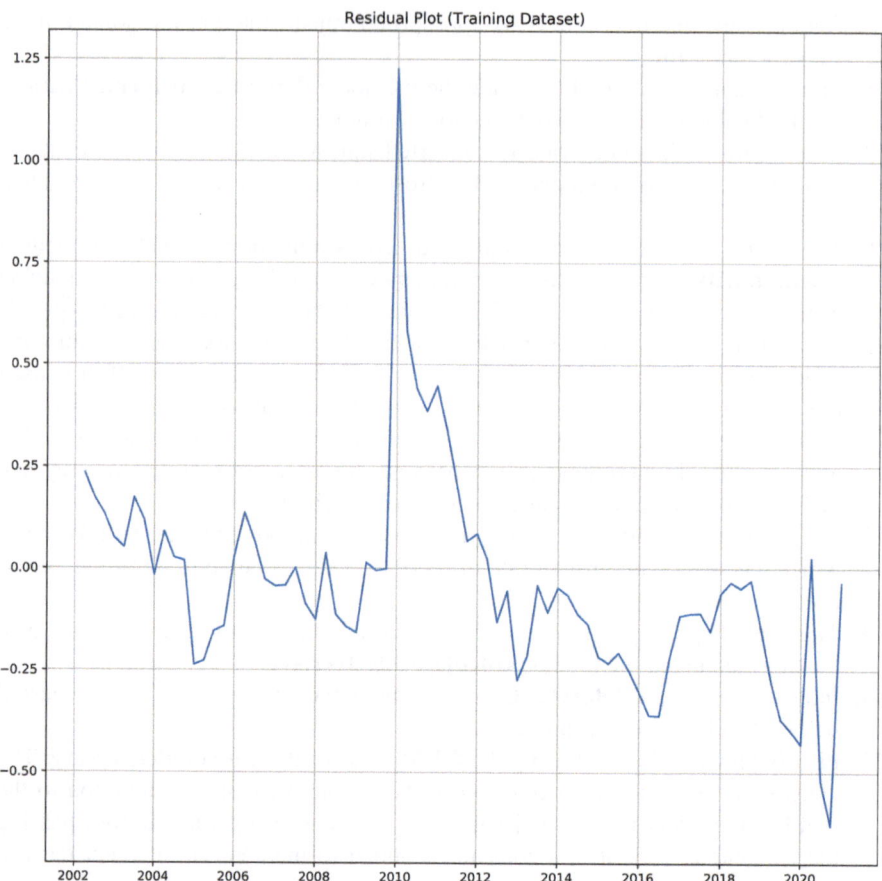

Fig. 3-5. Fitting Poisson GLM on Training Dataset

Reserve's sustained monetary tightening to curb inflation. These views are reflected in the macroeconomic variables used in the model. Toward the end of 2010, Federal Reserve officials began indicating their willingness to hold or even cut interest rates in the wake of easing inflation pressures. As economic conditions reverted to the ones observed in majority of the training data, prediction accuracy improved. This is an important feature of most statistical models – one must pay careful attention to whether test data is similar to training data. If it is significantly different, prediction accuracy will suffer. There are a few alternatives to dealing with this scenario, such as resorting to regime switching models or using a more representative training dataset. Regime switching models are discussed in a later chapter.

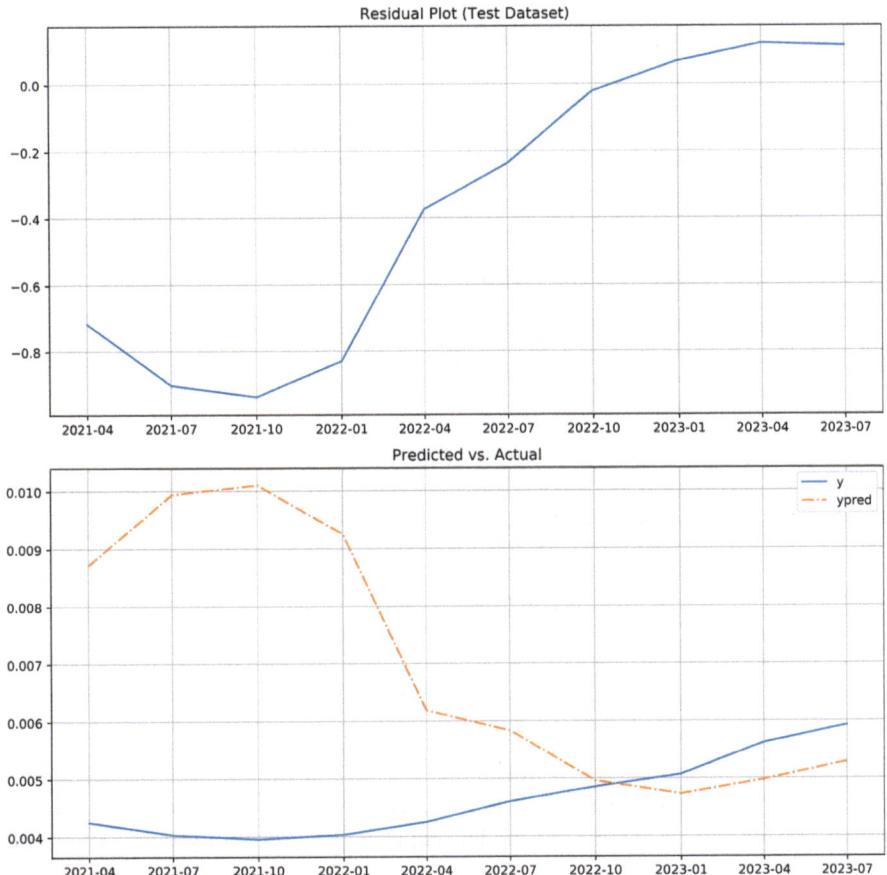

Fig. 3-6. Evaluating Poisson GLM on Test Dataset

3.2.3 Binomial Distribution

The binomial distribution is used to model variables that belong to one of two classes, for example, if a company will default on its debt or not in the next year or if a coin toss will yield heads or tails. Figure 3-7 shows the number of positive outcomes for a few examples of Binomial distribution with different mean values. The model output is a probability of belonging to one class, for example, probability that a company will default or a coin toss will give heads. Since the range of model output is [0, 1], the link function must transform the range $[-\infty, \infty]$ to [0, 1]. This can be seen by writing its probability mass function and comparing it with the exponential family, as shown in Equation 3-17.

$$f(y; \theta, \phi) = \binom{N}{y} p^y (1-p)^{N-y}$$

$$= \exp\left(y \log p + (N-y) \log(1-p) + \log N! - \right.$$

$$\log(N-y)! - \log y!)$$

$$l(\theta, \phi; y) = y \log \frac{p}{1-p} + N \log(1-p) + \log N! -$$

$$\log(N-y)! - \log y!$$

$$\therefore \theta = \log \frac{p}{1-p} \text{ or } p = \frac{1}{1+e^{-\theta}} \tag{3-17}$$

$$a(\phi) = 1$$

$$b(\theta) = -N \log(1-p) = N \log\left(1 + e^\theta\right)$$

$$\therefore E[y] = b'(\theta) = N\frac{1}{1+e^{-\theta}} = Np$$

$$\text{var}(y) = b''(\theta)a(\phi) = N\frac{e^{-\theta}}{\left(1+e^{-\theta}\right)^2}$$

$$= N\frac{1}{1+e^{-\theta}} \frac{e^{-\theta}}{1+e^{-\theta}} = Np(1-p)$$

Equation 3-17 shows that the mean and variance of the binomial distribution is Np and $Np(1-p)$, respectively. The canonical link function is obtained by setting $\eta = \theta = \log \frac{p}{1-p} = $ logit. The model output is the probability or p. Therefore, $\log \frac{y}{1-y} = \eta = \mathbf{X}\boldsymbol{\beta}$ implies that $y = \frac{1}{1+\exp -\mathbf{X}\boldsymbol{\beta}}$.

The **Bernoulli distribution** is a special case of binomial distribution with N = 1, i.e., an experiment with one trial. GLM with the Bernoulli distribution and logit link function is also known as logistic regression.

Application

Let us use the binomial distribution to predict whether weekly market return the following week (i.e., over the next five trading days) is positive or not. The endogenous variable is the probability of observing a positive weekly market return, denoted by $P(t)$ and defined as prob $\left(\frac{P(t+6)-P(t+1)}{P(t+1)} > 0\right)$. Let us use the following explanatory variables:

1. Indicator variable $I_r(t)$ that is 1 if the weekly return observed over the last week is greater than 2%, and 0 otherwise
2. Indicator variable $I_{MA}(t)$ that is 1 if the three-day moving average of closing price is greater than the ten-day moving average, and 0 otherwise

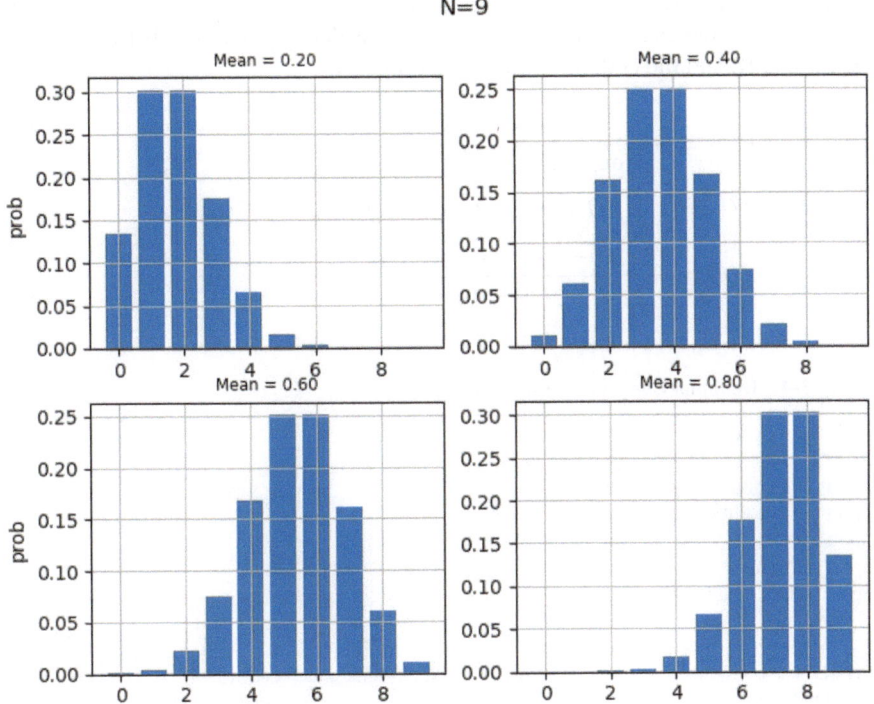

Fig. 3-7. Binomial Distribution

3. Indicator variable $I_{vol}(t)$ that is 1 if the one-month volatility is greater than one-year volatility, and 0 otherwise
4. Percentage change in the yield of one-month US government bill observed over the last one week, $\delta(t) = \frac{Y_{1Mo}(t-1)-Y_{1Mo}(t-6)}{Y_{1Mo}(t-6)}$

Being a prediction of market variable, we do not expect the fit to be very good because of the efficient market hypothesis. Market participants discount all available information, including forecasts from known macroeconomic variables, to assign market price. Therefore, any change in price from the opening price is due to unforeseen events – or random noise. Although the efficient market hypothesis is not true for prices of all securities and at all times, market tracking indices like S&P 500 generally follow this principle to a good extent.

Another shortcoming of the modeling methodology applied here is the constraint inherent in the GLM framework. The explanatory variables are combined with each other in a linear relationship, and this approach precludes non-linear interaction among the variables. There is also the possibility of not having included all requisite explanatory variables in the model. The model, as described above, is formulated in Equation 3-18 and the code is shown in Listing 3-4.

$$P(t) = \frac{1}{1 + \exp - Z(t)} + \epsilon(t)$$

$$Z(t) = \alpha + \beta_1 I_r(t) + \beta_2 I_{MA}(t) + \beta_3 I_{vol} + \beta_4 \delta(t) \qquad (3\text{-}18)$$

$$\delta(t) = \frac{Y_{1Mo}(t-1) - Y_{1Mo}(t-6)}{Y_{1Mo}(t-6)}$$

Listing 3-4. Binomial Generalized Linear Model

```
1   import numpy as np
2   import pandas as pd
3   import statsmodels.api as sm
4   import matplotlib.pyplot as plt
5   import os
6   import logging
7   import scipy.stats as ss
8
9
10  DATADIR = r"C:\prog\cygwin\home\samit_000\latex\book_stats\code\data"
11  PLOTDIR = r"C:\prog\cygwin\home\samit_000\latex\book_stats\code\plots"
12  logging.basicConfig(level=logging.DEBUG)
13
14
15  class BinomialGLM(object):
16      PERIOD = 5
17
18      def __init__(self, security, rates, trainPerc=0.9):
19          filename = os.path.join(DATADIR, security + ".csv")
20          y = pd.read_csv(filename, parse_dates=["DATE"])
21          self.convertColumnToFloat(y, security)
22          for xi in rates:
23              filename = os.path.join(DATADIR, xi + ".csv")
24              x = pd.read_csv(filename, parse_dates=["DATE"])
25              y = pd.merge(y, x, how="left", on=["DATE"])
26              self.convertRateToFloat(y, xi)
27          y = self.calculateEndogExogVars(y, security, rates)
28          self.df = y
29
30          self.testdata = int(trainPerc * self.df.shape[0]) - 1
31          self.logger = logging.getLogger(self.__class__.__name__)
32          self.model = None
33
34      def calculateEndogExogVars(self, y, col, rates):
35          vals = y.loc[:, col].values
36          ret = vals[1+self.PERIOD:]/vals[1:-self.PERIOD] - 1
37
```

```
38       y.loc [:,  " positive_ret "] = 0
39       self .endIndex = y.shape[0] − self .PERIOD − 2
40       y.loc [0: self .endIndex, " positive_ret "] = np.where(ret > 0, 1.0, 0.0)
41
42       self .endog = " positive_ret "
43       y.loc [:,  " lastret "] = 0
44       y.loc [2+ self .PERIOD:, " lastret "] = ret [0:−1]
45       y.loc [:,  " indicator "] = np.where(y. lastret . values > 0.02, 1.0, 0.0)
46
47       ma3 = self .movingAverage(vals, 3)
48       ma10 = self .movingAverage(vals, 10)
49       y.loc [:, "ma3_10"] = 0.0
50       y.loc [11:, "ma3_10"] = np.where(ma3[10:−1] > ma10[10:−1], 1.0, 0.0)
51
52       vol21day = self . volatility ( ret , 21)
53       vol1yr = self . volatility ( ret , 252)
54       y.loc [:, "vol21_252"] = 0
55       y.loc [253+ self .PERIOD:, "vol21_252"] = np.where(vol21day[252:] > vol1yr [252:],
             1.0, 0.0)
56
57       # percent change in  interest  rate  over  the  period
58       rate_change_cols = []
59       for  rate  in  rates :
60            col = rate + "_change"
61            rval = y.loc [:,  rate ]. values
62            rval = np.where(rval == 0, 1E−8, rval )
63            change = rval [ self .PERIOD:] / rval [0:− self .PERIOD] − 1
64            y.loc [:,  col] = 0
65            y.loc[1+ self .PERIOD:, col] = change[0:−1]
66            rate_change_cols .append(col)
67
68       self .exog = [" indicator ", "ma3_10", "vol21_252"] + rate_change_cols
69       self .nvars = len( self .exog)
70       self .beginIndex = 253
71       return  y
72
73  def movingAverage(self, arr , period):
74       res = np.zeros( arr .shape [0], dtype=np. float64 )
75       sumval = np.sum(arr [0: period ])
76       for i  in  range(period,  arr .shape [0]) :
77            res [i] = sumval / period
78            sumval += arr [i] − arr [i −period]
79
80       return  res
81
82  def  volatility ( self , arr , period):
83       res = np.zeros ( arr .shape [0], dtype=np. float64 )
84       sumval = np.sum(arr [0: period ])
85       sumsq = np.dot( arr [0: period ],  arr [0: period ])
86       for i  in  range(period,  arr .shape [0]) :
87            res [i] = np. sqrt (sumsq/period − (sumval/period)**2)
88            sumval += arr [i] − arr [i −period]
89            sumsq += arr [i]* arr [i] − arr [i −period]* arr [i −period]
```

```python
90          return res
91
92      def convertRateToFloat(self, df, col):
93          df.loc[:, col] = df.loc[:, col].replace(".", np.nan).astype(np.float64).ffill()
94          df.loc[:, col] = df.loc[:, col] / 100.0  # convert to decimal
95
96      def convertColumnToFloat(self, df, col):
97          if (df.loc[:, col] == ".").sum() > 0:
98              df.drop(np.where(df.loc[:, col] == ".")[0], inplace=True)
99              df.loc[:, col] = df.loc[:, col].astype(np.float64)
100             df.reset_index(drop=True, inplace=True)
101
102     def fit(self):
103         y = self.df.loc[self.beginIndex: self.testdata, self.endog].values
104         X = self.df.loc[self.beginIndex: self.testdata, self.exog].values
105         X = sm.add_constant(X, has_constant="add")
106         glm = sm.GLM(y, X, family=sm.families.Binomial(link=sm.families.links.logit()))
107         glm = glm.fit()
108         self.logger.info(glm.summary(xname=['constant'] + self.exog))
109         summaryfile = os.path.join(PLOTDIR, self.__class__.__name__ + ".txt")
110         with open(summaryfile, 'w') as fh:
111             fh.write(glm.summary(xname=['constant'] + self.exog).as_text())
112         self.model = glm
113
114     def plotResid(self):
115         fig, axs = plt.subplots(nrows=3, ncols=1, figsize=(10, 10))
116         yendog = self.model.model.endog
117         yhatv = self.model.predict(self.model.model.exog)
118         resid = yendog - yhatv
119
120         dates = self.df.loc[self.beginIndex: self.testdata, "DATE"].values
121
122         axs[0].plot(dates, resid)
123         axs[0].grid()
124         axs[0].set_title("Residual Plot (Training Dataset)")
125         axs[1].hist(yhatv, bins=40, density=True)
126         axs[1].grid()
127         axs[1].set_title("Histogram of Residuals")
128         axs[2].plot(dates, yendog, label="y")
129         axs[2].plot(dates, yhatv, "-.", label="ypred")
130         axs[2].grid()
131         axs[2].legend()
132         axs[2].set_title("Predicted vs. Actual")
133         plt.tight_layout()
134         plt.savefig(os.path.join(PLOTDIR, "trainResidBinomial.jpeg"),
135                     dpi=500)
136
137     def plotTestResults(self, y, ypred):
138         fig, axs = plt.subplots(nrows=2, ncols=1, figsize=(10, 10))
139         resid = (y - ypred)
140
141         dates = self.df.loc[self.testdata +1: self.endIndex, "DATE"].values
142         axs[0].hist(resid, bins=40, density=True)
```

```
143    axs [0]. grid ()
144    axs [0]. set_title ("Histogram of Residuals")
145    axs [1]. plot (dates, y, label="y")
146    axs [1]. plot (dates, ypred, "-.", label="ypred")
147    axs [1]. grid ()
148    axs [1]. legend ()
149    axs [1]. set_title ("Predicted vs. Actual")
150    plt . tight_layout ()
151    plt . savefig (os.path. join (PLOTDIR, "testResidBinomial.jpeg"),
152                   dpi=500)
153
154    def test ( self ):
155        testdata  = self . testdata  + 1
156        y = self .df. loc [ testdata : self .endIndex, self .endog]. values
157        X = self .df. loc [ testdata : self .endIndex, self .exog]. values
158        X = sm.add_constant(X, has_constant="add")
159        ypred  = self .model. predict (X)
160        self . plotTestResults (y, ypred)
161
162
163    if __name__ == "__main__":
164        glm = BinomialGLM("SP500", ["DGS1MO"])
165        # TERMCBCCALLNS: CB interest rate on credit cards (monthly)  100
166        # CCLACBW027SBOG: Loan on credit card and other revolving plans (weekly) 200
167        glm. fit ()
168        glm.plotResid ()
169        glm. test ()
```

Code Explanation

A code walk-through is presented below with the purpose of highlighting salient code features:

1. The code instantiates an object of class **BinomialGLM**, passing the constructor arguments specifying the endogenous and exogenous variables.
2. Daily end-of-day prices of S&P 500 are used to construct the endogenous variable. This dataset is available in the FRED database under the label **SP500** [14].
3. The **BinomialGLM** class reads the file containing S&P 500 daily closing prices.
4. It calls the method **calculateEndogExogVars** to calculate the final endogenous and exogenous variables.
5. The class defines a constant **PERIOD** to be five trading days (or approximately one week). It calculates five-day return on S&P500, with the value 5 coming from the definition of **PERIOD**.

6. It calculates column **lastret** as the trailing five-day return using the definition $lr(t) = \frac{P(t)}{P(t-5)} - 1$.

7. An indicator variable **indicator** is calculated which is defined to be 1 if **lastret** is greater than 2% or 0.02 and 0 otherwise.

8. It calculates the three-day and ten-day moving averages of closing price for S&P 500. It computes an exogenous variable **ma3_10** and the difference between three-day and ten-day moving averages of S&P 500 closing prices. The three-day moving average responds faster than the ten-day moving average to recent changes in price. Taking the difference between a short duration (three-day) and long duration (ten-day) moving average gives a measure of price momentum.

9. It calculates the 21-day and 252-day volatility of returns for S&P 500. 21 trading days correspond roughly to 1 month, while 252 trading days translate to 1 year. Following this, it computes an indicator variable **vol21_252** which takes a value of 1 if 21-day volatility is greater than 252-day volatility and 0 otherwise.

10. Because volatility values get populated after 252 days due to the definition of the 252-day volatility, data begins from the 252nd observation.

11. The class uses 90%–10% as the training-testing data split.

12. The endogenous variable is the following five-day return on S&P 500.

13. After adding a column of ones to the exogenous variables to include an intercept term, it fits the binomial GLM model using the **GLM** class from the **statsmodels** package and passing the endogenous variable and the exogenous variables including a constant and the **Binomial** family of functions with the **logit** link function.

14. Model fit statistics are printed.

15. It produces a plot of residuals observed for the training dataset.

16. It makes a prediction on the test dataset and plots the results against actual S&P 500 five-day returns.

The plot of the predicted probability of observing a positive weekly return against the ground truth if the actual weekly return was positive or not on the test dataset is shown in Figure 3-8. It can be seen from the plot that the predicted probability hovers around 0.6, which is the average probability of observing a positive weekly return in the training dataset. As expected for a market variable, the overall fit is poor. This can be improved by using non-linear models such as deep neural networks. It may also be possible to improve the fit by including additional exogenous variables.

3.2.4 Gamma Distribution

The gamma distribution is used to model the wait time for the αth occurrence of an event when the average time for an event to occur is β. While the Poisson distribution governs the number of event occurrences in a given time interval, the

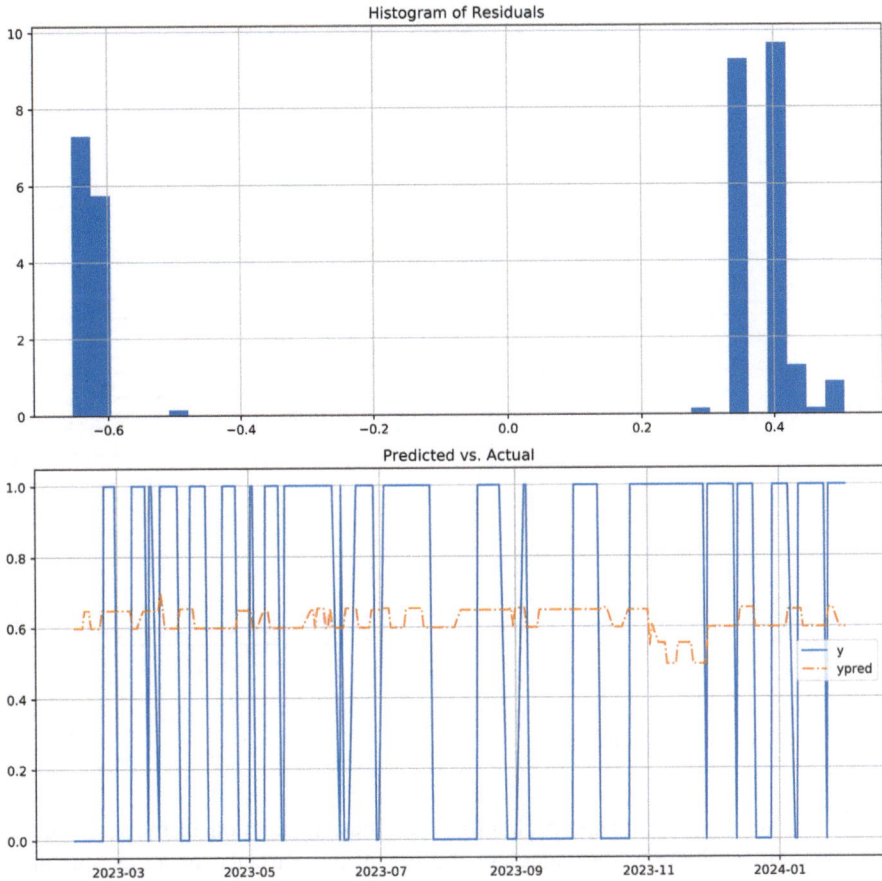

Fig. 3-8. Fitting Binomial GLM to Predict If Weekly Market Return Is Positive on Test Dataset

gamma distribution models the wait time, or the time for a certain number of event occurrences, as shown for a few sample gamma distributions in Figure 3-9. The distribution is given by Equation 3-19, where y denotes the wait time, α denotes the number of event occurrences, and β denotes the average time for an event to occur. The domain of the gamma distribution is $y \in [0, \infty)$.

$$f(y; \alpha, \beta) = \frac{e^{-\beta y} \beta^{\alpha} y^{\alpha-1}}{\Gamma(\alpha)}$$

$$\text{where } \Gamma(\alpha) = \int_0^{\infty} e^{-x} x^{\alpha-1} dx$$

(3-19)

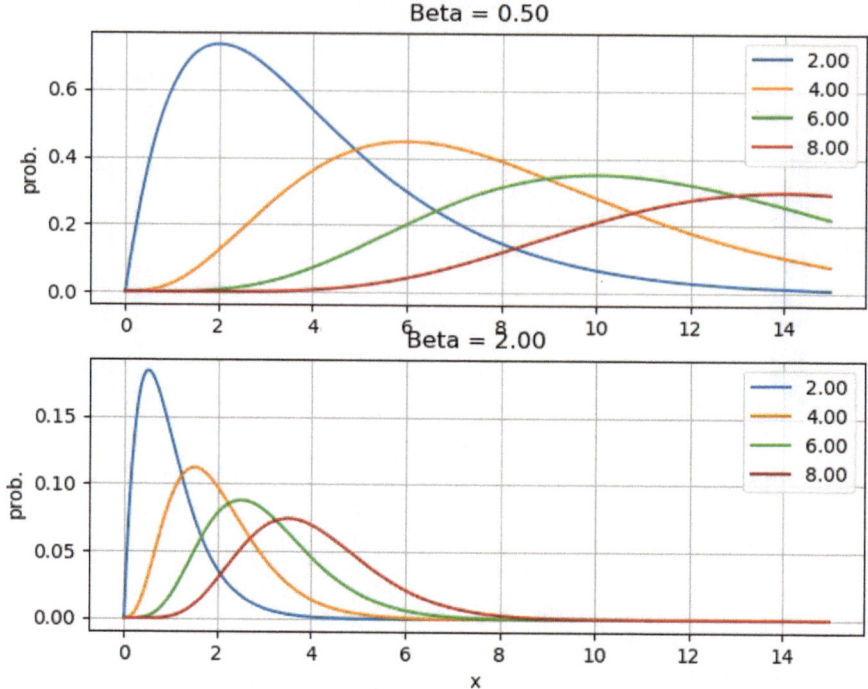

Fig. 3-9. Gamma Distribution

As before, let us write the probability distribution function in the form of the exponential distribution and calculate the mean and variance of y, as shown in Equation 3-20.

$$f(y; \alpha, \beta) = \exp\left(\frac{y\beta - \alpha \log \beta}{-1} + (\alpha - 1) \log y - \log \Gamma(\alpha)\right)$$

$$\therefore \theta = \beta$$

$$b(\theta) = \alpha \log \beta = \alpha \log \theta \text{ and } a(\phi) = -1 \qquad (3\text{-}20)$$

$$E[y] = b'(\theta) = \frac{\alpha}{\theta} = \frac{\alpha}{\beta}$$

$$\text{var}(y) = b''(\theta)a(\phi) = \frac{\alpha}{\theta^2} = \frac{\alpha}{\beta^2}$$

In order to get the canonical link function for the gamma distribution, set $\theta = \eta$. Recall that η is equal to the dot product of parameters and regressors, $\mathbf{X}$. β in Equation 3-20 must be distinguished from the parameters of the linear model, which was denoted using the same symbol. From Equation 3-20, $E[y] = \frac{\alpha}{\beta} = \frac{\alpha}{\eta}$ because

$\beta = \theta$. Therefore, $E[y] = \frac{\alpha}{\eta}$. We can absorb the constant α into model parameters in η to write $E[y] = \frac{1}{\eta}$. This shows that the canonical link function is the reciprocal function or $\eta = \frac{1}{\mu}$.

Application

Let us use the gamma distribution to predict the quarter in a year when the number of new non-farm payroll jobs created in the United States exceeds 300,000. Non-farm payroll is a key econometric measure tracked by economists, market participants, and the Federal Reserve to monitor the health of the economy and to predict where inflation is headed. Strong payroll numbers that exceed expectations are a reliable signal of the economy's strength and may presage higher inflation. This may spur the Federal Reserve to tighten monetary conditions. It is conceivable that economists may upgrade their assessment of the economy's health once the number of new jobs created crosses a certain threshold. Likewise, the central bank may need to recalibrate its interest rate strategy once the number of new jobs in the non-farm sector crosses a certain threshold. Let us see how the gamma distribution–based GLM can be used to predict when the number of new non-farm jobs created in a year crosses a threshold of 300,000.

The number of non-farm payroll jobs created monthly does not remain constant, as seen from the plot of this variable shown in Figure 3-10. This suggests that one of the exogenous variables should be the real GDP growth. Let us also use a second exogenous variable – growth rate in the market index (S&P 500). The model is shown in Equation 3-21. Real GDP is reported quarterly, S&P 500 closing prices are available daily, and the endogenous variable, non-farm payroll, is available monthly. Because the biggest time difference between two successive observations of a variable is a quarter, let us align all variables to the beginning of the quarter. S&P 500 closing price on the first trading day of each quarter is used. Non-farm payrolls reported for three months in a quarter are added to get the quarterly non-farm payroll. Finally, we would like to predict which quarter in a year would first witness the cumulative non-farm payrolls in that year venturing above the 300,000 level.

$$g\left(N(t)\right) = \alpha + \beta_1 \frac{GDP(t) - GDP(t - 3Yr)}{GDP(t - 3Yr)} +$$

$$\beta_2 \frac{SP500(t) - SP500(t - 3Yr)}{SP500(t - 3Yr)} + \epsilon(t) \tag{3-21}$$

$N(t) = $ Average quarterly non-farm payroll observed over a year

$$g(Z) = \frac{1}{Z}$$

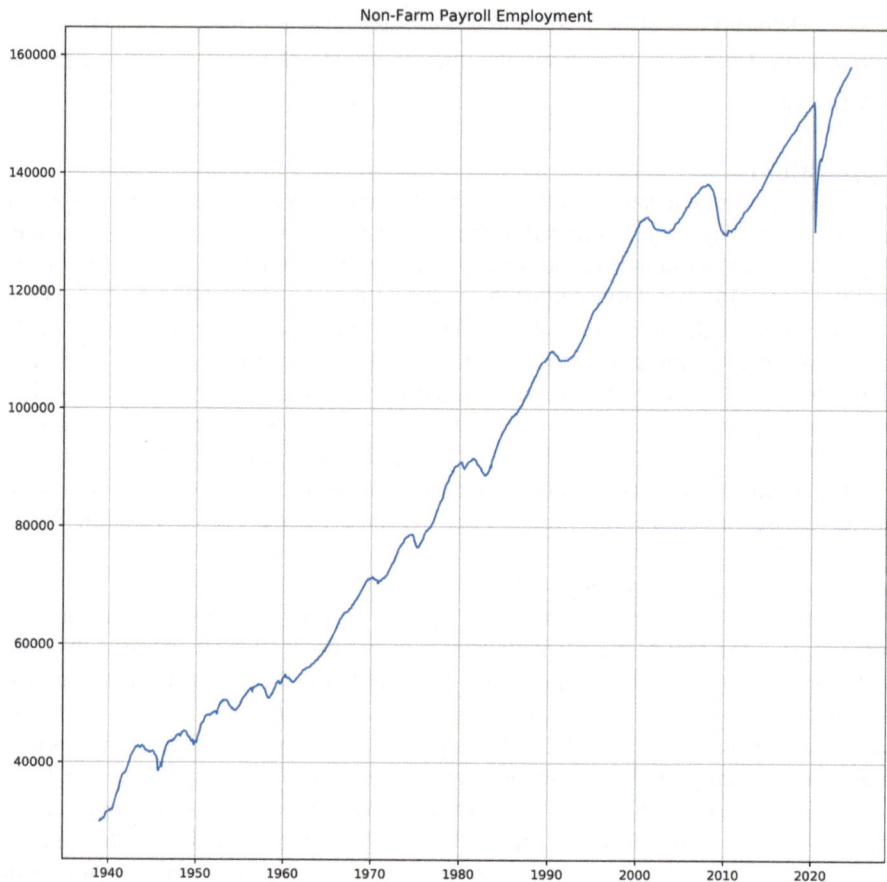

Fig. 3-10. Non-farm payroll

The model variables are tabulated below:

1. **Endogenous variable**: Average quarterly non-farm payroll. This variable is calculated by adding the non-farm payroll numbers for three months in each quarter, followed by taking an average of the quarterly non-farm payroll over the following four quarters (or one year). Seasonally adjusted non-farm payroll is published monthly by the US Bureau of Labor Statistics. It is denoted by $N(t)$.

2. **Exogenous variable 1**: Three-year growth rate in real GDP, denoted by $\frac{GDP(t)-GDP(t-3Yr)}{GDP(t-3Yr)}$.

3. **Exogenous variable 2**: Three-year growth rate in the S&P 500 index, denoted by $\frac{SP500(t)-SP500(t-3Yr)}{SP500(t-3Yr)}$.

Table 3-2. Testing Predictions of Quarterly Non-farm Payroll Using Gamma GLM Model

Date	Predicted Quarter	Actual Quarter	Pred Qtr Payroll	Actual QtrAvg Payroll	Perc Diff
2021-01-01	0	0.00	429443.89	438827.75	−0.02
2022-01-01	0	0.00	420511.25	457594.00	−0.08
2023-01-01	0	0.00	461238.15	468775.75	−0.02

Listing 3-5. Statistics of Training Gamma GLM Model

```
Generalized Linear Model Regression Results
==================================================================
Dep. Variable :     y   No. Observations :        11
Model:              GLM  Df Residuals :            8
Model Family:       Gamma  Df Model:               2
Link Function:      inverse_power   Scale :       0.00024260
Method:             IRLS  Log−Likelihood:          −114.38
Date:               Sun, 19 May 2024  Deviance:   0.0019295
Time:               12:05:38  Pearson chi2 :       0.00194
No. Iterations :    5  Pseudo R−squ. (CS):         0.6774
Covariance Type:    nonrobust
==================================================================
            coef    std err     z     P>|z|    [0.025    0.975]
------------------------------------------------------------------
constant  1.627e−06 4.17e−08 39.049  0.000  1.54e−06 1.71e−06
GDPC1_3year_perc_change −3.531e−07 3.07e−07 −1.151 0.250 −9.54e−07 2.48e−07
SP500_3year_perc_change 3.097e−07 9.78e−08 3.167 0.002 1.18e−07 5.01e−07
==================================================================
```

Statistics for fitting the gamma GLM model are shown in Listing 3-5. Pseudo-R^2 of the model is 0.68. Data begins from 2014 and is quarterly. 90% of the data is used for training, with the remaining 1% used for testing. Due to the quarterly sampling frequency, the number of data points is somewhat limited.

The test dataset contains three data points, and the fitted model is able to correctly predict that the number of non-farm payrolls will exceed 300,000 in the first quarter for all of them. Comparing the predicted quarterly non-farm payroll against the actual quarterly average non-farm payroll calculated over the following year, we observe the predictions to range between –2% and –8%, as shown in Table 3-2. For the quarter beginning on January 1, 2021, the model predicts quarterly non-farm payroll to be 429443.89, and the actual value was observed to be 438,827.75, showing that the model does a decent job at predicting quarterly non-farm payroll for this limited test dataset. The full code can be found in Listing 3-6.

Listing 3-6. Gamma Generalized Linear Model

```
import numpy as np
import pandas as pd
import statsmodels.api as sm
import matplotlib.pyplot as plt
import os
import logging


DATADIR = r"C:\prog\cygwin\home\samit_000\latex\book_stats\code\data"
PLOTDIR = r"C:\prog\cygwin\home\samit_000\latex\book_stats\code\plots"
logging.basicConfig(level=logging.DEBUG)



class GammaGLM(object):
    PERIOD = 5

    def __init__(self, endog, exog, trainPerc=0.9):
        filename = os.path.join(DATADIR, endog + ".csv")
        y = pd.read_csv(filename, parse_dates=["DATE"])
        self.convertColumnToFloat(y, endog)
        self.plotEndog(y, endog)
        y.loc[:, 'quarter'] = ((y.DATE.dt.month.values - 1) // 3)
        y.loc[:, 'year'] = y.DATE.dt.year
        ypart = y[['year', 'quarter', endog]]
        ypart = ypart.groupby(['year', 'quarter']).sum().reset_index(drop=False)
        y.drop(columns=[endog], inplace=True)
        y = pd.merge(y, ypart, on=['year', 'quarter'], how="left")
        self.beginIndex = None
        self.endIndex = None
        self.origEndog = endog
        for xi in exog:
            filename = os.path.join(DATADIR, xi + ".csv")
            x = pd.read_csv(filename, parse_dates=["DATE"])
            x.loc[:, "MonthBegin"] = x.DATE + pd.offsets.MonthBegin(0)
            x = x.groupby(["MonthBegin"]).first().reset_index(drop=False)
            x.DATE = x.MonthBegin
            y = pd.merge(y, x, how="inner", on=["DATE"])
            self.convertColumnToFloat(y, xi)
        y = self.calculateEndogExogVars(y, endog, exog)
        self.df = y

        self.testdata = int(trainPerc * (self.endIndex - self.beginIndex)) + self.
            beginIndex
        self.logger = logging.getLogger(self.__class__.__name__)
        self.model = None

    def plotEndog(self, df, endog):
        fig, axs = plt.subplots(nrows=1, ncols=1, figsize=(10, 10))
        y = df.loc[:, endog].values
        dates = df.DATE.values
        axs.plot(dates, y)
        axs.grid()
```

```python
        axs. set_title ("Non—Farm Payroll Employment")
        plt. tight_layout ()
        plt. savefig (os. path. join (PLOTDIR, "nonFarmPayroll.jpeg"),
                      dpi=500)

    def calculateEndogExogVars( self , df, endog, exog):
        vals = df.loc [:, endog].values
        yvar = "AvgNonFarmPayrollPerQtr"
        df.loc [:, yvar] = 0.0
        for i in range(3, df.shape [0]) :
            df.loc[i − 3, yvar] = vals [i−3:i+1].mean()
        self .endog = yvar

        # 3−year percent change in real GDP, SP500
        exog_cols = []
        for x in exog:
            col = x + "_3year_perc_change"
            rval = df.loc [:, x].values
            rval = np.where( rval == 0, 1E−8, rval)
            change = rval [11:] / rval [0:−11] − 1
            df.loc [:, col] = 0
            df.loc [11:, col] = change
            exog_cols.append(col)

        self .exog = exog_cols
        self .nvars = len( self .exog)
        self .beginIndex = 12
        self .endIndex = df.shape [0] − 12
        return df

    def convertColumnToFloat( self , df, col) :
        if (df.loc [:, col] == ".").sum() > 0:
            df.drop(np.where(df.loc [:, col] == ".") [0], inplace=True)
            df.loc [:, col] = df.loc [:, col ].astype (np. float64 )
            df. reset_index (drop=True, inplace=True)

    def fit ( self ):
        y = self .df.loc [ self .beginIndex: self . testdata , self .endog].values
        X = self .df.loc [ self .beginIndex: self . testdata , self .exog].values
        X = sm.add_constant(X, has_constant="add")
        glm = sm.GLM(y, X, family=sm.families.Gamma(link=sm.families.links .
            inverse_power ()))
        glm = glm. fit ()
        self . logger . info (glm.summary(xname=['constant'] + self .exog))
        summaryfile = os. path. join (PLOTDIR, self.__class__.__name__ + ".txt ")
        with open(summaryfile, 'w') as fh:
            fh. write (glm.summary(xname=['constant'] + self .exog). as_text ())
        self .model = glm

    def plotResid ( self ):
        fig , axs = plt. subplots (nrows=1, ncols=1, figsize =(10, 10))
        yendog = self .model.model.endog
        yhatv = self .model. predict ( self .model.model.exog)
```

```python
        dates = self.df.loc[self.beginIndex: self.testdata, "DATE"].values

        axs.plot(dates, yendog, label="y")
        axs.plot(dates, yhatv, "-.", label="ypred")
        axs.grid()
        axs.legend()
        axs.set_title("Predicted vs. Actual")
        plt.tight_layout()
        plt.savefig(os.path.join(PLOTDIR, f"trainResid{self.__class__.__name__}.jpeg"),
                    dpi=500)

    def tabulateTestResults(self):
        # calculate the month when non-farm payroll reaches or exceeds 300K
        yvar = "MonthPayRollGt300K"
        ypredvar = "PredMonthPayRollGt300K"
        df = self.df
        month = df.loc[:, "DATE"].dt.month
        begin_month = np.where(month.values == 1)
        df.loc[:, yvar] = 0.0
        df.loc[:, ypredvar] = 0
        endog = df.loc[:, self.origEndog].values
        month_indx = []

        for i in begin_month[0]:
            if (i > self.testdata) and (df.loc[i, self.endog] != 0.0):
                month_indx.append(i)
                total = np.cumsum(endog[i:i+4])
                df.loc[i, yvar] = np.where(total >= 300000)[0][0]
                df.loc[i, ypredvar] = int(300000 / df.loc[i, "ypred"])
                self.logger.info("Date: %s, predicted quarterly non-farm payroll: %f,
                    actual: %f", df.loc[i, "DATE"],
                                    df.loc[i, "ypred"], df.loc[i, self.endog])

        dates = df.loc[month_indx, "DATE"].values
        yval = df.loc[month_indx, yvar].values
        yvalhat = df.loc[month_indx, ypredvar].values
        prollhat = df.loc[month_indx, "ypred"].values
        proll = df.loc[month_indx, self.endog].values

        df = pd.DataFrame({"Date": dates, "Predicted Quarter": yvalhat,
                           "Actual Quarter": yval,
                           "Predicted Qtr Payroll": prollhat,
                           "Actual QtrAvg Payroll": proll})
        df.loc[:, "Perc Diff"] = df.loc[:, "Predicted Qtr Payroll"] / df.loc[:, "Actual
            QtrAvg Payroll"] - 1

        df.to_csv(os.path.join(PLOTDIR, f"{self.__class__.__name__}.csv"), index=False)
        self.logger.info(df.to_latex(index=False, float_format="{:.2f}".format))
        self.logger.info(df)

    def test(self):
        testdata = self.testdata + 1
```

```
155      X = self.df.loc[ testdata :,  self.exog]. values
156      X = sm.add_constant(X, has_constant="add")
157      ypred = self.model.predict (X)
158      self.df.loc [:,  "ypred"] = 0.0
159      self.df.loc[ testdata :,  "ypred"] = ypred
160      self . tabulateTestResults ()
161
162
163  if __name__ == "__main__":
164      glm = GammaGLM("PAYEMS", ["GDPC1", "SP500"])
165      # PAYEMS: Nonfarm payroll (monthly), seasonally  adjusted
166      # CCLACBW027SBOG: Loan on credit card and other revolving plans (weekly) 200
167      # PCEPI: PCE price index  (monthly) 100
168      # DSPIC96: Real disposable  income (monthly) 2000
169      # MORTGAGE30US: 30 year mortgage rate (weekly) 100
170
171      glm. fit ()
172      glm.plotResid ()
173      glm. test ()
```

Code Explanation
A code walk-through is presented below as an aid to understanding the code:

1. The code instantiates an object of class **GammaGLM**, passing endogenous and exogenous variables as constructor arguments.
2. The model uses seasonally adjusted non-farm payroll data available from the FRED database under the label **PAYEMS** [15]. This data is published monthly.
3. Exogenous variables include real gross domestic product **GDPC1** available from [16] and closing prices of S&P 500 [14]. While real GDP data is available quarterly, S&P 500 closing prices are available daily on trading days.
4. It reads the data files as comma-separated data using the pandas library.
5. It produces a plot of non-farm payroll to give a high-level overview of the data.
6. It extracts year quarter and month from the dates and joins the dataframes to obtain quarterly data. Since non-farm payroll data is available monthly, we only keep the values reported during months falling at the beginning of each quarter. The quarter is extracted from the date by performing an integer division of zero-index month with three because each quarter contains three months.
7. Final endogenous and exogenous variables are then computed inside the method **calculateEndogExogVars**. It calculates the average non-farm payroll per quarter using the three months falling in a quarter. This is the endogenous variable. Exogenous variables are three-year growth rates (or percent changes) observed in real GDP and S&P 500 closing prices. Since 3 years have 12 quarters, calculations are performed accordingly.
8. Gamma GLM is then fitted using the **Gamma** family of functions with the **inverse_power** link function. This link function takes the reciprocal of the argument.
9. Model fitting statistics are printed followed by testing results.

10. A table comparing the actual and predicted non-farm payroll for the test dataset is generated.

3.2.5 Inverse Gamma Distribution

The inverse gamma distribution is similar to the gamma distribution except that it involves the reciprocal of the input variable y. Its domain is $y \in [0, \infty)$. The probability density function for this distribution is shown in Equation 3-22. Parameter α denotes the average.

$$
\begin{aligned}
f(y; \alpha, \beta) &= \frac{\beta^\alpha}{\Gamma(\alpha)} \left(\frac{1}{y}\right)^{(\alpha+1)} \exp(-\frac{\beta}{y}) \\
&= \frac{\beta^\alpha}{\Gamma(\alpha)} (y)^{-(\alpha+1)} \exp(\frac{\beta}{y}) \\
&= f^{\text{gamma}}(\frac{1}{y}, \alpha, \beta)
\end{aligned}
\tag{3-22}
$$

This distribution can be transformed into a gamma distribution as shown in Equation 3-22. The canonical link for this distribution is $\eta = \frac{1}{\mu^2}$.

3.2.6 Multinomial Distribution

A multinomial distribution is often employed in classification tasks when a response variable can be one of K specified distinct values, where K is fixed and prespecified. This can be viewed as a classification problem, with the response variable y indicating the class to which the observation belongs. The multinomial distribution is a generic counterpart of the binomial distribution where the number of classes is 2. It is a discrete distribution.

Let p_i denote the probability of an observation belonging to class i. We have $\sum_{i=1}^{K} p_i = 1$. The probability mass function of the multinomial distribution gives the probability of observing y_i observations belonging to class i from a total of N classes. This probability mass function is shown in Equation 3-23.

$$
\text{prob}(y_1, y_2, \cdots, y_K | p_1, p_2, \cdots, p_K) = N! \prod_{i=1}^{K} \frac{p_i^{y_i}}{y_i!}
$$

$$
\sum_{i=1}^{K} y_i = N \text{ where } y_i \in [0, 1, \cdots, N]
\tag{3-23}
$$

$$
\sum_{i=1}^{K} p_i = 1
$$

For each class i in a set of N classes, the probability of belonging to class i can be written as shown in Equation 3-24. Being probabilities, they must all add to 1. This constraint is also shown in Equation 3-24. x_j denotes a set of $K+1$ exogenous variables including a constant, where $j \in (0, 1, 2, \cdots, K)$.

$$y_i = \exp\left(\beta_{0,i} + \beta_{1,i} x_1 + \cdots \beta_{K,i} x_K\right)$$

$$= \exp\left(\beta_{0,i} + \sum_{j=1}^{K} \beta_{j,i} x_j\right)$$

$$\tag{3-24}$$

$$i \in [1, 2, \cdots, N] \text{ number of classes}$$

$$p_i = \frac{y_i}{\sum_{a=1}^{N} y_a}$$

We can select one class as a pivot and express the probability of remaining classes as shown in Equation 3-25. In this equation, we have selected the first class as the pivot. As seen from this equation, the number of free parameters is $(N-1)*(K+1)$ if we include a constant in each equation. The free parameters are $\gamma_{j,i}$ where $j \in [0, 1, \cdots, K]$ and $i \in [2, 3, \cdots, N]$ where $\gamma_{j,i} = \beta_{j,i} - \beta_{j,1}$.

$$p_i = \frac{y_i}{\sum_{a=1}^{N} y_a} = \frac{\frac{y_i}{y_1}}{1 + \sum_{a=2}^{N} \frac{y_a}{y_1}}$$

$$\text{for } i \in [2, 3, \cdots, N]$$

$$p_1 = 1 - \sum_{a=2}^{N} p_a$$

$$p_i = \frac{\exp\left(\beta_{0,i} - \beta_{0,1} + \left(\beta_{1,i} - \beta_{1,1}\right) x_1 + \cdots + \left(\beta_{K,i} - \beta_{K,1}\right) x_K\right)}{1 + \sum_{a=2}^{N} \left[\beta_{0,a} - \beta_{0,1} + \left(\beta_{1,a} - \beta_{1,1}\right) x_1 + \cdots + \left(\beta_{K,a} - \beta_{K,1}\right) x_K\right]}$$

$$= \frac{\exp\left(\gamma_{0,i} + \gamma_{1,i} x_1 + \cdots + \gamma_{K,i} x_K\right)}{1 + \sum_{a=2}^{N} \left[\gamma_{0,a} + \gamma_{1,a} x_1 + \cdots + \gamma_{K,a} x_K\right]}$$

$$\tag{3-25}$$

Application
Volatility and return are important predictors used by portfolio managers to decide which stock to buy. Asset managers want to hold assets that have high return with low volatility or risk. Let us try to predict whether the return of S&P 500 over a five-day holding period will fall in one of the four buckets defined as follows. Let us partition the scale of returns into two distinct buckets signifying low or high

Table 3-3. Selecting Return
and Volatility Thresholds at
50% for S&P 500

Level	Return	Volatility
count	2262.000	2262.000
mean	0.002	0.011
std	0.023	0.009
min	−0.180	0.000
25%	−0.007	0.005
50%	0.004	0.008
75%	0.014	0.014
max	0.174	0.082

returns. The return partitions are $(-\infty, 0.004]$ and $(0.004, \infty)$ and represent low
and high returns, respectively. Similarly, let us partition volatility into two buckets
as $[0, 0.008]$ and $(0.008, \infty)$ corresponding to low and high volatility regimes,
respectively. The threshold values for return and volatility regimes are selected by
examining the distribution of return and volatility over the training dataset. Values
at 50% level are selected, as seen from Table 3-3. The final four buckets are then
defined by taking a cross product of the two return and volatility regimes; for
example, one would get high return high volatility, high return low volatility, etc.
Both return and volatility of returns are calculated over a five-day period. A plot of
S&P 500 closing value is shown in Figure 3-11.

Let us denote the endogenous variable as the probability that the five-day return
of S&P 500 falls in each of the four return-volatility buckets. Let $P_i(t)$ denote the
probability that the five-day return falls in bucket i, where $i \in [0, 1, 2, 3]$. Let us use
the following exogenous (predictor) variables:

1. Return for the past five days, $r(t - 5) = \frac{P(t)-P(t-5)}{P(t-5)}$.
2. Volatility of returns observed over the past five days, $V(t - 5)$.
3. Difference between five-day and ten-day historical moving averages of closing
   price, $ma_{5-10}(t)$. This feature is a momentum indicator because it shows how a
   shorter moving average (five-day) compares with a longer moving average (ten-
   day).
4. Change in the yield of a ten-year US treasury bond over the last five days,
   $\Delta Y_{10yr}(t)$.
5. Change in the yield of a one-month US treasury bill over the last five days,
   $\Delta Y_{1Month}(t)$.

The model is formulated in Equation 3-26. It has four equations with six
parameters each. There are five exogenous variables listed above and one constant
for each bucket. However, not all of these parameters are independent. The equation
for each bucket gives the probability of belonging to that bucket. Therefore, the sum

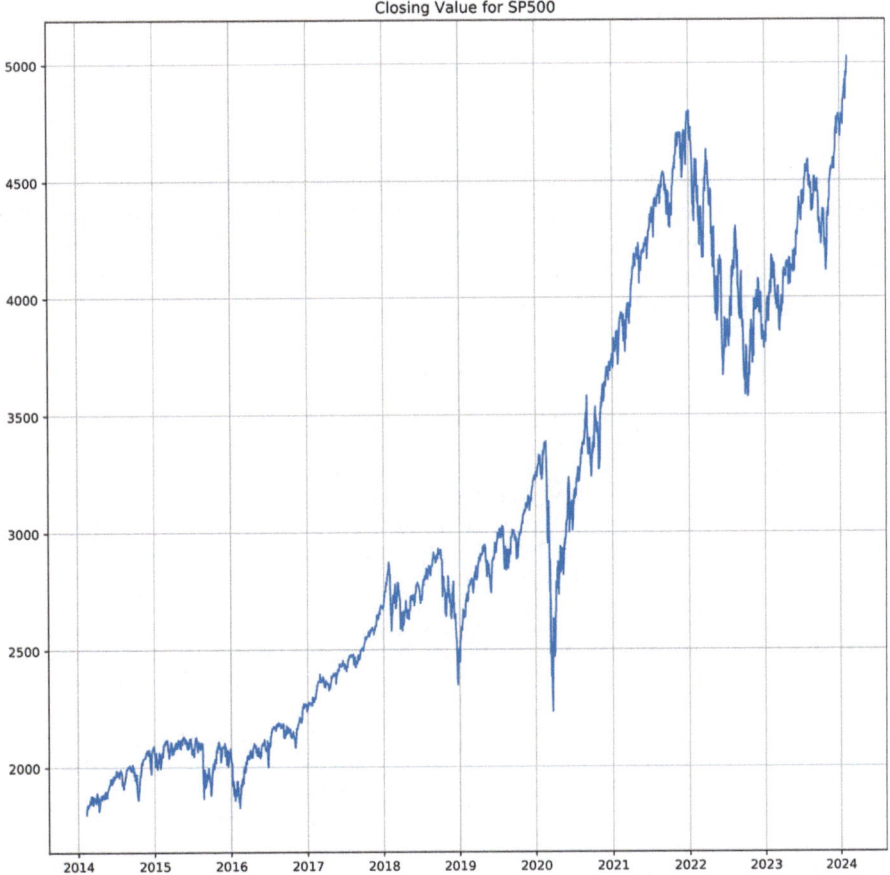

Fig. 3-11. S&P 500

of probabilities in all buckets should be one. As described in Equation 3-25, this constraint reduces the number of free parameters in the model to 18 ($=(4-1) \times 6$).

$$y_i(t) = \exp\left(\alpha_i + \beta_{i,1}r(t-5) + \beta_{i,2}V(t-5)+\right.$$

$$\beta_{i,3}ma_{5-10}(t) + \beta_{i,4}\Delta Y_{10yr}(t) + \beta_{i,5}\Delta Y_{1Month}(t)\left.\right)$$

$$i \in [0, 1, 2, 3, \cdots, 8]$$

$$p_i = \frac{y_i}{\sum_{a=1}^{4} y_a} = \frac{\frac{y_i}{y_1}}{1 + \sum_{a=2}^{4} \frac{y_a}{y_1}} \tag{3-26}$$

for i $\in [2, 3, 4]$

$$p_1 = 1 - \sum_{a=2}^{4} p_a$$

Fitting the model to data, we get the coefficients shown in Listing 3-7. The confusion matrix for training data is shown in Figure 3-12. Diagonal entries in the confusion matrix show the number of elements classified correctly for each class. We observe that the model does a better job in classifying observations belonging to low return low volatility and high return high volatility classes.

Listing 3-7. Fitting Multinomial Logistic GLM Model

MNLogit Regression Results						
Dep. Variable :		y	No. Observations :			2235
Model:		MNLogit	Df Residuals :			2217
Method:		MLE	Df Model:			15
Date:	Wed, 29 May 2024		Pseudo R−squ.:			0.1544
Time:		19:41:40	Log−Likelihood:			−2618.9
converged:		True	LL−Null:			−3097.1
Covariance Type:		nonrobust	LLR p−value:			2.954e−194
y=1	coef	std err	z	P>|z|	[0.025	0.975]
constant	−0.1205	0.180	−0.670	0.503	−0.473	0.232
Last5DayRet	24.9083	4.667	5.337	0.000	15.761	34.056
Last5DayVolat	79.3006	17.312	4.581	0.000	45.370	113.231
ma5_10	−0.5912	0.151	−3.921	0.000	−0.887	−0.296
DGS10_diff	−0.9804	0.685	−1.432	0.152	−2.323	0.362
DGS1MO_diff	−1.6608	1.223	−1.358	0.175	−4.058	0.737
y=2	coef	std err	z	P>|z|	[0.025	0.975]
constant	−1.5544	0.184	−8.449	0.000	−1.915	−1.194
Last5DayRet	0.5138	4.311	0.119	0.905	−7.936	8.964
Last5DayVolat	258.7038	16.662	15.526	0.000	226.046	291.362
ma5_10	−0.8325	0.153	−5.439	0.000	−1.133	−0.532
DGS10_diff	−2.6218	0.699	−3.749	0.000	−3.993	−1.251
DGS1MO_diff	1.1493	1.172	0.981	0.327	−1.147	3.446
y=3	coef	std err	z	P>|z|	[0.025	0.975]
constant	−1.7393	0.187	−9.280	0.000	−2.107	−1.372
Last5DayRet	−9.0519	4.361	−2.076	0.038	−17.599	−0.504
Last5DayVolat	273.0841	16.839	16.217	0.000	240.080	306.088
ma5_10	−1.1208	0.158	−7.100	0.000	−1.430	−0.811
DGS10_diff	−4.5161	0.725	−6.227	0.000	−5.938	−3.095
DGS1MO_diff	2.8562	1.178	2.425	0.015	0.548	5.164

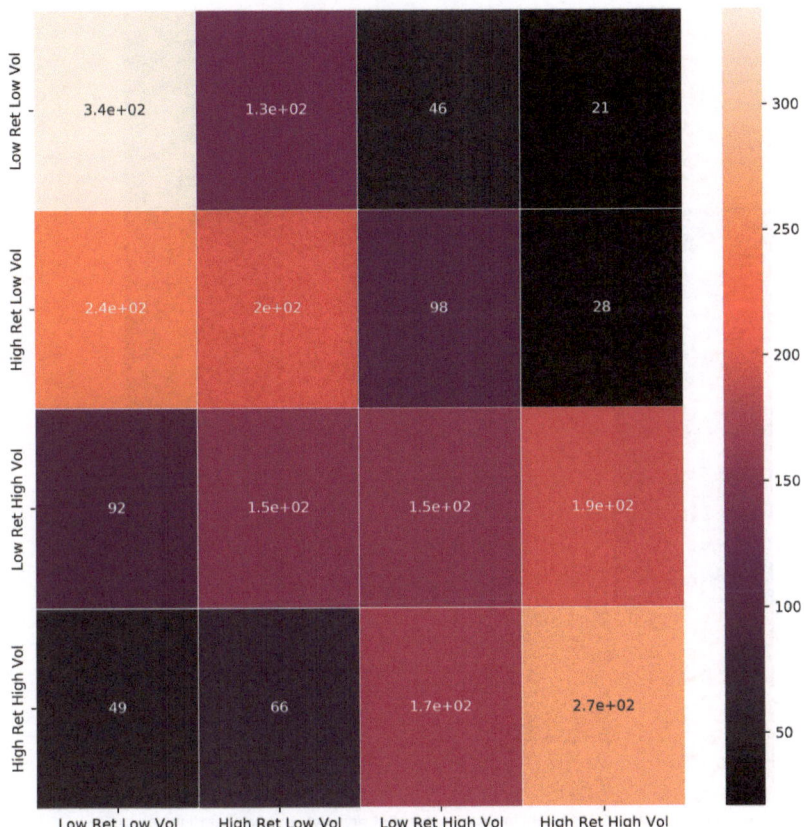

Fig. 3-12. Confusion Matrix of Multinomial GLM Model on Training Data

To evaluate the model performance, the confusion matrix of predictions on testing data is shown in Figure 3-13. On testing data, one observes that the model does a better job at predicting observations in high return high volatility and high return low volatility classes.

The complete code for fitting and testing the model is shown in Listing 3-8.

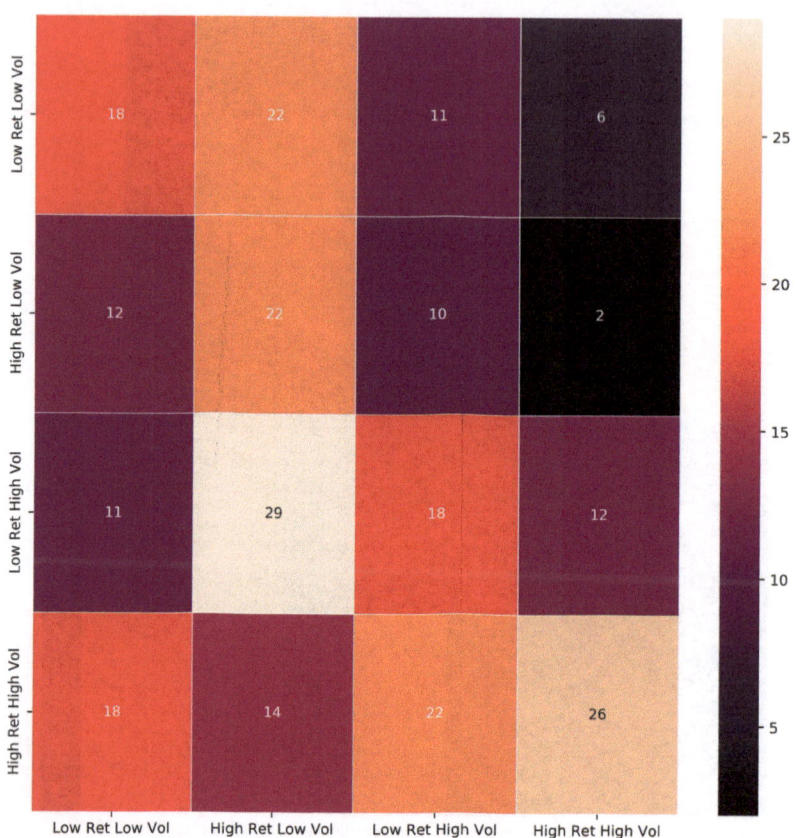

Fig. 3-13. Confusion Matrix of Multinomial GLM Model on Test Dataset

Listing 3-8. Multinomial Logistic Generalized Linear Model

```
1   import numpy as np
2   import pandas as pd
3   import statsmodels.api as sm
4   import matplotlib.pyplot as plt
5   import os
6   import logging
7   import bisect
8   from sklearn.metrics import confusion_matrix
9   import seaborn as sns
10
11
12  DATADIR = r"C:\prog\cygwin\home\samit_000\latex\book_stats\code\data"
13  PLOTDIR = r"C:\prog\cygwin\home\samit_000\latex\book_stats\code\plots"
14  logging.basicConfig(level=logging.DEBUG)
15
16
```

```
17  class MultinomialGLM(object):
18      PERIOD = 5
19
20      def __init__(self, endog, exog, trainPerc=0.9):
21          self.logger = logging.getLogger(self.__class__.__name__)
22          filename = os.path.join(DATADIR, endog + ".csv")
23          self.trainPerc = trainPerc
24          y = pd.read_csv(filename, parse_dates=["DATE"])
25          self.convertColumnToFloat(y, endog)
26          self.plotEndog(y, endog)
27          self.beginIndex = 0
28          self.retThresholds = [0.004]
29          self.volatThresholds = [0.008]
30          self.bucketNames = ["Low Ret Low Vol", "High Ret Low Vol",
31                              "Low Ret High Vol", "High Ret High Vol"]
32          for xi in exog:
33              filename = os.path.join(DATADIR, xi + ".csv")
34              x = pd.read_csv(filename, parse_dates=["DATE"])
35              y = pd.merge(y, x, how="inner", on=["DATE"])
36              self.convertColumnToFloat(y, xi)
37          self.endIndex = y.shape[0] - self.PERIOD
38          self.testdata = int(trainPerc * (self.endIndex - self.beginIndex)) + self.
                beginIndex
39          y = self.calculateEndogExogVars(y, endog, exog)
40          self.df = y
41          self.model = None
42
43      def getBucketNumber(self, retList, volatList):
44          b1 = np.array([bisect.bisect_left(self.retThresholds, ret) for ret in retList])
45          b2 = np.array([bisect.bisect_left(self.volatThresholds, volat) for volat in
                volatList])
46          return (len(self.retThresholds) + 1) * b2 + b1
47
48      def getBucketsAndName(self, num):
49          b2, b1 = divmod(num, len(self.retThresholds) + 1)
50          return b1, b2, self.bucketNames[num]
51
52      def plotEndog(self, df, endog):
53          fig, axs = plt.subplots(nrows=1, ncols=1, figsize=(10, 10))
54          y = df.loc[:, endog].values
55          dates = df.DATE.values
56          axs.plot(dates, y)
57          axs.grid()
58          axs.set_title(f"Closing Value for {endog}")
59          plt.tight_layout()
60          plt.savefig(os.path.join(PLOTDIR, f'{endog}.jpeg'),
61                      dpi=500)
62
63      def calculateEndogExogVars(self, df, endog, exog):
64          vals = df.loc[:, endog].values
65          ret = vals[self.PERIOD:] / vals[0:-self.PERIOD] - 1
66          volatility = self.volatility(ret, self.PERIOD)
67          df.loc[:, "ret"] = 0
```

```python
68          df.loc [:, " volatility "] = 0
69          df.loc [0: df.shape[0]− self.PERIOD−1, "ret"] = ret
70          df.loc [0: df.shape[0]− self.PERIOD−1, "volatility"] = volatility
71          self.logger.info (df.loc [0: self.testdata , ["ret", " volatility " ]].describe ())
72
73          bucketNums = self.getBucketNumber(ret, volatility )
74          df.loc [:, "bucket"] = 0
75          df.loc [0: df.shape[0]− self.PERIOD−1, "bucket"] = bucketNums
76          df.loc [:, "last5DayBucket"] = 0
77          df.loc [ self.PERIOD:, "last5DayBucket"] = bucketNums
78          self.beginIndex = max(self.PERIOD, 11)
79          self.endIndex = df.shape[0]− self.PERIOD
80          self.testdata = int ( self.trainPerc ∗ ( self.endIndex − self.beginIndex)) + self.
                beginIndex
81
82          bucketNameList = ["Bucket_" + str (i) for i in range((len( self.retThresholds )
                +1)∗(len( self.volatThresholds )+1))]
83          last5DayBucketNameList = ["Last5Day_" + b for b in bucketNameList]
84          for lnm, nm in zip(last5DayBucketNameList, bucketNameList):
85              df.loc [:, nm] = 0.0
86              df.loc [:, lnm] = 0.0
87
88          for i in range(df.shape[0]− self.PERIOD):
89              df.loc [i, "Bucket_%d"%bucketNums[i]] = 1.0
90              df.loc [i+ self.PERIOD, "Last5Day_Bucket_%d"%bucketNums[i]] = 1.0
91
92          df.loc [:, "Last5DayRet"] = 0
93          df.loc [ self.PERIOD:, "Last5DayRet"] = ret
94          df.loc [:, "Last5DayVolat"] = 0
95          df.loc [ self.PERIOD:, "Last5DayVolat"] = volatility
96
97          ma5 = self.movingAverage(vals, 5)
98          ma10 = self.movingAverage(vals, 10)
99          df.loc [:, "ma5_10"] = 0.0
100         df.loc [11:, "ma5_10"] = np.where(ma5[10:−1] > ma10[10:−1], 1.0, 0.0)
101
102         ecols_list = []
103         for exog1 in exog:
104             evals = df.loc [:, exog1].values
105             colnm = exog1 + "_diff"
106             df.loc [:, colnm] = 0
107             df.loc [ self.PERIOD:, colnm] = evals[ self.PERIOD:] − evals[0:−self.PERIOD]
108             ecols_list .append(colnm)
109
110         exog_cols = ["Last5DayRet", "Last5DayVolat", "ma5_10"] + ecols_list #
                last5DayBucketNameList + ["Last5DayRet", "Last5DayVolat", "ma5_10"]
111         self.exog = exog_cols
112         self.nvars = len( self.exog)
113         self.endog = bucketNameList
114         return df
115
116     def volatility ( self, arr, period):
117         res = np.zeros( arr.shape [0], dtype=np.float64 )
```

```python
118        sumval = np.sum(arr[0:period])
119        sumsq = np.dot(arr[0:period], arr[0:period])
120        for i in range(period, arr.shape[0]):
121            res[i] = np.sqrt(sumsq/period - (sumval/period)**2)
122            sumval += arr[i] - arr[i-period]
123            sumsq += arr[i]*arr[i] - arr[i-period]*arr[i-period]
124        return res
125
126    def movingAverage(self, arr, period):
127        res = np.zeros(arr.shape[0], dtype=np.float64)
128        sumval = np.sum(arr[0:period])
129        for i in range(period, arr.shape[0]):
130            res[i] = sumval / period
131            sumval += arr[i] - arr[i-period]
132        return res
133
134    def convertColumnToFloat(self, df, col):
135        if (df.loc[:, col] == ".").sum() > 0:
136            df.drop(np.where(df.loc[:, col] == ".")[0], inplace=True)
137            df.reset_index(drop=True, inplace=True)
138        df.loc[:, col] = df.loc[:, col].astype(np.float64)
139
140    def fit(self):
141        y = self.df.loc[self.beginIndex:self.testdata, self.endog].values
142        X = self.df.loc[self.beginIndex:self.testdata, self.exog].values
143        X = sm.add_constant(X, has_constant="add")
144        glm = sm.MNLogit(y, X)
145        glm = glm.fit()
146        glm.endog_names = self.endog
147        self.logger.info(glm.summary(xname=['constant'] + self.exog))
148        summaryfile = os.path.join(PLOTDIR, self.__class__.__name__ + ".txt")
149        with open(summaryfile, 'w') as fh:
150            fh.write(glm.summary(xname=['constant'] + self.exog).as_text())
151        self.model = glm
152
153    def plotTrainingConfusionMatrix(self):
154        yendog = self.model.model.endog
155        yhatv = self.model.predict(self.model.model.exog)
156        mostProbableBucket = np.argmax(yhatv, axis=1)
157        self.plotResults(yendog, mostProbableBucket, label="train")
158
159    def plotResults(self, y_true, y_pred, label="test"):
160        cm = confusion_matrix(y_true, y_pred)
161        df = pd.DataFrame(cm.astype(np.int32), index=self.bucketNames, columns=self.bucketNames)
162        plt.figure(figsize=(10, 10))
163        ax = sns.heatmap(df, annot=True, linewidths=0.5)
164        bottom, top = ax.get_ylim()
165        ax.set_ylim(bottom + 0.5, top - 0.5)
166        plt.savefig(os.path.join(PLOTDIR, f"confusionMatrix_{label}_{self.__class__.__name__}.jpeg"),
                    dpi=500)
168        self.logger.info(df)
```

```
169
170    def  test ( self ):
171        testdata  =  self . testdata  + 1
172        X = self . df . loc [ testdata :,  self .exog]. values
173        X = sm.add_constant(X,  has_constant="add")
174        ypred  =  self .model. predict (X)
175        mostProbableBucket = np.argmax(ypred,  axis=1)
176        actualBucket  = np.argmax( self . df . loc [ testdata :,  self .endog]. values ,  axis=1)
177        self . plotResults ( actualBucket ,  mostProbableBucket)
178
179
180 if  __name__ == "__main__":
181    glm = MultinomialGLM("SP500", ["DGS10", "DGS1MO"])
182    # PAYEMS: Nonfarm payroll (monthly), seasonally  adjusted
183    # CCLACBW027SBOG: Loan on credit card and other revolving plans (weekly) 200
184    # PCEPI: PCE price index  (monthly) 100
185    # DSPIC96: Real disposable  income (monthly) 2000
186    # MORTGAGE30US: 30 year mortgage rate (weekly) 100
187
188    glm. fit ()
189    glm. plotTrainingConfusionMatrix ()
190    glm. test ()
```

Code Explanation

Let us perform a code walk-through below:

1. An object of class **MultinomialGLM** is instantiated with endogenous and exogenous variables as arguments. Closing prices of S&P 500 are used to construct an endogenous variable, while yields on ten-year and one-month US government debt obligations are used as exogenous variables.
2. All the data is available daily, on each trading day. This aids in joining the datasets on date.
3. After reading the datasets, the code defines the four buckets used by the multinomial logistic model as low volatility low return, low volatility high return, high volatility low return, and high volatility high return.
4. Method **calculateEndogExogVars** computes the endogenous and exogenous variables. It first computes the five-day return of S&P 500 followed by the five-day return volatility.
5. The threshold separating low from high returns is set at 0.4%. Similarly, the threshold separating low from high volatility is set at 0.008. These thresholds are selected as 50 percentile points of return and volatility distributions observed in the training dataset, as explained earlier.

6. It assigns high or low return and volatility buckets to each data point inside method **getBucketNumber**.
7. It computes the remaining exogenous variables: last five-day return, last five-day volatility, and difference between five-day and ten-day moving average of S&P 500 prices.
8. The endogenous variable is the bucket to which the data belongs. There are four buckets: low volatility low return, low volatility high return, high volatility low return, and high volatility high return.
9. The multinomial model is fitted to the data using the training dataset using the **MNLogit** class of the **statsmodels** library, passing endogenous and exogenous variables as arguments. A column of ones is added to exogenous variables to include an intercept.
10. Model fit statistics as well as testing results are generated.

3.3   Maximum Likelihood

The objective function for generalized linear models is obtained using the log-likelihood function. The maximum likelihood principle involves maximizing the log-likelihood or minimizing the negative of log-likelihood with respect to model parameters. Since the model parameters are encapsulated within parameter θ in the log-likelihood expression of the exponential family of distributions, we only consider terms involving θ. Using the canonical link function, $\theta = \eta = \mathbf{X}\boldsymbol{\beta}$, we can write the objective function for optimization for each of the distributions considered earlier.

For the normal distribution, the maximum likelihood involves minimizing the function in Equation 3-27, which can be obtained by taking the negative of the log-likelihood function in Equation 3-12 and retaining only the terms involving θ, which is equal to $\mathbf{X}\boldsymbol{\beta}$. For the canonical link function, this is also equal to the predicted output, $\hat{y}$. Equation 3-27 can be solved using OLS.

$$\min_{\beta} \sum_{i=1}^{N} (y_i - \mathbf{X}\boldsymbol{\beta})^2 \text{ normal distribution,}$$

$$\min_{\beta} \sum_{i=1}^{N} \left(y_i - \hat{y}_i\right)^2 \tag{3-27}$$

$$\therefore \sum_{i=1}^{N} (y_i - \mathbf{X}_i \beta)\, \mathbf{X}_i = 0$$

Similarly, for the Poisson distribution, the objective function can be written using the negative of the log-likelihood function from Equation 3-15, as shown in Equation 3-28. $\hat{y}_i$ denotes the output of the model. Equation 3-28 is a non-linear equation that needs to be solved for β using Newton-Raphson or one of the secant methods.

$$\min_{\beta} \sum_{i=1}^{N} y_i \log \mu_i - \mu_i \text{ Poisson distribution}$$

$$\min_{\beta} \sum_{i=1}^{N} y_i \log \hat{y}_i - \hat{y}_i$$

$$\min_{\beta} \sum_{i=1}^{N} y_i \mathbf{X}_i \beta - \exp(\mathbf{X}_i \beta) \text{ using canonical link} \qquad (3\text{-}28)$$

$$\therefore \sum_{i=1}^{N} y_i \mathbf{X}_i - \exp(\mathbf{X}_i \beta) \mathbf{X}_i = 0$$

For the binomial distribution, each observation has $N = 1$. Setting this in Equation 3-17, we get the objective function as shown in Equation 3-29.

$$\min_{\beta} \sum_{i=1}^{N} y_i \log \frac{\hat{y}_i}{1 - \hat{y}_i} + \log\left(1 - \hat{y}_i\right) \text{ binomial distribution}$$

$$\min_{\beta} \sum_{i=1}^{N} y_i \log \hat{y}_i + (1 - y_i) \log\left(1 - \hat{y}_i\right) \qquad (3\text{-}29)$$

$$\min_{\beta} \sum_{i=1}^{N} y_i \log \frac{1}{1 + \exp(-\mathbf{X}_i \beta)} + (1 - y_i) \log\left(1 - \frac{1}{1 + \exp(-\mathbf{X}_i \beta)}\right)$$

For the gamma distribution, note that $E[y] = \hat{y} = \frac{\alpha}{\beta}$ as shown in Equation 3-20. Taking the negative of the likelihood expression from Equation 3-20, we get the objective function shown in Equation 3-30.

$$\min_{\beta} \sum_{i=1}^{N} \frac{y_i}{\hat{y}_i} - \log \hat{y}_i \text{ gamma distribution}$$

$$\min_{\beta} \sum_{i=1}^{N} y_i \mathbf{X}_i \beta + \log(\mathbf{X}_i \beta) \qquad (3\text{-}30)$$

3.4 Binary Data

Binary data involves the output belonging to one of two classes, and it is customary to denote the output as 0 or 1. Within the framework of generalized linear models, there are a few alternative link functions available to handle binary data depending upon the data distribution. A few of these alternatives are shown in Equation 3-31.

$$g(\mu) = \log\left(\frac{\mu}{1-\mu}\right) \text{ logistic regression}$$

$$g(\mu) = \Phi^{-1}(\mu) \text{ probit regression} \tag{3-31}$$

$$g(\mu) = \log\left(-\log\left(1-\mu\right)\right)$$

$$g(\mu) = -\log\left(-\log\mu\right)$$

The first alternative shown in Equation 3-31 corresponds to logistic regression using the sigmoid link function. This is the canonical link function when the response variable y follows a Bernoulli distribution, i.e., $P(y|\mathbf{X}) \sim$ bernoulli. The second alternative in Equation 3-31 corresponds to probit regression. Probit regression is often employed in the Heckman two-step correction procedure for censored regression, described in a later chapter. The third and fourth alternatives are less often used in practice.

Kernel Regression

4

Kernel regression is an elegant method of extending generalized linear regression to infinite-dimensional subspaces by leveraging the tools developed for linear regression. In generalized linear regression, the exogenous variables are linear. By including higher powers and non-linear combinations of exogenous variables, we can take one step toward generalization. This enhancement enables us to improve the accuracy of the fitted model. As an example, let us try to fit a logistic regression model (from the class of generalized linear models) to data shown in Figure 4-1. Logistic regression would attempt to fit a model shown in Equation 4-1. However, it only involves linear exogenous variables. If one augments the space of exogenous variables by adding $\frac{x^2}{a^2} + \frac{y^2}{b^2}$, one can use logistic regression to classify the points. This modified equation is shown in Equation 4-2.

$$y = \frac{1}{1 + \exp - (\alpha + \beta_1 x + \beta_2 y)}$$

(4-1)

$y \in [0, 1]$ probability of belonging to class 1

$$y = \frac{1}{1 + \exp - \left(\alpha + \beta_1 x + \beta_2 y + \beta_3 \left(\frac{x^2}{a^2} + \frac{y^2}{b^2}\right)\right)}$$

(4-2)

$y \in [0, 1]$ probability of belonging to class 1

© Samit Ahlawat 2025
S. Ahlawat, *Statistical Quantitative Methods in Finance*,
https://doi.org/10.1007/979-8-8688-0962-0_4

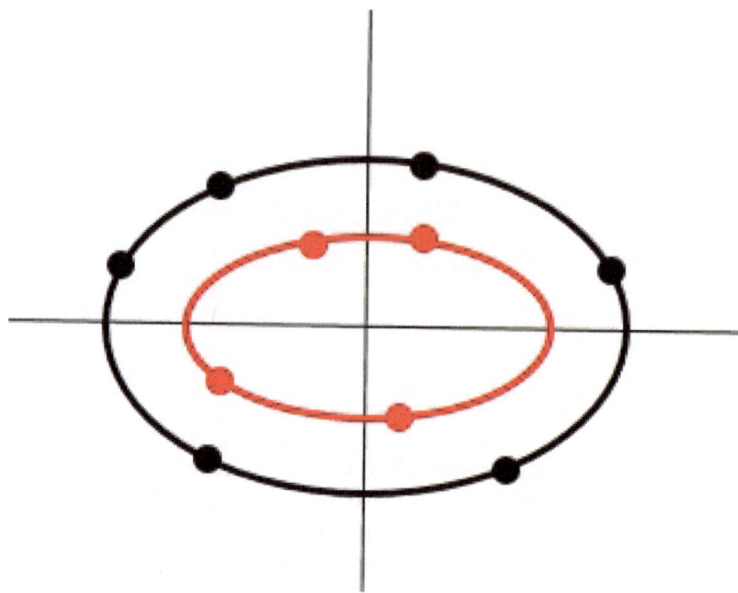

Figure 4-1. Data That Cannot Be Classified Using Logistic Regression with Linear Regressors

Kernel regression generalizes this approach by introducing kernel functions to include infinite-dimensional spaces spanned by exogenous variables. In order to understand how kernel methods achieve this, let us begin with linear regression but include $\psi(\mathbf{X})$ as a vector of exogenous variables. Function ψ may involve raising exogenous variables to powers different from one or taking products with other exogenous variables. We can write the loss function in Equation 4-3. Assuming that the space of transformed exogenous variables $\psi(\mathbf{X})$ span the subspace of model coefficients $\boldsymbol{\beta}$, we can write the weights β_k as a linear combination of transformed exogenous variables $\psi(\mathbf{X})$, as shown in Equation 4-4.

$$L(\boldsymbol{\beta}) = \frac{1}{2}\sum_{i=1}^{N}\left(y_i - \sum_{j=1}^{k}\beta_j\psi_j(\mathbf{X_i})\right)^2 \qquad (4\text{-}3)$$

$$\beta_j = \sum_{k=1}^{K} \gamma_k \psi_k(\mathbf{X})$$

$$L(\boldsymbol{\beta}) = \frac{1}{2} \sum_{i=1}^{N} \left(y_i - \sum_{j=1}^{K} \sum_{k=1}^{K} \gamma_k \psi_k(\mathbf{X_i}) \psi_j(\mathbf{X_i}) \right)^2$$

$$= \frac{1}{2} \sum_{i=1}^{N} \left(y_i - \sum_{j=1}^{K} \sum_{k=1}^{K} \gamma_k K \left(\psi_k(X_i), \psi_j(X_i) \right) \right)^2$$

$$= \frac{1}{2} \sum_{i=1}^{N} \left(y_i - \sum_{k=1}^{K} \gamma_k \sum_{j=1}^{K} K \left(\psi_k(X_i), \psi_j(X_i) \right) \right)^2 \qquad (4\text{-}4)$$

$$= \frac{1}{2} \sum_{i=1}^{N} \left(y_i - \sum_{k=1}^{K} \gamma_k \sum_{j=1}^{K} K_{j,k} \right)^2$$

$$= \frac{1}{2} \sum_{i=1}^{N} \left(y_i - [\gamma_1 \gamma_2 \cdots \gamma_K] \begin{bmatrix} K_{1,1} & K_{1,2} & \cdots & K_{1,K} \\ K_{2,1} & K_{2,2} & \cdots & K_{2,K} \\ & & \cdots & \\ K_{K,1} & K_{K,2} & \cdots & K_{K,K} \end{bmatrix} \right)^2$$

In Equation 4-4, function $K_{i,j}$ is called the kernel function. If we can evaluate the kernel functions, the problem in Equation 4-4 can be solved using OLS. The kernel matrix K needs to be positive semi-definite, i.e., $\mathbf{y}^T K \mathbf{y} \geq 0$ for any vector $\mathbf{y}$. A few examples of kernel functions are listed below:

1. **Linear kernel**: $K(i, j) = \mathbf{X}\mathbf{X}^T$, where $\mathbf{X}^T$ denotes the transpose of vector $\mathbf{X}$. This kernel can also be written as $K_{i,j} = x_i x_j$, where x_i is a component of vector $\mathbf{X}$.
2. **Gaussian or radial kernel**: $K(i, j) = \exp\left(-\frac{\|x_i - x_j\|^2}{2\sigma^2}\right)$.
3. **Laplace kernel**: $K(i, j) = \exp\left(-L \|x_i - x_j\|\right)$.

4.1 Nadaraya-Watson Kernel Regression

Nadaraya [17] and Watson [18] proposed a methodology for kernel regression that circumvents the necessity to solve a linear system. This methodology has been widely adopted for kernel regression in practice because it speeds up computation. By normalizing the kernels as shown in Equation 4-5, one can simply multiply the kernel value with the endogenous variable and obtain a sum for all endogenous variable to make a prediction. N denotes the number of data points or observations.

Equation 4-5 formulates a value of endogenous variable without resorting to OLS for solving a linear system of equations.

$$\hat{y} = \frac{\sum_{i=1}^{N} y_i K_h(x, x_i)}{\sum_{i=1}^{N} K_h(x, x_i)}$$

$$K_h(x, x_i) = \frac{1}{h} K\left(\frac{x}{h}, \frac{x_i}{h}\right) \tag{4-5}$$

$$K_h(\text{gaussian})(x, x_i) = \frac{1}{h\sqrt{2\pi}} \exp\left(-\frac{\|x - x_i\|^2}{2h^2}\right)$$

In Equation 4-5, $\hat{y}$ is the endogenous variable we want to predict when exogenous variables have value x.

For categorical variables, Aitchison and Aitken [19] proposed a kernel shown in Equation 4-6.

$$K_{\text{categorical}}(x, x_d) = \begin{cases} 1 - \lambda & \text{if } x = x_d \\ \frac{\lambda}{M-1} & \text{if } x \neq x_d \end{cases}$$

$$x \in (x_1, x_2, \cdots, x_d, \cdots, x_M) \tag{4-6}$$

There are M categories

λ is a smoothing parameter

For applications with mixed categorical and numerical data, we can combine kernels for categorical and numerical exogenous variables, as shown in Equation 4-7.

$$K_{\text{mixed}}(x, x_i) = \frac{1}{Mh} \sum_{i=1}^{N} K_{\text{categorical}}(x_{\text{categorical}}, x_{d,\text{categorical}})$$

$$K_{\text{numerical}}(x_{\text{numerical}}, x_{d,\text{numerical}}) \tag{4-7}$$

4.2   Application

Let us use kernel regression to predict sector returns using market returns as the predictor (exogenous) variable. The S&P 500 index can be decomposed into 11 different sectors, based upon the nature of business and industry type of component firms. Eleven sectors are listed below. Each of these sectors is represented by an ETF (exchange-traded fund), so that investors can gain investment exposure to individual sectors. Ticker symbols for sectors are listed alongside sector names:

1. Communication services (XLC)
2. Consumer discretionary (XLY)
3. Consumer staples (XLP)

4. Energy (XLE)
5. Financials (XLF)
6. Health care (XLV)
7. Industrials (XLI)
8. Materials (XLB)
9. Real estate (XLRE)
10. Technology (XLK)
11. Utilities (XLU)

Let us predict the daily return of each sector ETF using daily market returns for the last ten days. The model we are using to predict sector return is shown in Equation 4-8.

$$r_{sector}(t) = \beta_1 r_{mkt}(t - 1) + \cdots + \beta_{10} r_{mkt}(t - 10)$$

Predict $r_{sector}(t)$ when $r_{mkt}(t)$ is known

(4-8)

Let us use a Gaussian kernel with σ set as the ten-day standard deviation of market returns. Kernel regression is used to predict daily sector returns for a period of around 4.5 years, from January 2000 to July 2024. Plots of predicted and actual returns for each of the sectors are shown in Figures 4-2 to 4-12.

The root-mean-square error of predicted vs. actual returns is computed using Equation 4-9 and is shown in Table 4-1.

$$\text{RMSE} = \sqrt{\frac{\sum_{i=1}^{N} (y_i - \hat{y}_i)^2}{N}}$$

$\hat{y}_i$ = Predicted sector return

y_i = Actual sector return

(4-9)

Adjusted R^2 is computed using Equation 4-10. In this equation, N denotes the number of observations, $M = 10$ because we use ten kernels, y is the actual return, and $\hat{y}$ is the predicted return.

$$\text{Adj. } R^2 = 1 - \frac{N - 1}{N - M - 1} \frac{\sum_{i=1}^{N} (y_i - \hat{y}_i)^2}{\sum_{i=1}^{N} (y_i - \overline{y})^2}$$

$$\overline{y} = \frac{\sum_{i=1}^{N} y_i}{N}$$

(4-10)

As seen from Table 4-1, the model does a good job of predicting daily returns. Among all the sectors, highest adjusted R^2 of 0.91 is observed for the technology sector, while lowest adjusted R^2 of 0.79 is observed for the energy sector. The root-mean-square error of the energy sector is the largest at around 0.01.

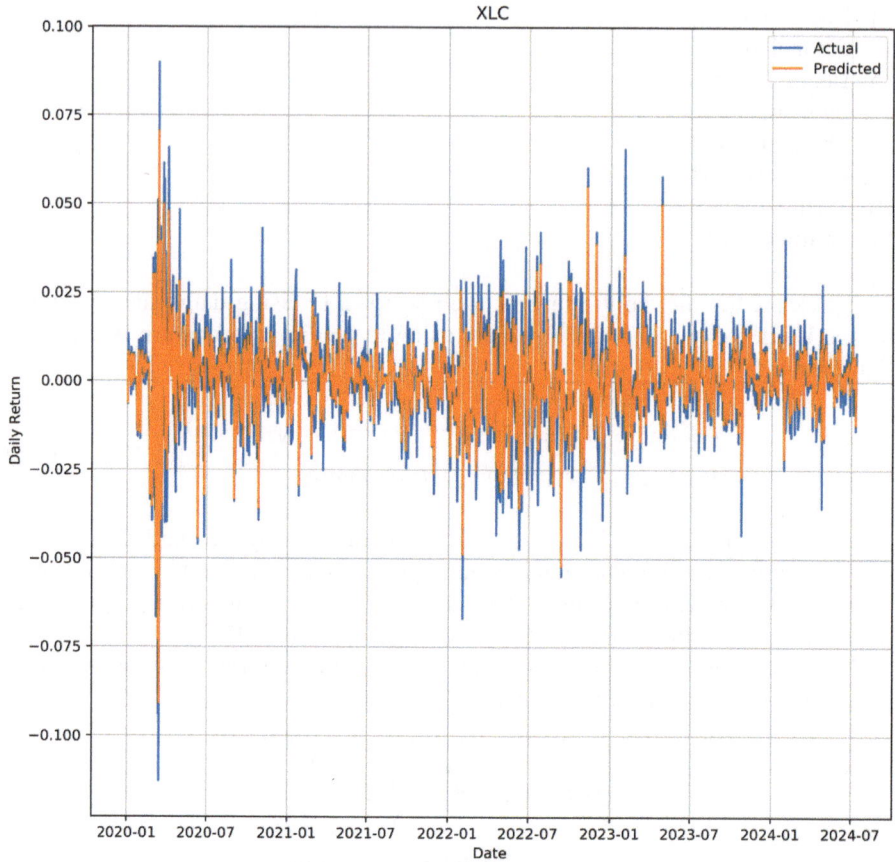

Figure 4-2. Predicted Daily Returns for Communication Services (XLC) vs. Actual Returns

We have used contemporaneous market return as one of the explanatory variables. In practice, we would need a model to predict this quantity because it is not known on day t.

The code for fitting the kernel regression model is shown in Listing 4-1.

Listing 4-1. Using Kernel Regression to Predict Daily Sector Returns

```
1  import numpy as np
2  import pandas as pd
3  import matplotlib.pyplot as plt
4  import logging
5  import os
6
7  logging.basicConfig(level=logging.DEBUG)
8
9
```

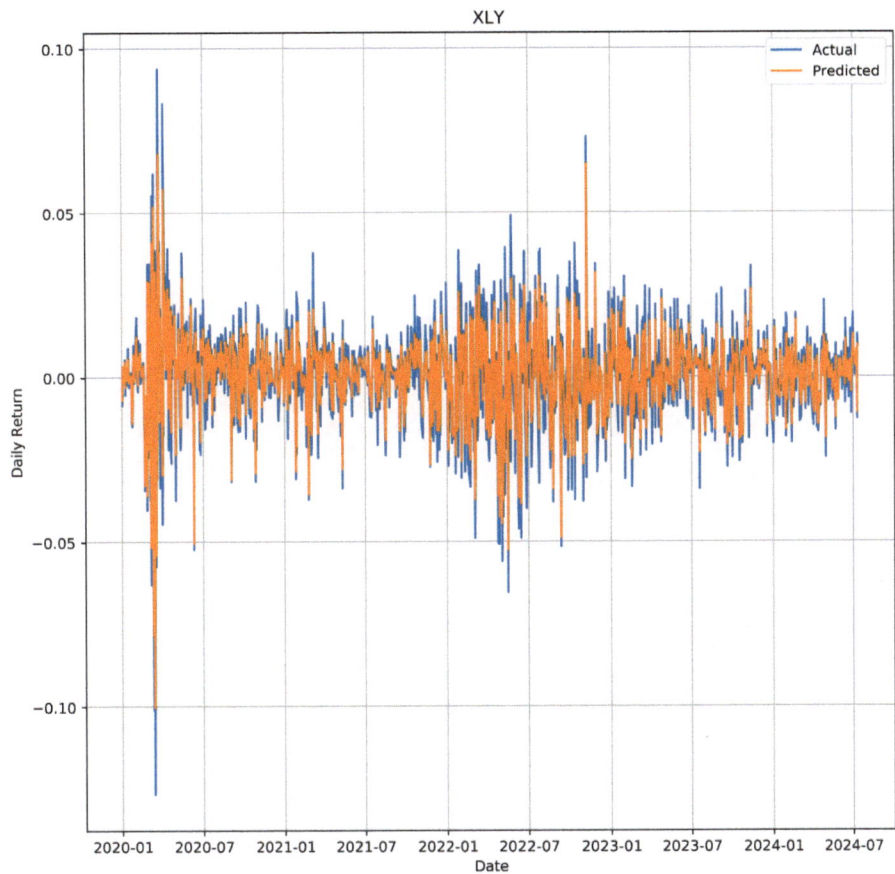

Figure 4-3. Predicted Daily Returns for Consumer Discretionary (XLY) vs. Actual Returns

```
10   class  KernelRegression :
11       def  __init__ ( self , dirname , mktFile="SPY"):
12           self .dirname = dirname
13           self . sectors  = {  'Communication services': 'XLC',
14                               'Consumer discretionary': 'XLY',
15                               'Consumer staples': 'XLP',
16                               'Energy': 'XLE',
17                               ' Financials ': 'XLF',
18                               'Health  care': 'XLV',
19                               ' Industrials ': 'XLI',
20                               ' Materials ': 'XLB',
21                               'Real  estate ': 'XLRE',
22                               'Technology': 'XLK',
23                               ' Utilities ': 'XLU'
24                              }
25           self .mktFile = mktFile
26           self . logger  = logging .getLogger( self .__class__.__name__)
```

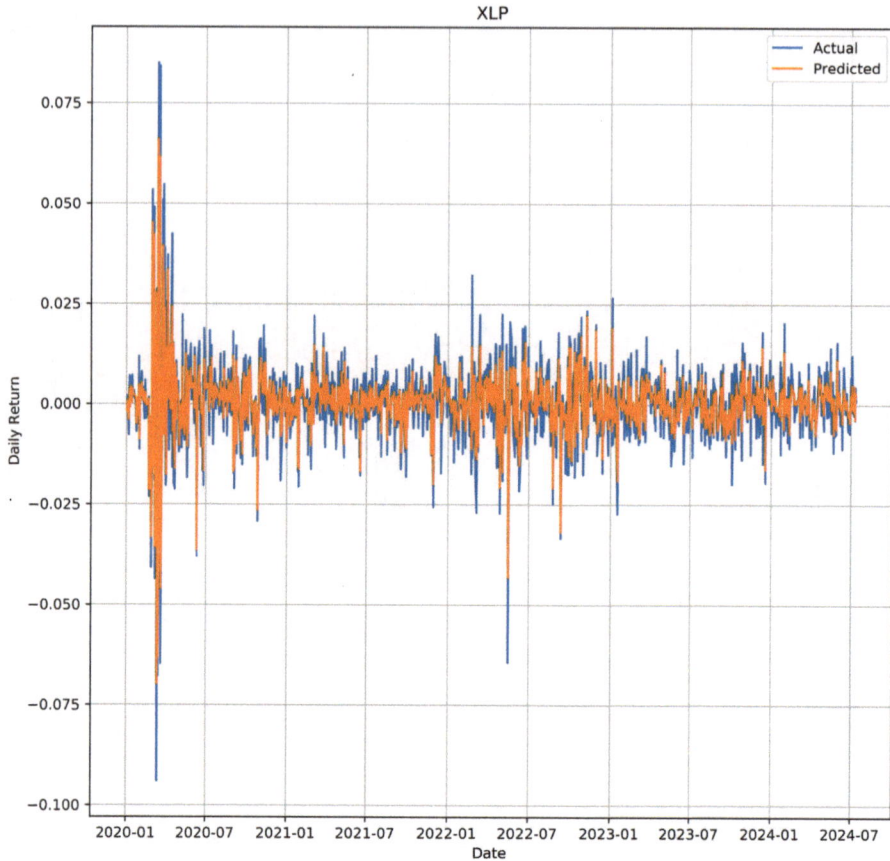

Figure 4-4. Predicted Daily Returns for Consumer Staples (XLP) vs. Actual Returns

```
27          self .symbolToEtf = {v:k  for  k,  v  in  self . sectors . items () }
28          self . dfs  = {}
29          self . mktDf = None
30          self . variance  = None
31          self . readFiles ()
32          self . calculateEndogExogVars()
33
34      def  readFiles ( self ):
35          for  symbol in  self .symbolToEtf.keys () :
36              self . dfs [symbol] = pd.read_csv(os. path . join ( self .dirname, f"{symbol}.csv"),
                        parse_dates =["Date"])
37          self . mktDf = pd.read_csv(os. path . join ( self .dirname, f"{ self .mktFile }.csv"),
                    parse_dates =["Date"])
38
39      def  calculateEndogExogVars( self ):
40          dfs  = [ self .mktDf] +  list ( self . dfs . values () )
41          for  df  in  dfs :
```

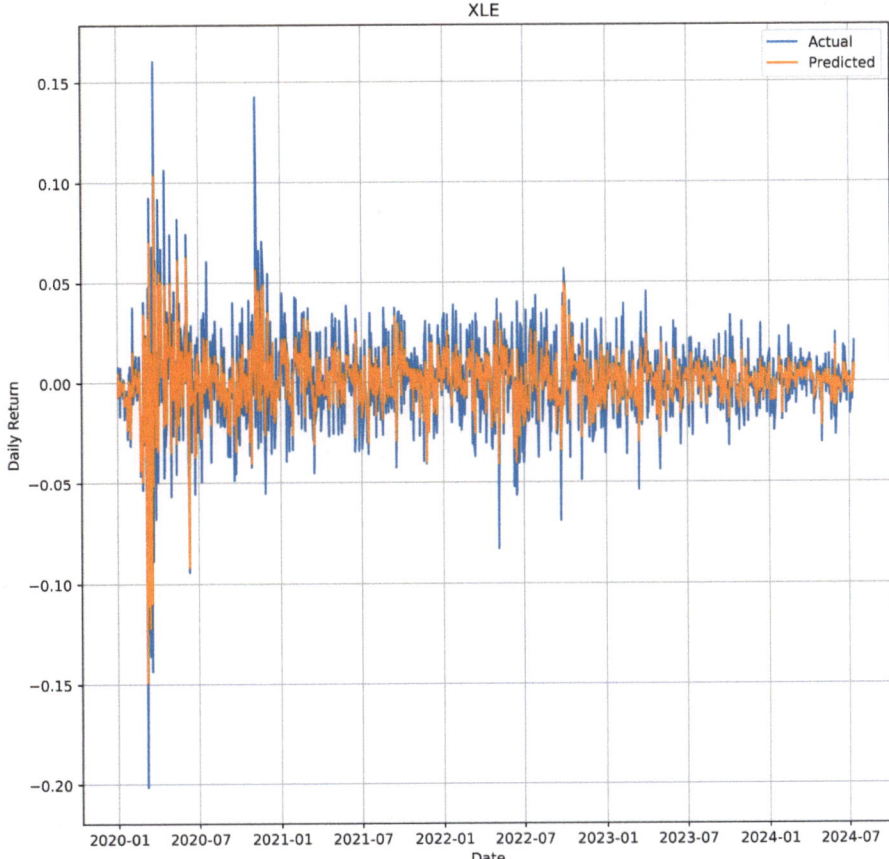

Figure 4-5. Predicted Daily Returns for Energy (XLE) vs. Actual Returns

```
42            df.loc [:, "returns"] = 0
43            price = df.loc [:, "Close"]. values
44            returns = price [1:] / price[0:−1] − 1
45            df.loc [0: df.shape[0]−2, "returns"] = returns
46
47      def calculateKernels (self, x, xi):
48            multipliers = np.array ([1.0/h for h in range(len(xi), 0, −1)])
49            kernels = ( multipliers / self.variance) * np.exp(−(((x − xi) / self.variance)
                    **2)/2.0)
50            normalizedKernels = kernels / kernels.sum()
51            return normalizedKernels
52
53      def calculateRMSE(self, actual, predicted):
54            diff = ( actual − predicted)
55            return np.sqrt (np.sum(diff ** 2) / diff.shape[0])
56
57      def calculateAdjustedR2 ( self, actual, predicted):
```

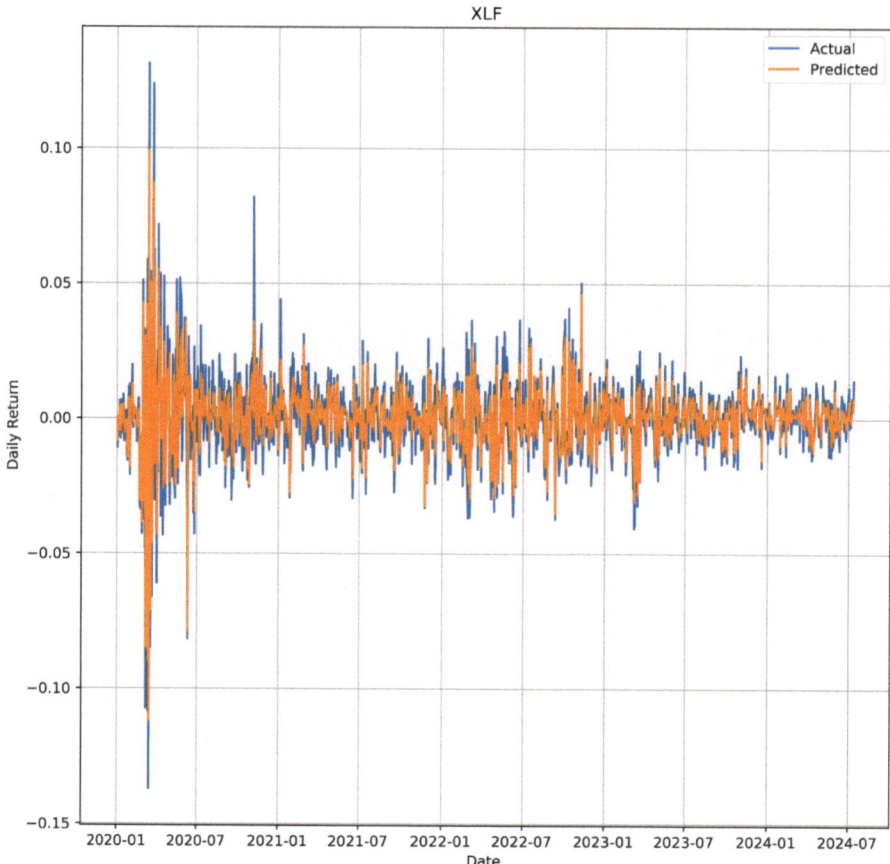

Figure 4-6. Predicted Daily Returns for Financials (XLF) vs. Actual Returns

```
58          diff  = ( actual  − predicted )
59          ssModel = np.sum( diff  ∗∗ 2)
60          avg = np.mean( actual )
61          ssTotal  = np.sum(( actual  − avg) ∗∗ 2)
62          n  =  actual . shape [0]
63          adjR2 = 1  −  ((n−1)/(n−10−1)) ∗ ssModel/ssTotal
64          return  adjR2
65
66      def  plot ( self ,  actual ,  predicted ,  sector ,  begin, end):
67          fig ,  axs  =  plt . subplots (1,  1,  figsize =(10, 10))
68          df  =  self . dfs [ sector ]
69          dates  = df . loc [ begin:end, "Date"]. values
70          axs . plot ( dates ,  actual ,  label ="Actual")
71          axs . plot ( dates ,  predicted ,  label ="Predicted ")
72          axs . grid ()
73          axs . legend ()
74          axs . set_xlabel ("Date")
```

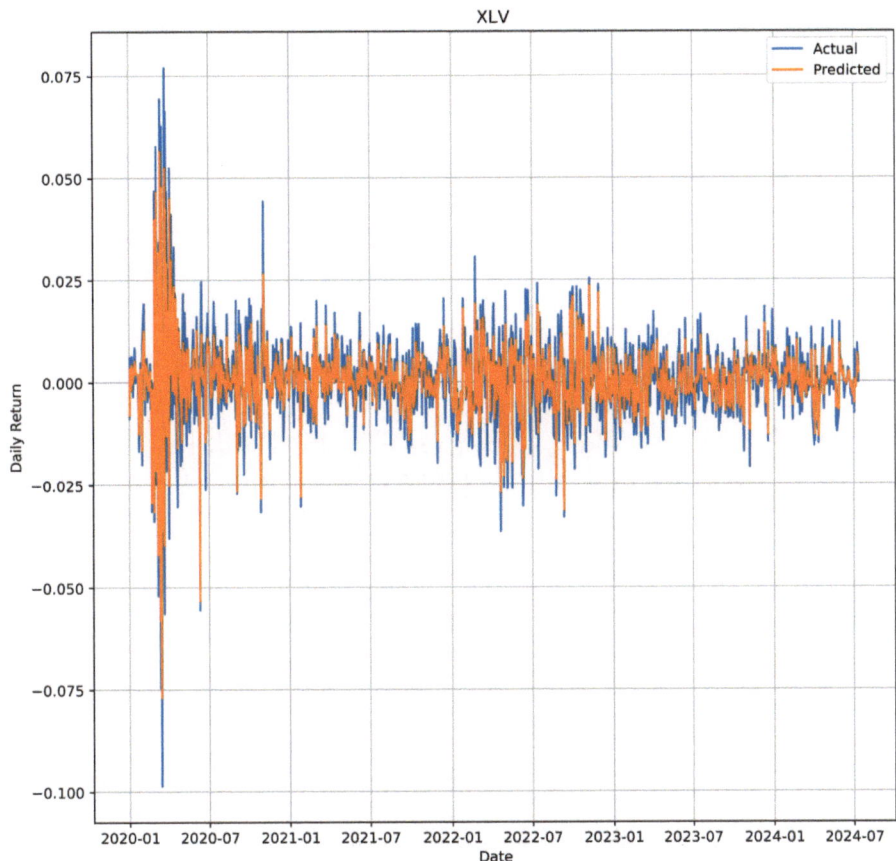

Figure 4-7. Predicted Daily Returns for Health Care (XLV) vs. Actual Returns

```
75        axs.set_ylabel("Daily Return")
76        axs.set(title=sector)
77        plt.savefig(os.path.join(self.dirname, f"kernel_{sector}.jpeg"),
78                    dpi=500)
79        plt.show()
80
81    def predict(self, beginDate, endDate):
82        beginDate = pd.to_datetime(beginDate)
83        endDate = pd.to_datetime(endDate)
84        rmseList = []
85        sectorList = []
86        symbolList = []
87        adjR2List = []
88        for sector in self.symbolToEtf.keys():
89            df = self.dfs[sector]
90            begin = df.loc[df.Date == beginDate, :].index[0]
91            end = df.loc[df.Date == endDate, :].index[0]
```

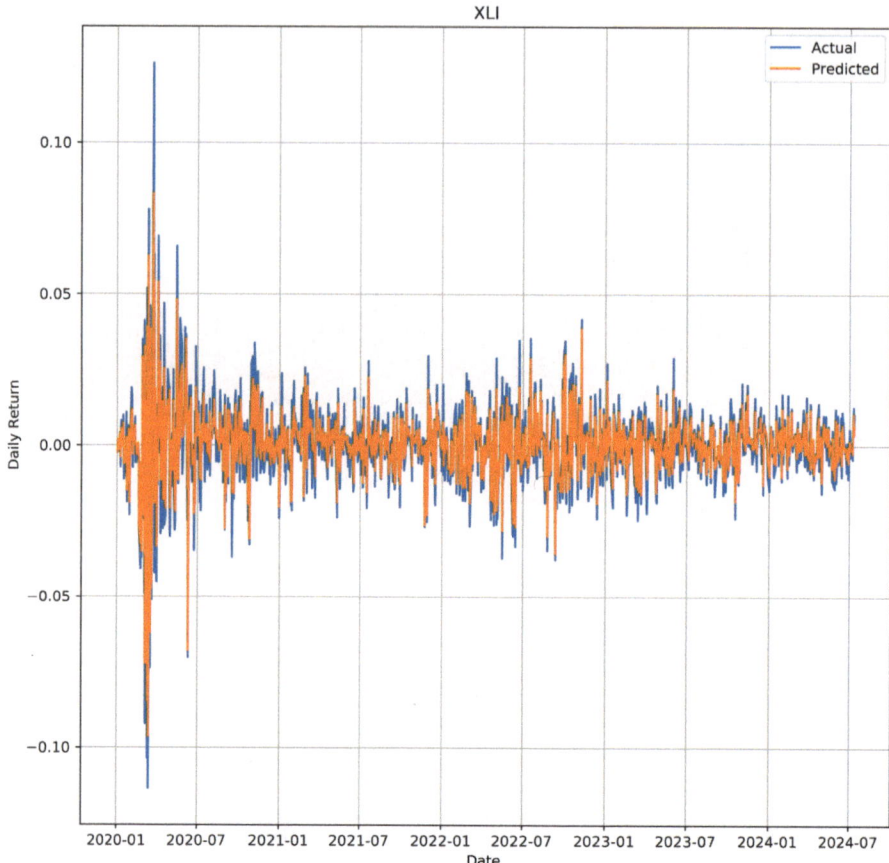

Figure 4-8. Predicted Daily Returns for Industrials (XLI) vs. Actual Returns

```
92         beginMkt = self.mktDf.loc[self.mktDf.Date == beginDate, :].index[0]
93         actual = df.loc[begin:end, "returns"].values
94         predicted = np.zeros(actual.shape[0], dtype=np.float32)
95         for j in range(begin, end+1, 1):
96             beginIdx = beginMkt + j − begin
97             mktRet = self.mktDf.loc[beginIdx, "returns"]
98             self.variance = np.std(self.mktDf.loc[beginIdx−10:beginIdx, "returns"].
                   values)
99             prevMktReturns = self.mktDf.loc[beginIdx−10:beginIdx, "returns"].values
100            prevSectorReturns = df.loc[j − 10:j, "returns"].values
101            kernels = self.calculateKernels(mktRet, prevMktReturns)
102            predicted[j−begin] = np.dot(prevSectorReturns, kernels)
103
104        rmse = self.calculateRMSE(actual, predicted)
105        adjr2 = self.calculateAdjustedR2(actual, predicted)
106        self.plot(actual, predicted, sector, begin, end)
```

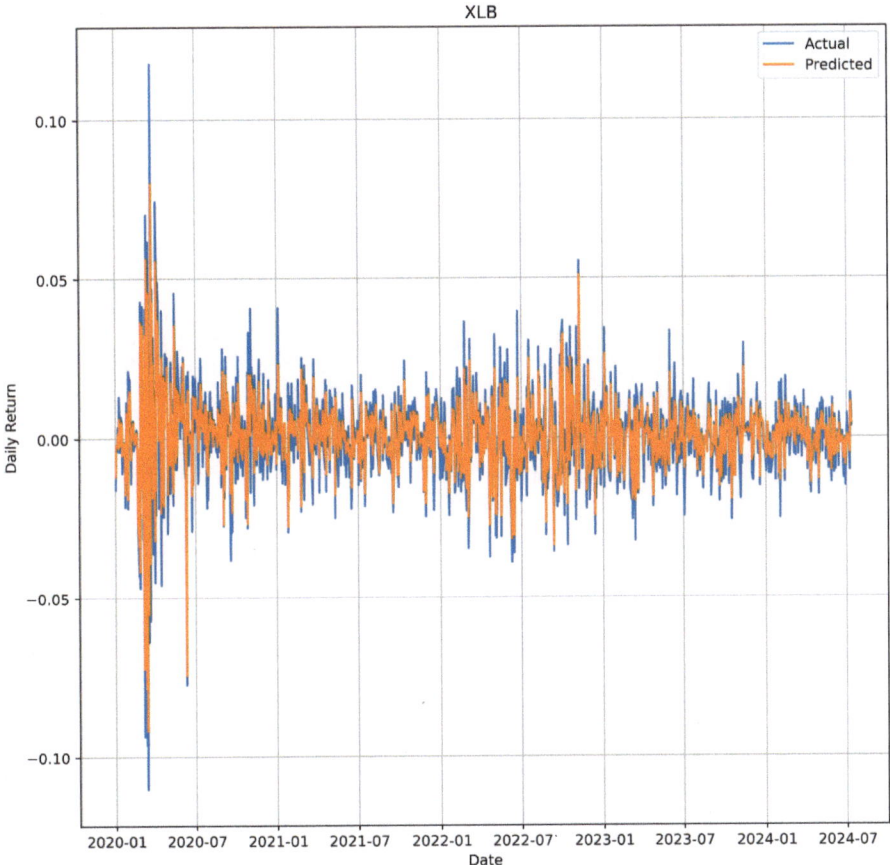

Figure 4-9. Predicted Daily Returns for Materials (XLB) vs. Actual Returns

```
107           self.logger.info("RMSE for sector %s: %f, adj R^2: %f", sector, rmse, adjr2
                  )
108          rmseList.append(rmse)
109          sectorList.append(self.symbolToEtf[sector])
110          symbolList.append(sector)
111          adjR2List.append(adjr2)
112
113     df = pd.DataFrame({"Sector": sectorList,
114                         "Symbol": symbolList,
115                         "RMSE": rmseList,
116                         "Adj. R2": adjR2List})
117     self.logger.info(df)
118
119
120 if __name__ == "__main__":
121     dirname = r"C:\prog\cygwin\home\samit_000\RLPy\data_merged\sectors"
122     kernelReg = KernelRegression(dirname)
```

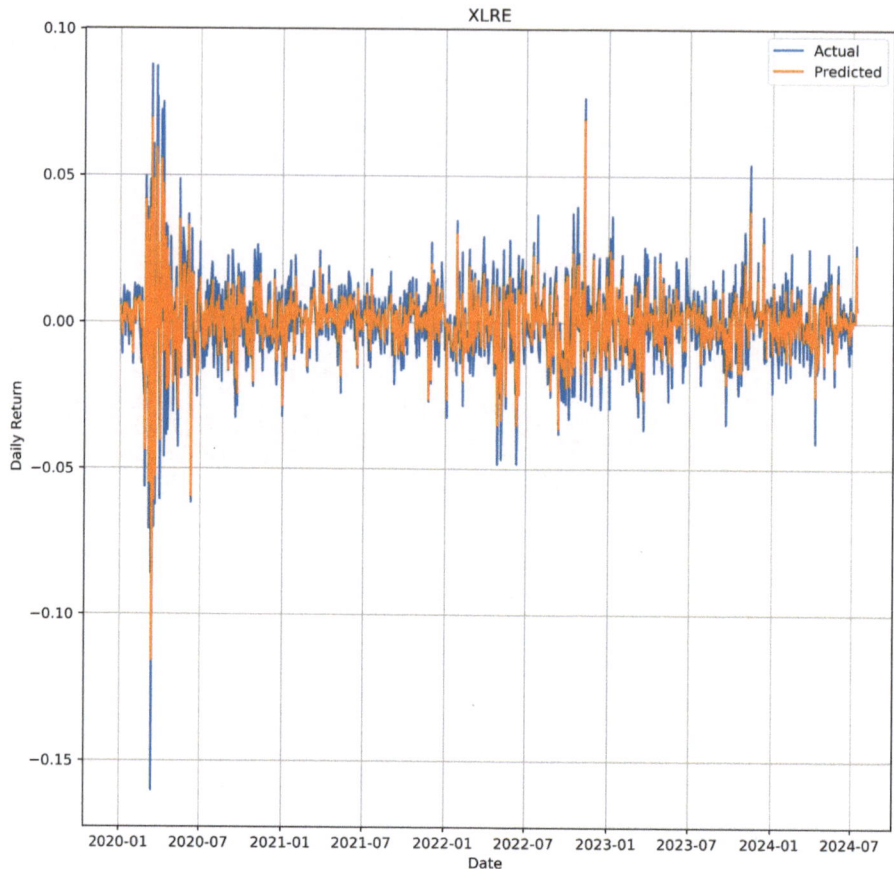

Figure 4-10. Predicted Daily Returns for Real Estate (XLRE) vs. Actual Returns

```
123    beginDate = "2020−01−02"
124    endDate = "2024−07−12"
125    kernelReg. predict (beginDate,  endDate)
```

Code Explanation

Let us do a code walk-through to explain the salient features of the code:

1. The code begins with instantiating an object of class **KernelRegression**, passing the directory name containing data files as an argument. The constructor (inside method **__init__**) has one additional argument with a default value:

   - **mktFile**: Name of file containing S&P 500 end-of-day prices

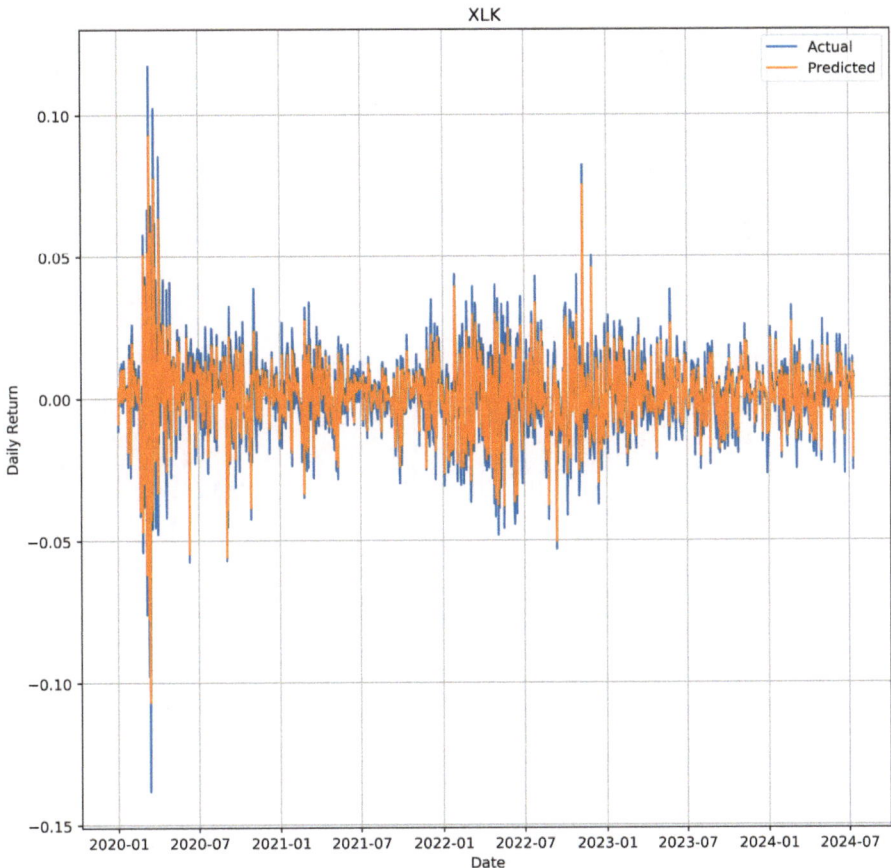

Figure 4-11. Predicted Daily Returns for Technology (XLK) vs. Actual Returns

2. Inside the constructor, data files are read as pandas dataframes. One data file for each sector and one file for S&P 500 prices is read. This happens inside method **readFiles**.
3. Endogenous and exogenous variables are calculated inside method **calculateEn-dogExogVars**. The endogenous variable is a one-day return, calculated using Equation 4-11.

$$y_{\text{sector}}(t) = \frac{P_{\text{sector}}(t)}{P_{\text{sector}}(t-1)} - 1 \qquad (4\text{-}11)$$

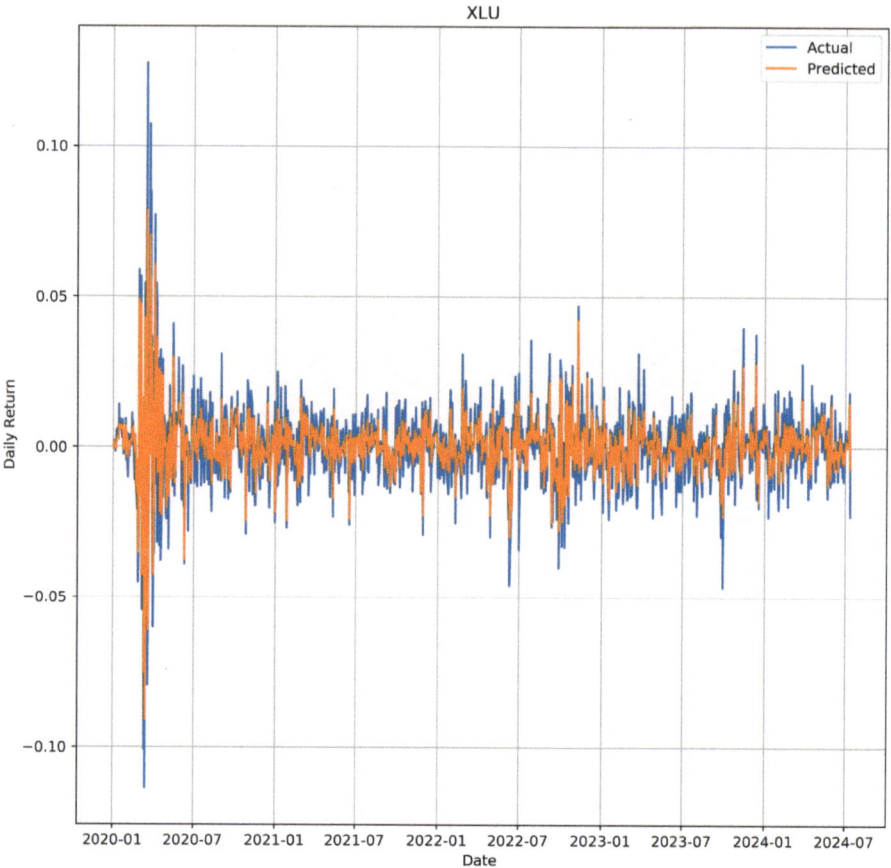

Figure 4-12. Predicted Daily Returns for Utilities (XLU) vs. Actual Returns

Exogenous variables are ten lagged daily returns, calculated using Equation 4-12.

$$x_{i,\mathrm{mkt}}(t) = \frac{P_{\mathrm{mkt}}(t-i)}{P_{\mathrm{mkt}}(t-i-1)} - 1$$

$$i \in (1, 2, \cdots, 10)$$

$$(4\text{-}12)$$

4. After this, the predict method is called with begin date (January 2, 2000) and end date (July 12, 2024) as arguments.
5. Note that kernel regression does not need a model fitting step. We may nevertheless choose to tune some hyper-parameters specific to the kernel in the training phase. For the radial basis function (RBF) kernel, the standard deviation is a hyper-parameter. This is set to the standard deviation of ten-day returns for

Table 4-1. Kernel regression for predicting daily sector returns

	Sector	Symbol	RMSE	Adj. R^2
0	Communication services	XLC	0.005306	0.885886
1	Consumer discretionary	XLY	0.005345	0.896089
2	Consumer staples	XLP	0.004090	0.861334
3	Energy	XLE	0.010848	0.791136
4	Financials	XLF	0.005855	0.882100
5	Health care	XLV	0.004194	0.874688
6	Industrials	XLI	0.004840	0.895615
7	Materials	XLB	0.005450	0.876916
8	Real estate	XLRE	0.006115	0.858806
9	Technology	XLK	0.005154	0.914948
10	Utilities	XLU	0.006254	0.828313

each day we predict a return. Hence, the choice of RBF kernel obviates the need to train the model.

6. Inside the **predict** method, the code steps through each of the 11 sectors. For each sector, it computes the begin and end rows corresponding to **beginDate** and **endDate** arguments passed to this method.

7. For each day in the prediction period, i.e., from begin date to end date, it does the following steps:

- It computes the hyper-parameter for the RBF kernel, h, for each day in the forecast as the standard deviation of daily returns for S&P 500 for the last ten days.
- It obtains the S&P 500 return on the prediction day, $r_{mkt}(t)$, and for ten previous days, $r_{mkt}(t-1)$, $r_{mkt}(t-2)$, ..., $r_{mkt}(t-10)$.
- It computes the ten kernels for each day. It scales the kernels harmonically. This is because, intuitively, we would expect the last day's observation to have a greater impact on prediction than the value ten days ago. The kernel is shown in Equation 4-13.

$$\tilde{K}(i, j) = \frac{1}{i} \frac{1}{h} \exp\left(-\frac{(r_{mkt}(t-i) - r_{mkt}(t-j))^2}{2h^2}\right)$$

$$K(i, j) = \frac{\tilde{K}(i, j)}{\sum_{l=1}^{10} \tilde{K}(i, l)} \quad \text{for } j \in (1, 2, \cdots, 10)$$

(4-13)

- Sector return is predicted using kernel regression and kernels computed earlier, as shown in Equation 4-14.

$$\hat{r_{sector}}(t) = \sum_{j=1}^{10} r_{sector}(t - j)K(t, j) \qquad (4\text{-}14)$$

8. Adjusted R^2 and RMS errors are computed for each sector.
9. It plots the adjusted vs. actual returns for each sector and reports statistics to measure prediction accuracy.

Dynamic Regime Switching Models

5

Regression models we have seen so far involve an equation describing a relationship between endogenous (dependent) and exogenous (independent) variables with coefficients or weights that remain constant. A natural extension of the model involves cases where the structural form of relationship between endogenous and exogenous variables remains the same, but coefficients and error term variance change depending upon the state of some variable. The variable that determines the regime could be latent, i.e., not observed explicitly, or an observable variable. Dynamic regime switching models compute the probability of a particular regime manifesting itself when an observation is recorded. For simplicity, it is customary to use the Markov model for the regime switching process. This assumption implies that the probability of occurrence of a regime is dependent only upon the prior period's regime and other observable variables. For this reason, models analyzed in this chapter will belong to the class of Markov dynamic regime switching models.

Dynamic regime switching models use a set of modeling equations based on manifestation of a particular regime. The individual equations share a set of exogenous variables but have different coefficients for those variables. The error terms of individual equations may have different variances to account for heteroskedasticity across regimes. However, their correlation with each other is zero. Simplifications of regime switching models involve making simplifying assumptions such as restricting the individual equations to share common coefficients for certain exogenous variables or restrictions of common variance of error term across regimes. The model is estimated using the maximum likelihood method.

5.1   Model Formulation

Let us formulate the model mathematically. Let us denote the number of regimes by N and each individual regime by i. This implies there are N distinct governing equations between dependent and independent variables. Let each equation have k

© Samit Ahlawat 2025
S. Ahlawat, *Statistical Quantitative Methods in Finance*,
https://doi.org/10.1007/979-8-8688-0962-0_5

exogenous (explanatory) variables, $\beta_{1,i}, \beta_{2,i}, \cdots, \beta_{k,i}$, with a constant $\beta_{0,i}$. This is shown in Equation 5-1. Linear regime switching models, such as the one shown in Equation 5-1, have linear coefficients with no cross product or powers different from unity involving the coefficients. Let ϵ_i denote the error term in the equation corresponding to regime i.

$$y = \beta_{0,i} + \beta_{1,i}x_1 + \beta_{2,i}x_2 + \cdots + \beta_{k,i}x_k + \epsilon_i$$

$$i \in \{1, 2, \cdots, N\}$$

$$P(s(t)) = \begin{bmatrix} p_{1,1} & p_{1,2} & \cdots & p_{1,N} \\ p_{2,1} & p_{2,2} & \cdots & p_{2,N} \\ & & \cdots & \\ p_{N,1} & p_{N,2} & \cdots & p_{N,N} \end{bmatrix} \begin{bmatrix} P_{s(t-1)=1} \\ P_{s(t-1)=2} \\ \cdots \\ P_{s(t-1)=N} \end{bmatrix}$$

$$\sum_{i=1}^{N} p_{j,i} = 1 \forall j$$

(5-1)

$$p_{j,i} \geq 0 \ \forall i \text{ and } \forall j$$

$$\epsilon_i \sim N(0, \sigma_i^2)$$

$$\text{Variance}(\epsilon_i) = \sigma_i^2$$

$$\text{Covariance}(\epsilon_i, \epsilon_j) = 0$$

Equation 5-1 is the most general formulation of a linear, first-order Markov dynamic regime switching model with constant state transition probabilities. The correlation between error terms of different regimes is assumed to be zero. This assumption is essential to establish the independence of equation between dependent and independent variables across regimes conditional on the Markov assumption. Model parameters in Equation 5-1 are the model coefficients $\beta_{0,i}, \beta_{1,i}, \cdots, \beta_{k,i}$ for $i \in \{1, 2, \cdots, N\}$, error variances σ_i^2, regime transition probabilities $p_{j,i}$, and initial regime probabilities $P(s(0) = i)$, subject to the constraint $\sum_{i=1}^{N} p_{j,i} = 1$ for all regimes j. Regime transition probability $p_{j,i}$ denotes the probability of transitioning from regime i to regime j in the next time step. They are estimated using the maximum likelihood method. Let us look at the method of computing the parameters in the next section.

5.2   Model Estimation Using Maximum Likelihood

In order to estimate model parameters using the maximum likelihood method, we must write the probability of observing a set of observations under the assumption that the model is true. In a regime switching model, each regime has a probability of manifesting itself at a given time step. If we have T observations of exogenous and endogenous variables, we can think of time going from $t = 1, 2, \cdots, T$. At time step t, the probability of manifestation of regime i is $P(s(t) = i)$. According to the Markov assumption, this is determined by the regime transition probabilities $p_{i,j}$ and the regime probability at previous time step $t - 1$. Accordingly, probability $P(s(t) = i)$ can be written as $\sum_{j=1}^{N} P(s(t) = i | s(t - 1) = j) P(s(t - 1) = j)$. Using matrix multiplications and iterating the time steps backward toward the initial observation at $t = 0$, we can write the probability of manifestation of regime i, $P(s(t) = i)$, at time t using Equation 5-2.

$$
\begin{aligned}
P\left(s(t)\right) &= \begin{bmatrix} p_{1,1} & p_{1,2} & \cdots & p_{1,N} \\ p_{2,1} & p_{2,2} & \cdots & p_{2,N} \\ & & \cdots & \\ p_{N,1} & p_{N,2} & \cdots & p_{N,N} \end{bmatrix} \begin{bmatrix} P_{s(t-1)=1|s(t-2)} \\ P_{s(t-1)=2|s(t-2)} \\ \cdots \\ P_{s(t-1)=N|s(t-2)} \end{bmatrix} \\[2em]
&= \begin{bmatrix} p_{1,1} & p_{1,2} & \cdots & p_{1,N} \\ p_{2,1} & p_{2,2} & \cdots & p_{2,N} \\ & & \cdots & \\ p_{N,1} & p_{N,2} & \cdots & p_{N,N} \end{bmatrix}^2 \begin{bmatrix} P_{s(t-2)=1|s(t-3)} \\ P_{s(t-2)=2|s(t-3)} \\ \cdots \\ P_{s(t-2)=N|s(t-3)} \end{bmatrix} \\[2em]
&= \begin{bmatrix} p_{1,1} & p_{1,2} & \cdots & p_{1,N} \\ p_{2,1} & p_{2,2} & \cdots & p_{2,N} \\ & & \cdots & \\ p_{N,1} & p_{N,2} & \cdots & p_{N,N} \end{bmatrix}^t \begin{bmatrix} P_{s(0)=1} \\ P_{s(0)=2} \\ \cdots \\ P_{s(0)=N} \end{bmatrix}
\end{aligned}
\tag{5-2}
$$

If regime i is applicable, the probability of observation at time step t is given by Equation 5-3.

$$
p\left(y(t), x(t) | s(t) = i\right) = \frac{1}{\sqrt{2\pi \sigma_i^2}} \exp\left(-\frac{\epsilon_i(t)^2}{2\sigma_i^2}\right)
\tag{5-3}
$$

$$
\epsilon_i(t) = y(t) - \beta_{0,i} + \beta_{1,i} x_1(t) + \beta_{2,i} x_2(t) + \cdots + \beta_{k,i} x_k(t)
$$

However, each regime has a probability of occurrence at time step t. Therefore, the combined probability of observing the data $y(t), x_1(t), \cdots, x_k(t)$ at time step t is given by Equation 5-4.

$$
p\left(y(t), x(t)\right) = \sum_{i=1}^{N} p\left(y(t), x(t) | s(t) = i\right) P\left(s(t) = i\right)
\tag{5-4}
$$

Finally, the combined probability of observing the dataset across all time steps from $t = 1$ to T is given by Equation 5-5. It is customary to use log-likelihood instead of raw likelihood. This alleviates the problem of numerical underflow in computation when multiplying a large number of observations less than one. The transformation does not change the maximum of likelihood equation because the logarithm is a monotonic function. Taking the logarithm of Equation 5-5, we get the final expression for log-likelihood as shown in Equation 5-6.

$$P(\text{data}) = \prod_{t=1}^{T} p\left(y(t), x(t)\right) \tag{5-5}$$

$$\log\left(P(\text{data})\right) = \sum_{t=1}^{T} \log\left(p\left(y(t), x(t)\right)\right) \tag{5-6}$$

Substituting Equations 5-2, 5-3, and 5-4 in Equation 5-6 yields the final expression of log-likelihood of observing the data conditional on all model parameters. In order to find the maximum of this expression with respect to model parameters, we take the derivative and set it to zero as shown in Equation 5-7. The evaluation of resulting equations must be done numerically. For example, one could use Newton's iterations shown in Equation 5-8, with derivatives evaluated numerically at the current parameter values, in order to obtain updated parameter values. θ denotes the vector of model parameters in Equation 5-8. $\theta(m)$ denotes the model parameters at iteration m. We can observe that Equation 5-8 involves gradient ∇_θ and inverse of Hessian $\nabla_\theta \nabla_\theta$ matrices.

$$\max_{\theta} \log\left(P(\text{data})\right)$$

$$\nabla_\theta \log\left(P(\text{data})\right) = 0 \tag{5-7}$$

$$\equiv f(y, x, \theta) = 0$$

$$\theta(m) = \theta(m-1) - \left(\nabla_\theta f(y, x, \theta)\right)^{-1} f(y, x, \theta)$$
$$= \theta(m-1) - \left(\nabla_\theta \nabla_\theta \log\left(P(\text{data})\right)\right)^{-1} \nabla_\theta \log\left(P(\text{data})\right) \tag{5-8}$$

Iterations begin with an initial guess for parameter values and proceed using Newton's iterations. Gradient and Hessian matrices can be calculated analytically or estimated numerically.

In order to assess the goodness of fit, one uses an information criterion such as the Akaike Information Criterion (AIC) or the Bayesian Information Criterion (BIC). The aforementioned information criteria are described in greater detail in Subsection 2.8.4. Intuitively, the information criteria account for the goodness of fit by including a term proportional to negative log-likelihood and penalize complex models—those with greater number of free parameters—by including a term that scales with the number of model parameters. Better models have lower information criteria.

5.3 Dynamic Transition Probabilities

We can generalize the model presented in the last section by equipping it with changing transition probabilities. It may be recalled that in the last section, regime transition probabilities conditional on prior state were constant because the transition matrix with probabilities $p_{i,j}$ in Equation 5-1 had constant values. Diebold et al. (1993) proposed a generalization to the Markov dynamic regime switching model to incorporate time-varying conditional regime transition probabilities. This additional capability is apposite for applications where it is reasonable to believe that conditional transition probability will depend on exogenous variables in addition to current and prior regimes. It is instructive to note that in the last section, conditional transition probability $p_{i,j}$ was dependent only on current regime i and prior regime j.

Let us write the conditional transition probability matrix as shown in Equation 5-9. In the most general form, this method adds $(k-1) \times N \times$ additional free parameters to the model where $k-1$ is the number of exogenous variables in the expression for transition probability and N is the number of regimes.

$$p_{i,j}(t) = \frac{\exp\left(\gamma_{0,i,j} + \gamma_{1,i,j}x_1(t-1) + \cdots + \gamma_{k-1,i,j}x_{k-1}(t-1)\right)}{1 + \exp\left(\gamma_{0,i,j} + \gamma_{1,i,j}x_1(t-1) + \cdots + \gamma_{k-1,i,j}x_{k-1}(t-1)\right)}$$

$$p_{i,i}(t) = 1 - \sum_{j=1}^{k-1} \frac{\exp\left(\gamma_{0,i,j} + \gamma_{1,i,j}x_1(t-1) + \cdots + \gamma_{k-1,i,j}x_{k-1}(t-1)\right)}{1 + \exp\left(\gamma_{0,i,j} + \gamma_{1,i,j}x_1(t-1) + \cdots + \gamma_{k-1,i,j}x_{k-1}(t-1)\right)}$$

$$(5-9)$$

5.4 EM Algorithm

The expectation-maximization, or EM, algorithm is widely used in machine learning and statistics. It is used to obtain maximum likelihood estimates when the log-likelihood expression involves latent (missing or unobserved) variables, and the likelihood expression can only be evaluated after knowing the latent variables. This necessitates estimating the latent variables first followed by maximization of the resulting log-likelihood expression. Because estimation of latent variables requires specification of model parameter values, the latent variables must be updated once model parameters have been recalculated following maximization of the log-likelihood expression. This yields an iterative process.

Let us formulate the EM algorithm in a generic fashion before applying it to estimate the parameters of the dynamic Markov regime switching model with varying transition probabilities. Let Z denote the latent variables of a model and β denote the set of other model parameters, i.e., those not related to latent state. Latent state Z may depend upon exogenous variables X and some additional parameters, θ. To illustrate with an example, in the Markov dynamic regime switching model with varying transition probabilities, $p_{i,j}$ are the latent variables. They depend upon exogenous variables X and parameters shown in Equation 5-9. Let $p(y, X, Z, \beta)$ denote the likelihood of observing the data (X, y), where y is the endogenous and X is the exogenous variable. Z belongs to a continuum of values but could belong to a discrete set of values in certain applications. Equation 5-10 shows the likelihood expression. This expression can only be evaluated when we know the value of latent state Z. The evaluation of the likelihood expression requires writing it as a probabilistic sum of likelihood values for each realization of latent state, as shown in Equation 5-10. The inequality in Equation 5-10 is a consequence of the concavity of the logarithmic function and the application of Jensen's inequality.

$$p(y, X, Z, \beta) = \int_\theta p(y, X, Z, \beta | \theta) p(\theta) d\theta$$

$$\log p(y, X, Z, \beta) = \log \int_\theta p(y, X, Z, \beta | \theta) p(\theta) d\theta$$

$$\geq \int_\theta \log \left(p(y, X, Z, \beta | \theta) \right) p(\theta) d\theta$$

$$= \int_\theta \log \frac{\left(p(y, X, Z, \beta | \theta) p(y, X, \beta | Z, \theta_0) \right)}{p(y, X, \beta | Z, \theta_0} p(\theta) d\theta$$

$$= \int_\theta \log \left(p(y, X, Z, \beta | \theta) p(y, X, \beta | Z, \theta_0) \right) p(\theta) d\theta -$$

$$\int_\theta \log p(y, X, \beta | Z, \theta_0) p(\theta) d\theta$$

$$= Q(\theta, \theta_0) - E \left[\log p(y, X, \beta | Z, \theta_0) \right]$$

$$\geq Q(\theta, \theta_0)$$

$$\text{because } E \left[\log p(.) \right] < 0$$

$$(5\text{-}10)$$

Equation 5-10 implies that any θ that increases the right-hand side must also increase the log-likelihood by virtue of the inequality. θ_0 denotes an initial value. Furthermore, quantity $Q(\theta, \theta_0)$ is tractable and can be computed. This principle yields the iterative procedure followed by the EM algorithm outlined below:

1. Assume a value of parameters β and θ.
2. **E step**: Compute the latent states using θ and any required exogenous variable.
3. **M step**: Write the log-likelihood expression using the assumed value of θ. Maximize this expression to get the value of parameters β and θ. Go back to the previous step and recompute the latent states.
4. Iterations converge when there is little change in the values of β and θ. At this point, we have the parameter values that give maximum likelihood.

5.4.1   K-Means Algorithm

The K-means algorithm is perhaps one of the most simplistic applications of the expectation-maximization (EM) algorithm. K-means is a clustering algorithm that can operate in multidimensional space and produces hard, i.e., non-probabilistic, assignment of points to one of the clusters.

Let us begin with a set of M points, represented as $\mathbf{X_i}$, where $\mathbf{X}$ represents multidimensional coordinates of a point i drawn from N-dimensional space. If we suspect the data to be clustered and have an intuition about the number of clusters, we can use the K-means algorithm to classify the points into one of those clusters.

The algorithm needs the number of clusters, K, as an input. We are trying to minimize the function shown in Equation 5-11 with respect to the assignment of points to clusters, I, and cluster centroids, C. In this formulation, I is the latent variable and C are the model parameters. We cannot evaluate the objective function without knowing the assignment of points to clusters, I. Assuming we have cluster centroids fixed, we can show that the objective function in Equation 5-11 is minimized by the assignment of points shown in Equation 5-12.

$$\min_{I,C} \sum_{i=1}^{M} \sum_{j=1}^{K} I_{i,j} \left\| \mathbf{X_i} - \mathbf{C_j} \right\|^2 \tag{5-11}$$

$I_{i,j} = 1$ if point i belongs to cluster j and 0 otherwise

Assign point i to cluster j

$$\underset{j}{\operatorname{argmin}} \left\| \mathbf{X_i} - \mathbf{C_j} \right\|^2 \tag{5-12}$$

Once we know the assignment of points to clusters, I, the objective function in Equation 5-11 is minimized when cluster centroids are set to the values shown in Equation 5-13.

$$C_j = \frac{\sum_{i=1}^{M} I_{i,j} \mathbf{X_i}}{\sum_{i=1}^{M} I_{i,j}} \tag{5-13}$$

$I_{i,j} = 1$ if point i belongs to cluster j and 0 otherwise

The algorithm is sketched in pseudo-code 3.

Algorithm 3 K-Means Algorithm

Require: A set of points with coordinates X_l and a number K representing the number of clusters to define. Let us assume the points are drawn from N dimensional space.

1: Set threshold ϵ to a small value. This threshold is used to assess if the algorithm has converged.
2: Select K points representing the centroid (center) of each of the K clusters. This can be assigned arbitrarily. Let us denote the coordinates of the jth cluster centroid by $C(j)$.
3: **for** t = 1, 2, $\cdots$, until convergence **do**
4:     **E-step**: Assign each point to the cluster with the centroid closest to that point, as shown in Equation 5-12. We could use the Euclidean distance to compute the distance as $s = \sqrt{\sum_{i=1}^{N} (x_i - c(j)_i)^2}$ or use another definition more appropriate for the problem.
5:     **M-step**: Recompute the cluster centroid coordinates using the points assigned to each cluster as shown in Equation 5-13. This step changes the coordinates of cluster centroids.
6:     Calculate the distance by which each cluster centroid moved. Repeat until the total change falls below the threshold ϵ.
7: **end for**
8: Report the assignment of points to clusters and cluster centroids.

As an example, let us verify if high volatility of S&P 500 co-occurs with low future returns. If high historical volatility precedes low subsequent returns, we should expect the points to cluster. We can define the problem, including coordinates of the points as shown below:

1. Let us consider daily end-of-day prices for the S&P 500 index from 2000 to 2024.
2. Daily returns are calculated using the equation $r(t) = \frac{P(t)}{P(t-1)} - 1$, where $P(t)$ represents the closing price on day t.
3. We calculate the trailing one-month volatility of daily returns using Equation 5-14. This equation uses 21 observations because there are approximately 21 trading days in a month.
4. Calculate the forward-looking five-day return using Equation 5-15.
5. Define the coordinates of a point as (one-month volatility, next five-day return). A scatter plot of the points can be seen in Figure 5-1. In most problems, the points will lie in a multidimensional space, and it may not be so convenient to visualize them. But in this problem, the points belong to a two-dimensional space and are amenable to visualization using a two-dimensional plot.
6. Using $K = 4$, assign the points to one of four clusters using the K-means algorithm.
7. The assigned points are shown in Figure 5-2.

$$\text{volatility}(t) = \sqrt{\frac{\sum_{i=1}^{21} (r(t-i) - \bar{r}(t))^2}{21}}$$

$$\bar{r}(t) = \frac{\sum_{i=1}^{21} r(t-i)}{21}$$

$$(5\text{-}14)$$

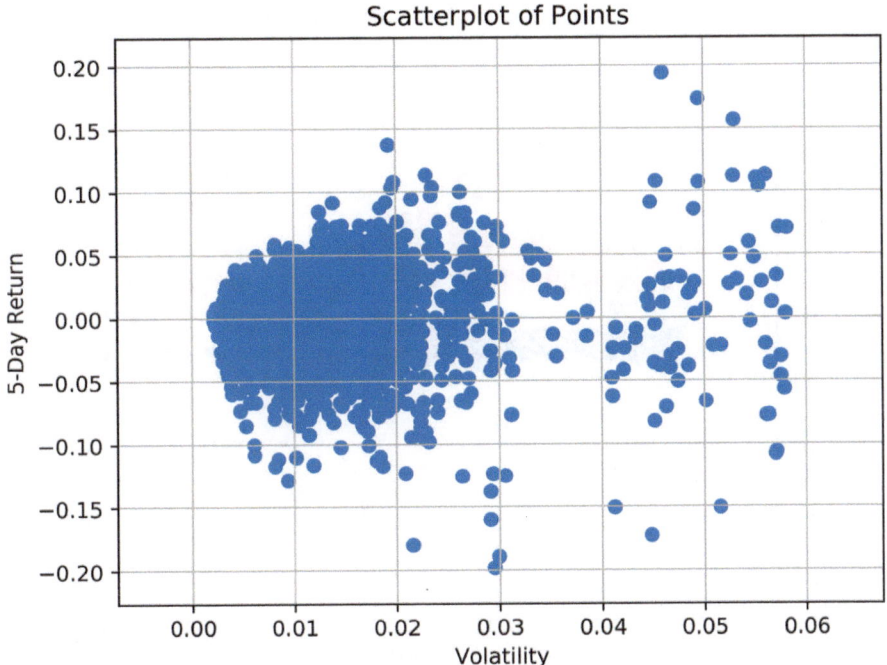

Figure 5-1. Scatter Plot of Return-Volatility Observations of S&P 500 to Be Clustered into Groups

$$R_{5\,\mathrm{day}}(t) = \frac{P(t+5)}{P(t)} - 1 \qquad (5\text{-}15)$$

Figure 5-2 shows the points classified into one of four clusters. Cluster centers are shown in black. From the plot, one can discern weak evidence for the ubiquitous notion that increased volatility is associated with low subsequent returns. While it is true that the cluster with the lowest return (cluster 3) has the highest volatility, the cluster with the highest return (cluster 2) also has high volatility which is only marginally lower than that of cluster 3. The remaining two clusters with low volatilities have low subsequent returns.

The code for fitting the k-means model and plotting the results can be found in Listing 5-1. The code uses the implementation of the K-means algorithm from the scipy library.

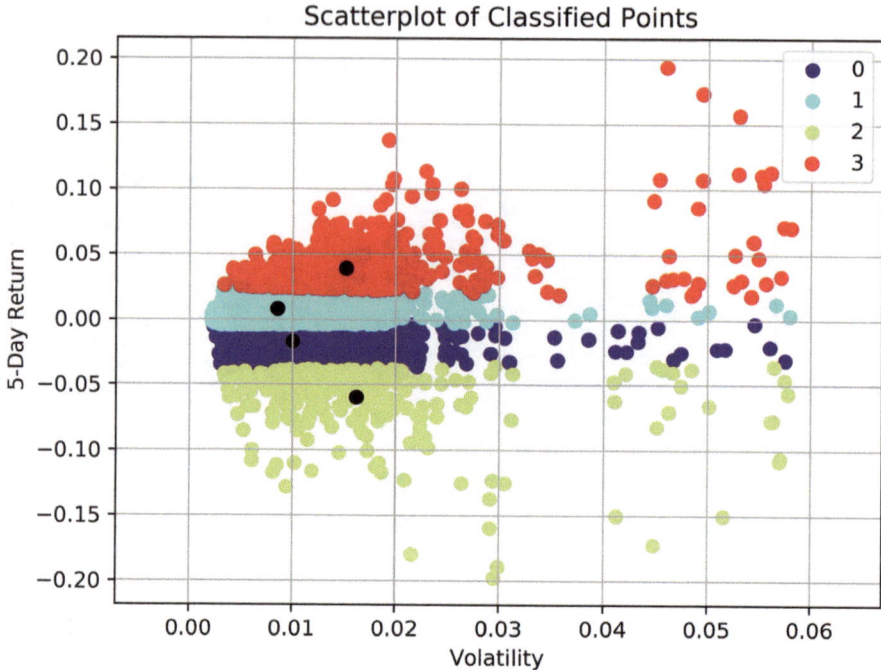

Figure 5-2. Clustered Return-Volatility Observations of S&P 500

Listing 5-1. Fitting Volatility-Return Observations of S&P 500 Using K-Means Algorithm

```
1   import numpy as np
2   import pandas as pd
3   import os
4   import logging
5   import matplotlib.pyplot as plt
6   from sklearn.cluster import KMeans
7   from matplotlib.pyplot import cm
8
9   logging.basicConfig(level=logging.DEBUG)
10
11
12  class KMeansCluster:
13      PERIOD = 5
14      VOLAT_PERIOD = 21
15
16      def __init__(self, dirname, spy, K=4):
17          self.logger = logging.getLogger(self.__class__.__name__)
18          filename = os.path.join(dirname, f"{spy}.csv")
19          self.df = pd.read_csv(filename, parse_dates=["Date"])
20          self.priceCol = "Close"
21          self.dirname = dirname
22          self.calculateVolatAndReturns()
```

```
23          self.model = KMeans(n_clusters=K)
24
25      def calculateVolatAndReturns(self):
26          prices = self.df.loc[:, self.priceCol].values
27          returns1Day = prices[1:]/prices[0:-1] - 1
28          returnPeriod = prices[self.PERIOD:]/prices[0:-self.PERIOD] - 1
29
30          self.df.loc[:, "volat"] = 0
31          for i in range(self.VOLAT_PERIOD, self.df.shape[0], 1):
32              self.df.loc[i, "volat"] = np.std(returns1Day[i-self.VOLAT_PERIOD:i])
33
34          self.df.loc[:, "return"] = 0
35          self.df.loc[0: self.df.shape[0]-self.PERIOD-1, "return"] = returnPeriod
36
37          x = self.df.loc[self.VOLAT_PERIOD:self.df.shape[0]-self.PERIOD-1, "volat"].
                    values
38          y = self.df.loc[self.VOLAT_PERIOD:self.df.shape[0]-self.PERIOD-1, "return"].
                    values
39          plt.scatter(x, y)
40          plt.grid()
41          plt.xlabel("Volatility")
42          plt.ylabel("5-Day Return")
43          plt.title("Scatterplot of Points")
44          plt.savefig(os.path.join(self.dirname, f"scatterplot_{self.__class__.__name__
                    }.jpeg"),
45                      dpi=500)
46          plt.show()
47
48      def fit(self):
49          x = self.df.loc[self.VOLAT_PERIOD:self.df.shape[0] - self.PERIOD - 1, "volat"].
                    values
50          y = self.df.loc[self.VOLAT_PERIOD:self.df.shape[0] - self.PERIOD - 1, "return"
                    ].values
51          X = np.vstack((x, y)).T
52          self.model.fit(X)
53          labels = self.model.labels_
54          clusterCenters = self.model.cluster_centers_
55          colors = cm.rainbow(np.linspace(0, 1, self.model.n_clusters))
56
57          self.logger.info("Cluster centeres")
58          self.logger.info(clusterCenters)
59
60          for i in range(self.model.n_clusters):
61              xlab = x[labels == i]
62              ylab = y[labels == i]
63              plt.scatter(xlab, ylab, c=colors[i], label=str(i))
64              plt.scatter([clusterCenters[i, 0]], [clusterCenters[i, 1]], c='black')
65
66          plt.grid()
67          plt.xlabel("Volatility")
68          plt.ylabel("5-Day Return")
69          plt.title("Scatterplot of Classified Points")
70          plt.legend()
```

```
71    plt . savefig (os . path . join ( self .dirname,  f"  scatterplot_classified_  { self . __class__
         . __name__ }.jpeg"),
72                    dpi=500)
73    plt . show()
74
75
76
77 if __name__ == "__main__":
78    dirname = r"C:\prog\cygwin\home\samit_000\latex\ book_stats \code\data"
79    kmeans = KMeansCluster(dirname, "SPY")
80    kmeans. fit ()
```

Code Explanation

Let us walk through the code on using the K-means algorithm:

1. The code begins with instantiating an object of class **KMeansCluster**. The constructor of this class takes the directory name containing data files and the name of the file with S&P 500 end-of-day closing prices as arguments.
2. Inside the constructor (**__init__** method), a comma-separated file containing S&P 500 end-of-day prices is read using the pandas library.
3. Five-day returns and volatility of returns are calculated inside the method **calculateVolatAndReturns**. Five-day returns are calculated using $r(t) = \frac{P(t+5)}{P(t)} - 1$. The code defines a class-level variable **PERIOD** that controls the number of days used in return calculation. Five trading days roughly correspond to one week. The method also computes the volatility of one-day returns using 21 days as the look-back period. 21 trading days roughly correspond to one month. Since one-day returns are forward-looking, i.e., use the next day's price, volatility for day t uses one-day returns from $t - 21$ to $t - 1$, excluding day t. This avoids in-sampling bias. This task being focused on clustering does not presuppose that there is no in-sample bias. Volatility is computed using the **std** method from the **numpy** library.
4. It constructs data points using volatility and five-day returns and fits them using the K-means algorithm inside the **fit** method. Four clusters are used for this exercise. This value is passed as a default argument to the constructor. The code uses the **fit** method of the **KMeans** class available in the **sklearn** library.
5. The code prints the cluster centers and plots the classified points using a scatter plot. K-means assigns a particular class to each point that can be obtained using the **labels_** attribute of the **KMeans** class.

5.4.2   Gaussian Mixture Model

The Gaussian mixture model can be conceptualized as a "soft" clustering algorithm that assigns probabilities to each data point belonging to a Gaussian distribution centered at a cluster centroid and having a cluster-specific variance. By assigning probabilities to each data point belonging to a cluster, the Gaussian mixture model can handle cluster assignment of data points that are not linearly separable. The K-means algorithm, being a distance-based method from a centroid, may fail to converge for cases with non-linearly separable data. Generalizing the definition of distance beyond the Euclidean distance does not cure this defect entirely. The Gaussian mixture model is a generative model, i.e., it can generate new data points following the same probability distribution learned by the model during training.

Let us formulate the model mathematically. Let us assume there are K clusters or groups. The probability of observing a point $\mathbf{X_i}$ conditional on the point belonging to cluster j is modeled using a Gaussian density function, as shown in Equation 5-16. To gain an intuitive understanding of this equation, we observe that moving the point $\mathbf{X_i}$ closer to cluster center $\mu_\mathbf{j}$ increases the probability of observing the point under the assumption that the point belongs to cluster j. Likewise, increasing the cluster variance $\mathbf{\Sigma_j}$ has the impact of increasing the probability of observing the point $\mathbf{X_i}$, everything else remaining the same. In the most general case, $\mathbf{\Sigma_j}$ is an $M \times M$ matrix where M is the number of dimensions of the point $\mathbf{X_i}$.

$$P(\mathbf{X_i}|j) = N\left(\mathbf{X_i} - \mu_\mathbf{j}, \mathbf{\Sigma_j}\right)$$

$$= \frac{1}{\sqrt{(2\pi)^M \det\left(\mathbf{\Sigma_j}\right)}} \exp\left(-\frac{1}{2}\left(\mathbf{X_i} - \mu_\mathbf{j}\right)' \mathbf{\Sigma_j}^{-1}\left(\mathbf{X_i} - \mu_\mathbf{j}\right)\right) \qquad (5\text{-}16)$$

The probability of selecting a cluster is also called prior probability and is denoted by $P(j)$. This could be an uninformative prior where we set probabilities of selecting each cluster to be equal, i.e., $\frac{1}{K}$, or could be something more informative that conveys our prior belief about selecting each cluster. This value denotes the probability of selecting a cluster without using any information about the training data. Being a probability, it must obey the constraint that $\sum_{j=1}^{K} P(j) = 1$. The unconditional probability of observing a data point $\mathbf{X_i}$ can now be written as shown in Equation 5-17.

$$P(\mathbf{X_i}) = \sum_{j=1}^{K} P(\mathbf{X_i}|j)P(j) \qquad (5\text{-}17)$$

Using Equation 5-17, we can write the combined probability of observing all N data points as shown in Equation 5-21. Taking the logarithm, we obtain the log probability of observing the data. In order to fit the model and evaluate $\mu_\mathbf{j}$ and $\mathbf{\Sigma_j}$ for $j \in (1, 2, \cdots, K)$, we maximize the log probability with respect to these parameters. The probability expression in Equation 5-17 can only be evaluated once we assume initial values for $\mu_\mathbf{j}$ and $\mathbf{\Sigma_j}$. This yields the EM algorithm–based model fitting approach for the Gaussian mixture model, as shown in pseudo-code 4.

For making an inference, i.e., classifying a new point among one of the clusters, we select a cluster that maximizes the posterior probability using Bayes' rule as shown in Equation 5-20.

Algorithm 4 Gaussian Mixture Model

Require: A set of N points with coordinates $\mathbf{X_i}$ and a number K representing the number of clusters to define. Let us assume the points are drawn from M dimensional space.

1: Set threshold ϵ to a small value. This threshold is used to assess if the algorithm has converged.
2: Select K points representing the centroid (center) of each of the K clusters. This can be assigned arbitrarily. Let us denote the coordinates of the jth cluster centroid by $\mu_j(0)$.
3: Select initial variance-covariance matrices for each of the j clusters, $\mathbf{\Sigma_j}(0)$. For the first iteration, this can be set to identity matrix.
4: Assume an initial prior probability for each cluster, $P_0(j)$.
5: **for** t = 1, 2, $\cdots$, until convergence **do**
6:     **E-step:** Evaluate the probability of each data point i belonging to a cluster j, $P(\mathbf{X_i} \in j)$, using the values for μ, $\mathbf{\Sigma}$, and $P(j)$ from the previous iteration, as shown in Equation 5-18.

$$P(\mathbf{X_i}|j) = N\left(\mathbf{X_i} - \mu_j(t-1), \mathbf{\Sigma_j}(t-1)\right)$$

$$P(\mathbf{X_i} \in j) = \frac{P(\mathbf{X_i}|j)P_{t-1}(j)}{P(\mathbf{X_i})} \tag{5-18}$$

$$= \frac{P(\mathbf{X_i} \in j|j)P_{t-1}(j)}{\sum_{l=1}^{K} P(\mathbf{X_i} \in l|l)P_{t-1}(l)}$$

7:     **M-step:** Update the values of μ, $\mathbf{\Sigma}$, and $P(j)$ using Equation 5-19 and values computed in E-step. Superscript $'$ indicates a transpose.

$$\mu_j(t) = \frac{\sum_{i=1}^{N} P(\mathbf{X_i} \in j)\mathbf{X_i}}{\sum_{i=1}^{N} P(\mathbf{X_i} \in j)}$$

$$\mathbf{\Sigma}(t) = \frac{\sum_{i=1}^{N} P(\mathbf{X_i} \in j)\left(\mathbf{X_i} - \mu_j(t-1)\right)\left(\mathbf{X_i} - \mu_j(t-1)\right)'}{\sum_{i=1}^{N} P(\mathbf{X_i} \in j)} \tag{5-19}$$

$$P_t(j) = \frac{\sum_{i=1}^{N} P(\mathbf{X_i} \in j)}{N}$$

8:     Calculate the difference $\|\mu(t) - \mu(t-1)\|^2$, $\|\Sigma(t) - \Sigma(t-1)\|^2$, and $\sum_{j=1}^{K}(P_t(j) - P_{t-1}(j))^2$. Repeat until the total change falls below threshold ϵ.
9: **end for**
10: Report the cluster centers μ, variance matrices $\mathbf{\Sigma}$, and cluster probabilities $P(j)$.

$$P(\mathbf{X_i} \in j|\mathbf{X_i}) = \frac{P(\mathbf{X_i}|\mathbf{X_i} \in j)P(\mathbf{X_i} \in j)}{P(\mathbf{X_i})}$$

$$= \frac{P(\mathbf{X_i}|j)P(j)}{P(\mathbf{X_i})} \tag{5-20}$$

$$= \frac{P(\mathbf{X_i}|j)P(j)}{\sum_{k=1}^{K} P(\mathbf{X_i}|k)P(k)}$$

Select j that maximizes posterior probability $P(\mathbf{X_i} \in j|\mathbf{X_i})$

$$P(\mathbf{X}) = \prod_{i=1}^{N} P(\mathbf{X_i})$$

$$\log P(\mathbf{X}) = \sum_{i=1}^{N} \log P(\mathbf{X_i})$$

(5-21)

We can derive the equations for updating parameters in M-step by differentiating the log-likelihood function with respect to the parameters.

Let us apply the Gaussian mixture model to cluster the trailing one-month volatility and ensuing five-day return for S&P 500 examined in the last section. We keep four clusters, with equal prior probability assigned to each cluster. For making an inference on the training dataset, the model uses Equation 5-20. Clustered points along with cluster centers (black dot) are shown in Figure 5-3. The Gaussian mixture model gives a very different result as compared with K-means, identifying different clusters. We can observe that points close to a cluster center are not necessarily classified in that cluster. This is due to the fact that the Gaussian mixture model can employ a non-diagonal variance-covariance matrix, with principal axis aligned in directions not necessarily along the dimensions of the data space. The matrix is learned during training.

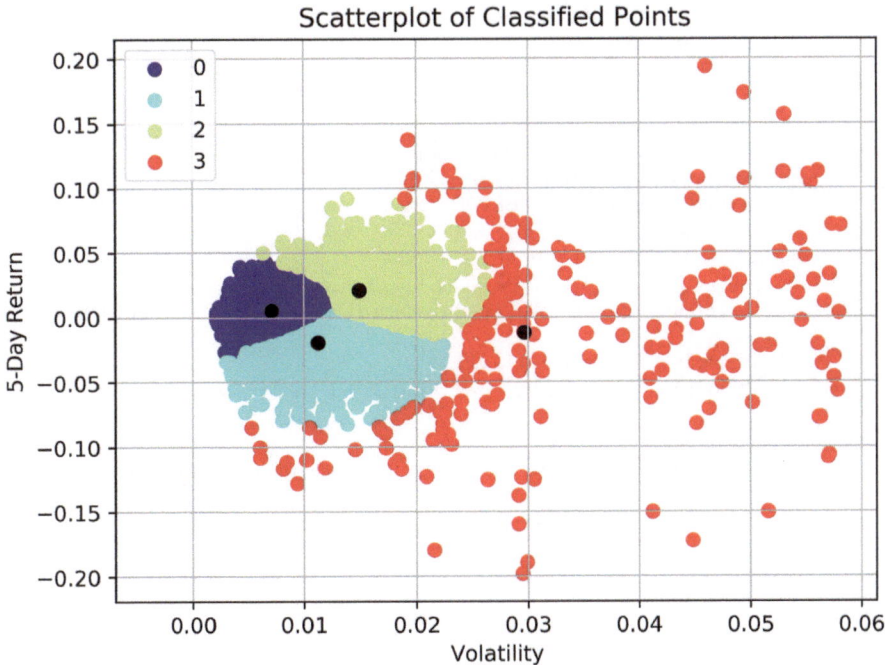

Figure 5-3. Clustered Return-Volatility Observations of S&P 500

The code for fitting a Gaussian mixture model can be found in Listing 5-2. We use the GaussianMixture class from the sklearn library.

Listing 5-2. Fitting Gaussian Mixture Model to Volatility-Return Observations of S&P 500

```python
import numpy as np
import pandas as pd
import os
import logging
import matplotlib.pyplot as plt
from sklearn.mixture import GaussianMixture
from matplotlib.pyplot import cm

logging.basicConfig(level=logging.DEBUG)


class GaussianMixtureModel:
    PERIOD = 5
    VOLAT_PERIOD = 21

    def __init__(self, dirname, spy, K=4):
        self.logger = logging.getLogger(self.__class__.__name__)
        filename = os.path.join(dirname, f"{spy}.csv")
        self.df = pd.read_csv(filename, parse_dates=["Date"])
        self.priceCol = "Close"
        self.dirname = dirname
        self.calculateVolatAndReturns()
        self.nClusters = K
        self.model = GaussianMixture(n_components=K, random_state=0)

    def calculateVolatAndReturns(self):
        prices = self.df.loc[:, self.priceCol].values
        returns1Day = prices[1:]/prices[0:-1] - 1
        returnPeriod = prices[self.PERIOD:]/prices[0:-self.PERIOD] - 1

        self.df.loc[:, "volat"] = 0
        for i in range(self.VOLAT_PERIOD, self.df.shape[0], 1):
            self.df.loc[i, "volat"] = np.std(returns1Day[i-self.VOLAT_PERIOD:i])

        self.df.loc[:, "return"] = 0
        self.df.loc[0:self.df.shape[0]-self.PERIOD-1, "return"] = returnPeriod

    def fit(self):
        x = self.df.loc[self.VOLAT_PERIOD:self.df.shape[0] - self.PERIOD - 1, "volat"].values
        y = self.df.loc[self.VOLAT_PERIOD:self.df.shape[0] - self.PERIOD - 1, "return"].values
        X = np.vstack((x, y)).T
        self.model.fit(X)
        clusterCenters = self.model.means_
        labels = self.model.predict(X)

        colors = cm.rainbow(np.linspace(0, 1, self.nClusters))
```

```
47
48          self . logger . info (" Cluster  centeres ")
49          self . logger . info ( clusterCenters )
50
51          for  i  in  range( self . nClusters ):
52              xlab  = x[ labels  == i]
53              ylab  = y[ labels  == i]
54              plt . scatter (xlab,  ylab,  c=colors[i],  label=str (i))
55              plt . scatter ([ clusterCenters [i,  0]],  [ clusterCenters [i,  1]],  c='black')
56
57          plt . grid ()
58          plt . xlabel (" Volatility ")
59          plt . ylabel ("5−Day Return")
60          plt . title (" Scatterplot  of  Classified  Points")
61          plt . legend ()
62          plt . savefig (os. path . join ( self .dirname,  f"  scatterplot_classified_  { self . __class__
                    . __name__}.jpeg"),
63                      dpi=500)
64          plt . show()
65
66
67
68  if  __name__ == "__main__":
69      dirname = r"C:\prog\cygwin\home\samit_000\latex \ book_stats \code\data"
70      gmm = GaussianMixtureModel(dirname, "SPY")
71      gmm.fit ()
```

Code Explanation

A code walk-through is presented below:

1. An object of class **GaussianMixtureModel** is instantiated, with the constructor receiving the directory name containing data files and the name of the file with S&P 500 end-of-day closing prices as arguments.
2. Inside the constructor (**__init__** method), a comma-separated file containing S&P 500 end-of-day prices is read using the pandas library.
3. Five-day returns and volatility of returns are calculated inside the method **calculateVolatAndReturns**. Five-day returns are calculated using $r(t) = \frac{P(t+5)}{P(t)} - 1$. The code defines a class-level variable **PERIOD** that controls the number of days used in return calculation. Five trading days roughly correspond to one week. The method also computes the volatility of one-day returns using 21 days as the look-back period. 21 trading days roughly correspond to one month. Since one-day returns are forward-looking, i.e., use the next day's price, volatility for day t uses one-day returns from $t - 21$ to $t - 1$, excluding day t. This avoids in-sampling bias. This task being focused on clustering does not presuppose that there is no in-sample bias. Volatility is computed using the **std** method from the **numpy** library.
4. It constructs data points using volatility and five-day returns and fits them using the Gaussian mixture model inside the **fit** method. Four clusters are used for

this exercise. This value is passed as a default argument to the constructor. The code uses the **fit** method of the **GaussianMixture** class available in the **sklearn** library. The constructor of the **GaussianMixture** class takes the number of clusters to define and an optional random state which should be passed for reproducibility of results. Random state is used as a seed in random number generation used for picking the cluster centers for first iteration.

5. The code prints the cluster centers and plots the classified points using a scatter plot. The Gaussian mixture model does not assign a particular class to each point, it produces a probability of each point belonging to a cluster. The code uses the **predict** method of the **GaussianMixture** class to predict the cluster to which the points belong. This method assigns the cluster with the highest probability among all clusters to contain a point. Class labels are produced as output by the **predict** method.

6. Labeled points along with clusters are plotted on a scatter plot with color chosen to indicate cluster membership.

5.5   Estimating Dynamic Transition Probabilities

After understanding the EM algorithm, we are in a position to apply it for obtaining the time-varying transition probabilities in a Markov dynamic regime switching model. Time-varying transition probabilities (TVTP) were assumed to belong to the parametric family shown in Equation 5-9. This expression can be used at each time step to obtain the unconditional probability of observing a state, as shown in Equation 5-22. We assume initial values of parameters γ in Equation 5-9 in order to calculate values of $p_{i,j}(t)$.

$$
P(s(t)) = \begin{bmatrix} p_{1,1}(t) & p_{1,2}(t) & \cdots & p_{1,N}(t) \\ p_{2,1}(t) & p_{2,2}(t) & \cdots & p_{2,N}(t) \\ & & \cdots & \\ p_{N,1}(t) & p_{N,2}(t) & \cdots & p_{N,N}(t) \end{bmatrix} \begin{bmatrix} P_{s(t-1)=1|s(t-2)} \\ P_{s(t-1)=2|s(t-2)} \\ \cdots \\ P_{s(t-1)=N|s(t-2)} \end{bmatrix}
$$

$$
= \begin{bmatrix} p_{1,1}(t) & p_{1,2}(t) & \cdots & p_{1,N}(t) \\ p_{2,1}(t) & p_{2,2}(t) & \cdots & p_{2,N}(t) \\ & & \cdots & \\ p_{N,1}(t) & p_{N,2}(t) & \cdots & p_{N,N}(t) \end{bmatrix} \begin{bmatrix} p_{1,1}(t-1) & \cdots & p_{1,N}(t-1) \\ p_{2,1}(t-1) & \cdots & p_{2,N}(t-1) \\ & \cdots & \\ p_{N,1}(t-1) & \cdots & p_{N,N}(t-1) \end{bmatrix}
$$

$$
\begin{bmatrix} P_{s(t-2)=1|s(t-3)} \\ P_{s(t-2)=2|s(t-3)} \\ \cdots \\ P_{s(t-2)=N|s(t-3)} \end{bmatrix}
$$

$$\tag{5-22}$$

Using the unconditional state probabilities at any given time t, we can write the pseudo-code for fitting the parameters of the Markov dynamic regime switching model with time-varying transition probabilities (TVTP) as shown in pseudo-code 5.

Algorithm 5 Fitting Markov Dynamic Regime Switching Model with Time-Varying Transition Probabilities Using EM Algorithm

---

Require: Exogenous variables X, endogenous variables y, and number of regimes.
1: Set threshold ϵ to a small value. This threshold is used to assess if the algorithm has converged.
2: Assume initial values for parameters γ in the parametric expression for transition probabilities (TVTP) in Equation 5-9.
3: **for** n = 1, 2, $\cdots$, until convergence **do**
4:      **E-step**: Calculate conditional transition probabilities (TVTP) using Equation 5-9 and the value of parameters $\gamma(t-1)$ from the last iteration. Evaluate the unconditional state probabilities at each time step t using Equation 5-22 and computed time-varying transition probabilities (TVTP).
5:      **M-step**: Using state probabilities found in E-step, write the maximum log-likelihood function as was done for the case with constant state transition probabilities and find the parameter values by maximizing the expression with respect to each parameter. Also, find the updated values of parameters γ required in Equation 5-9 by performing a first-order expansion around the values of γ from the previous iteration.
6:      Calculate the sum of square difference in value of parameters from the last iteration. Repeat until the total change falls below threshold ϵ.
7: **end for**
8: Use the fitted parameter values to make predictions.

5.6   Application

In this section, let us look at a few applications of Markov dynamic regime switching models.

5.6.1   Taylor Rule

The Taylor rule is a cornerstone of monetary policy. It prescribes the level of short-term interest rate, also known as federal funds rate in the United States, in relation to inflation and unemployment for the conduct of stable monetary policy. Monetary policy is stable if it sets short-term interest rates that foster actual inflation rate equal to target inflation rate and GDP growth rate equal to the potential rate of GDP growth. Potential GDP growth occurs when all capital and available manpower is used for production of goods and services. This corresponds to a state of full employment. American economist John B. Taylor proposed the rule in 1993 in a paper titled "Discretion versus policy rules in practice." The rule is stated in Equation 5-23. β_0, β_1, and β_2 are the model parameters. The term $y_t - \bar{y}_t$ represents the output gap and refers to the amount of GDP growth below its full potential. In his original paper, Taylor argued that both β_1 and β_2 should be close to 0.5.

$$i_t = \pi_t + \beta_0 + \beta_1 \left(\pi_t - \pi^* \right) + \beta_2 \left(y_t - \bar{y}_t \right)$$

i_t : Short-term interest rate

π_t : Inflation rate

π^* : Target rate of inflation   (5-23)

y_t : GDP growth rate

$\bar{y}_t$: Natural GDP growth rate consistent with inflation at target rate

In the original paper, Taylor proposed setting the inflation target, π^*, to 2%. This is also the level adopted by the Federal Reserve, so let us set the inflation target to 2%. In practice, the output gap is computed by first estimating a natural GDP growth trend and applying the trend over a period to obtain potential GDP growth. The output gap is then calculated as the difference between the current quarter's GDP growth and potential GDP growth. There are several methods to compute the natural GDP growth trend, such as smoothing or applying the HP filter. The Federal Reserve calculates this trend and uses it to compute the output gap. Let us use the values for the output gap published by the Federal Reserve Bank of St. Louis under its FRED data [20].

Using data from 1980 to 2024, let us first estimate the model coefficients using OLS. Inflation is measured using the GDP price deflator published by the Bureau of Economic Analysis (BEA). The short-term interest rate in the Taylor rule is the effective federal funds rate (EFFR) for the United States. This is the overnight rate charged by depositary institutions for lending funds to each other. The trading desk at the Federal Reserve Bank of New York ensures that the prevailing effective federal funds rate is close to the target set by the Federal Reserve using market transactions of borrowing or lending short-term funds to depositary institutions. This data is available in the FRED database made available by the Federal Reserve Bank of St. Louis. All data variables except EFFR are available at quarterly frequency. EFFR is available daily; therefore, we take the average of all EFFR observations in a quarter to get quarterly EFFR.

Listing 5-3 provides the summary statistics of fitting the OLS model using data from 1980 to 2024. The model has a low adjusted R^2 Of 0.001. The Durbin-Watson statistic of 0.205 indicates the presence of serial correlation in residuals.

Listing 5-3. Fitting Taylor Rule Using OLS Model

```
OLS Regression Results
==================================================================
Dep. Variable :                   y   R−squared:        0.012
Model:                          OLS   Adj. R−squared: 0.001
Method:               Least  Squares  F− statistic :     1.097
Date:             Sun,  14  Jul  2024  Prob (F− statistic ):0.336
Time:                      00:10:05   Log−Likelihood: 357.53
```

	coef	std err	t	P>|t|	[0.025	0.975]
const	0.0160	0.003	5.331	0.000	0.010	0.022
x1	0.1607	0.122	1.314	0.190	−0.081	0.402
x2	0.0801	0.121	0.664	0.507	−0.158	0.318

No. Observations: 177   AIC: −709.1
Df Residuals: 174   BIC: −699.5
Df Model: 2
Covariance Type: nonrobust

Omnibus: 11.897   Durbin−Watson: 0.205
Prob(Omnibus): 0.003   Jarque−Bera (JB):14.209
Skew: −0.485   Prob(JB): 0.000821
Kurtosis: 3.992   Cond. No. 50.4

Notes:
[1] Standard Errors assume that the covariance matrix of the errors is correctly specified .

Let us fit a Markov dynamic regime switching model to the same data. Let us use three regimes with different coefficients, intercept, and error variances in the regimes. Summary statistics from fitting the model are shown in Listing 5-4. Statistics show that AIC of the fitted model using the Markov dynamic regime switching model (−1046.7) is better than that obtained by using OLS (−709.1). Lower AIC value indicates a model is better. Certain error intervals in regime 3 in Listing 5-4 appear as nan (not a number) because the variance is too small, at $6.096e − 11$, indicating a high degree of confidence in calculated coefficients. From the results, it can be observed that regime 2 has coefficient $\beta_2 = 0.5962$ – close to Taylor's recommendation of 0.5. This is the coefficient for the output gap exogenous variable. It can be concluded that the traditional form of the Taylor rule is more apt as an accurate predictor of EFFR in moderate real interest rate regimes (regime 2).

We can interpret the three regimes as low, medium, and high EFFR regimes. Computed transition probabilities are shown in Listing 5-4. Regime occurrence probabilities are plotted in Figure 5-4. As observed in Figure 5-4, from 1981 to 1986, we were in a period of high interest rates when the Fed under Paul Volcker aggressively hiked interest rates to fight inflation. From 1990 to 2002, we were in a period of medium interest rates, with the Federal Reserve setting effective federal funds rate a little above prevailing inflation. After the Great Financial Crisis of 2008–2009 up until 2022, we were in a period of low interest rates. Only after 2022, we transitioned to a regime of high interest rates with the Federal Reserve raising rates to fight persistent inflation. However, the interpretation of regimes may be more nuanced. This is because the regime switching model assigning regime probabilities also defines them in relation to the levels of exogenous and endogenous variables. The exogenous variables are plotted in Figure 5-5.

Listing 5-4. Fitting Taylor Rule Using OLS Model

Markov Switching Model Results
==

Dep. Variable :		y	No. Observations :	177
Model:	MarkovRegression		Log Likelihood	541.329
Date:	Sun, 14 Jul 2024		AIC	−1046.658
Time:		00:10:08	BIC	−989.487
Sample:		0	HQIC	−1023.472
Covariance Type:		approx		

Regime 0 parameters
==

	coef	std err	z	P>|z|	[0.025	0.975]
const	−0.0060	0.001	−4.856	0.000	−0.008	−0.004
x1	−1.1565	0.045	−25.557	0.000	−1.245	−1.068
x2	0.3295	0.046	7.135	0.000	0.239	0.420
sigma2	3.456e−05	6.27e−06	5.509	0.000	2.23e−05	4.69e−05

Regime 1 parameters
==

	coef	std err	z	P>|z|	[0.025	0.975]
const	0.0320	0.002	17.914	0.000	0.029	0.036
x1	0.2627	0.074	3.544	0.000	0.117	0.408
x2	0.5962	0.103	5.807	0.000	0.395	0.797
sigma2	0.0001	2.3e−05	5.426	0.000	7.96e−05	0.000

Regime 2 parameters
==

	coef	std err	z	P>|z|	[0.025	0.975]
const	0.0299	0.004	6.772	0.000	0.021	0.039
x1	0.3774	0.211	1.786	0.074	−0.037	0.791
x2	−0.5413	0.143	−3.789	0.000	−0.821	−0.261
sigma2	0.0004	9.46e−05	4.231	0.000	0.000	0.001

Regime transition parameters
==

	coef	std err	z	P>|z|	[0.025	0.975]
p[0−>0]	0.9674	0.020	47.535	0.000	0.927	1.007
p[1−>0]	0.0126	0.014	0.897	0.370	−0.015	0.040
p[2−>0]	0.0172	0.021	0.832	0.405	−0.023	0.058
p[0−>1]	6.096e−11	nan	nan	nan	nan	nan
p[1−>1]	0.9728	0.022	44.294	0.000	0.930	1.016
p[2−>1]	0.0420	0.034	1.244	0.213	−0.024	0.108

==

Warnings:
[1] Covariance matrix calculated using numerical (complex−step) differentiation .

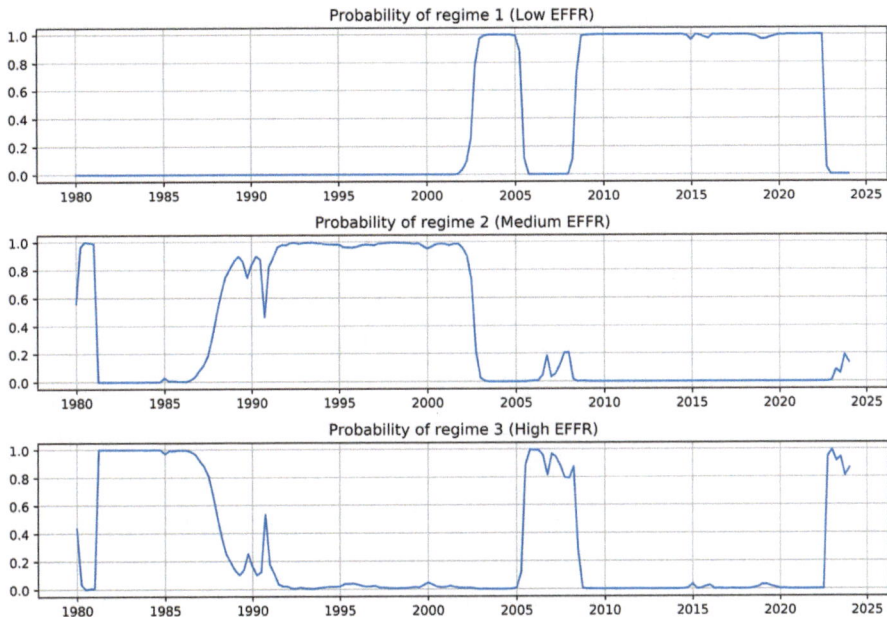

Figure 5-4. Regime Probabilities for Fitting Taylor Rule Using Dynamic Regime Switching Model

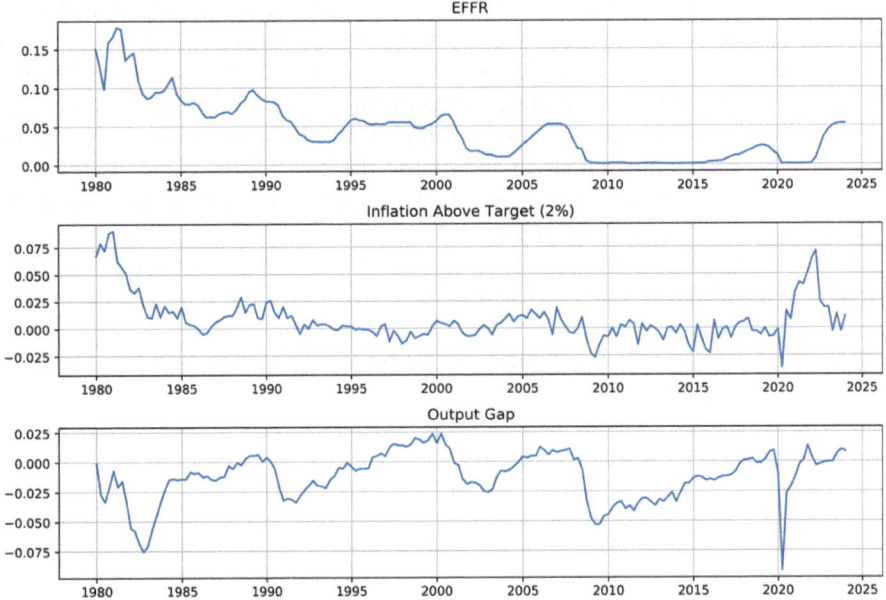

Figure 5-5. Exogenous and Endogenous Variable Plots

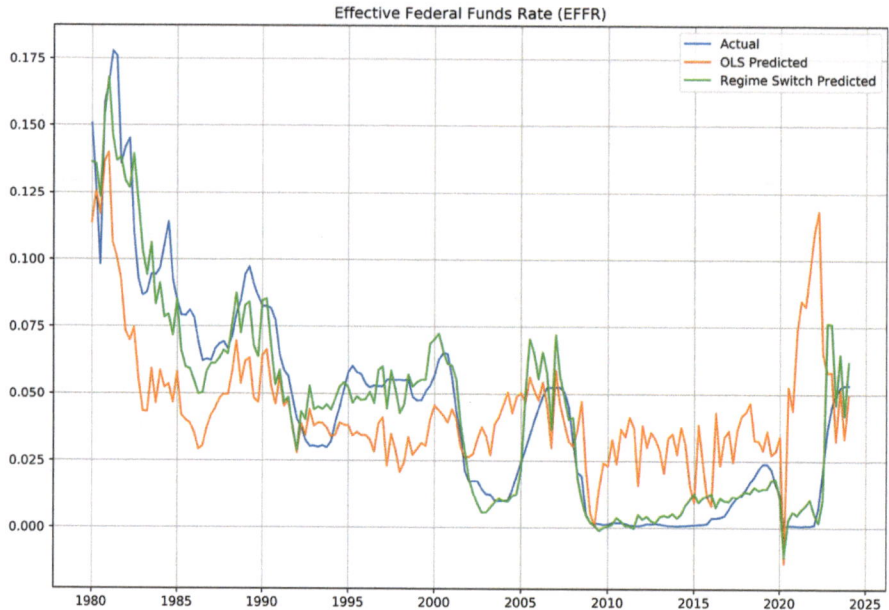

Figure 5-6. Comparison of Taylor Rule Fitted Using OLS and Dynamic Regime Switching Model Against EFFR Set by the Federal Reserve

Finally, let us compare the predicted values of EFFR using OLS and Markov dynamic regime switching models to fit the data to the Taylor rule and compare it against the actual value of the effective federal funds rate (EFFR) set by the Federal Reserve. Because the Federal Reserve takes into account a range of economic variables and considers the forecast of inflation and unemployment derived from models and expert's projections in addition to the Taylor rule, this fit will not be exact. As seen from the plot in Figure 5-6, the degree of fit is much closer using the Markov dynamic regime switching model.

The code for fitting the Taylor rule using OLS and Markov dynamic regime switching models is presented in Listing 5-5.

Listing 5-5. Fitting Taylor Rule Using OLS and Markov Dynamic Regime Switching Model

```
import numpy as np
import pandas as pd
import statsmodels.api as sm
import logging
import os
import matplotlib.pyplot as plt


logging.basicConfig(level=logging.DEBUG)



class TaylorRule:
```

```
12    def __init__(self, dirname, effrFile, inflationFile, outputGapFile, inflationTarget
          =0.02,
13                      beginDate="1980−01−01", trainTestSplit=0.9):
14        self.dirname = dirname
15        self.trainTestSplit = trainTestSplit
16        self.logger = logging.getLogger(self.__class__.__name__)
17        effr = pd.read_csv(os.path.join(dirname, f"{ effrFile }.csv"), parse_dates=["
              DATE"])
18        effr.rename(columns={"DFF": "intRate"}, inplace=True)
19        self.convertColToFloat(effr, "intRate", divideBy=100)
20        effr = self.convertToQuarterly(effr, ["intRate"])
21        self.beginDate = pd.to_datetime(beginDate)
22        self.endog = "intRate"
23
24        inflation = pd.read_csv(os.path.join(dirname, f"{ inflationFile }.csv"),
              parse_dates=["DATE"])
25        inflation.rename(columns={ inflationFile : "pi"}, inplace=True)
26        self.convertColToFloat(inflation, "pi", divideBy=100)
27        inflation.loc[:, "pi_m_target"] = inflation.loc[:, "pi"] − inflationTarget
28
29        df = pd.merge(effr, inflation, on=["DATE"], how="inner")
30
31        outputGap = pd.read_csv(os.path.join(dirname, f"{outputGapFile}.csv"),
              parse_dates=["DATE"])
32        outputGap.rename(columns={"GDPC1_GDPPOT": "output_gap"}, inplace=True)
33        self.convertColToFloat(outputGap, "output_gap", divideBy=100)
34        df = pd.merge(df, outputGap, on=["DATE"], how="inner")
35
36        df = df.loc[df.DATE >= self.beginDate, :].reset_index(drop=True)
37
38        self.exog = ["pi_m_target", "output_gap"]
39        self.df = df
40
41    def convertColToFloat(self, df, col, divideBy=1.0):
42        if (df.loc[:, col] == ".").sum() > 0:
43            df.drop(np.where(df.loc[:, col] == ".")[0], inplace=True)
44        df.loc[:, col] = df.loc[:, col].astype(np.float64) / divideBy
45        df.reset_index(drop=True, inplace=True)
46
47    def convertToQuarterly(self, df, cols):
48        df.loc[:, 'quarter'] = ((df.DATE.dt.month.values − 1) // 3)
49        df.loc[:, 'year'] = df.DATE.dt.year
50        ypart = df[['year', 'quarter'] + cols]
51        ypart = ypart.groupby(['year', 'quarter']).mean().reset_index(drop=False)
52        ydate = df[["DATE", "year", "quarter"]].groupby(["year", "quarter"]).first().
              reset_index(drop=False)
53        ydate.loc[:, "DATE"] = ydate.DATE + pd.offsets.MonthEnd(0) + pd.offsets.
              MonthBegin(−1)
54        df = pd.merge(ydate, ypart, on=["year", "quarter"], how="inner")
55        df.drop(columns=["year", "quarter"], inplace=True)
56        return df
57
58    def trainData(self):
```

```python
ntrain = int ( self .df.shape[0] * self . trainTestSplit )
y = self .df. loc [0: ntrain ,  self .endog]. values − self .df. loc [0: ntrain ,  "pi" ].
    values
X = self .df. loc [: ntrain ,  self .exog]. values
return y, X

def fitOLS( self ):
    y, X = self . trainData ()
    X = sm.add_constant(X, has_constant ="add")
    self .olsModel = sm.OLS(y, X)
    self .olsModel = self .olsModel. fit ()
    self . logger . info ( self .olsModel.summary())
    summaryfile = os. path . join ( self .dirname, self .__class__.__name__ + "_ols. txt ")
    with open(summaryfile, 'w') as fh:
        fh. write ( self .olsModel.summary().as_text ())


def  plotTrainingFit ( self ):
    y, X = self . trainData ()
    fig , ax = plt . subplots (1,  figsize =(10, 7))
    ntrain = int ( self .df.shape[0] * self . trainTestSplit )
    date = self .df. loc [0: ntrain ,  "DATE"].values
    ax. plot ( date , y + self .df. loc [0: ntrain ,  "pi" ]. values ,  label ="Actual")
    yPred = self .olsModel. fittedvalues
    ax. plot ( date ,  yPred + self .df. loc [0: ntrain ,  "pi" ]. values ,  label ="OLS Predicted
        ")
    yPred = self .markovModel. fittedvalues
    ax. plot ( date ,  yPred + self .df. loc [0: ntrain ,  "pi" ]. values ,  label ="Regime Switch
        Predicted")
    ax. legend ()
    ax. set ( title =" Effective  Federal  Funds  Rate  (EFFR)")
    ax. grid ()
    fig . tight_layout ()
    plt . savefig (os. path . join ( self .dirname,  f" effr_ { self .__class__.__name__}.jpeg"),
        dpi=500)
    plt .show()


def fitRegimeSwitch( self ):
    y, X = self . trainData ()
    ntrain = int ( self .df.shape[0] * self . trainTestSplit )
    date = self .df. loc [0: ntrain ,  "DATE"].values
    np.random.seed(1024)
    self .markovModel = sm.tsa.MarkovRegression(endog=y, k_regimes=3, trend='c',
        exog=X,
                                                switching_trend =True,
                                                switching_exog=True,
                                                switching_variance =True)
    self .markovModel = self .markovModel.fit ()

    self . logger . info ( self .markovModel.summary())
    summaryfile = os. path . join ( self .dirname,  self .__class__.__name__ + "
        _regimeSwitch.txt")
    with open(summaryfile, 'w') as fh:
        fh. write ( self .markovModel.summary().as_text())
```

```
107
108    fig, axes = plt.subplots(3, figsize =(10, 7))
109
110    ax = axes[0]
111    ax.plot(date, self.markovModel.filtered_marginal_probabilities[:, 0])
112    ax.set(title ="Probability of regime 1 (Low EFFR)")
113    ax.grid()
114
115    ax = axes[1]
116    ax.plot(date, self.markovModel.filtered_marginal_probabilities[:, 1])
117    ax.set(title ="Probability of regime 2 (Medium EFFR)")
118    ax.grid()
119
120    ax = axes[2]
121    ax.plot(date, self.markovModel.filtered_marginal_probabilities[:, 2])
122    ax.set(title ="Probability of regime 3 (High EFFR)")
123    ax.grid()
124
125    fig.tight_layout()
126    plt.savefig(os.path.join(self.dirname, f"regime_prob_{self.__class__.__name__
            }.jpeg"),
127                    dpi=500)
128    plt.show()
129
130    def fit(self):
131        self.fitOLS()
132        self.fitRegimeSwitch()
133        self.plotTrainingFit()
134
135
136 if __name__ == "__main__":
137     dirname = r"C:\prog\cygwin\home\samit_000\latex\book_stats\code\data"
138     trule = TaylorRule(dirname, "DFF", "A191RI1Q225SBEA", "fredgraph_OutputGap",
            trainTestSplit=1.0)
139     trule.fit()
```

To aid explanation of the code in Listing 5-5, let us look at pseudo-code 6 elucidating the code logic.

Code Explanation

A code walk-through is presented below:

1. An object of class **TaylorRule** is instantiated with five arguments passed to the constructor. The first argument is the directory name containing data files. The next argument indicates the name of the file containing effective federal funds rate (EFFR). The following argument indicates the name of the inflation file, and the one after that contains the output gap data. The data files are all comma separated, with identical name of column containing the data as the file name. For example, file **A191RI1Q225SBEA** contains the inflation data reported by BEA, with inflation data falling under column with name **A191RI1Q225SBEA**.

Algorithm 6 Using OLS and Markov Dynamic Regime Switching Model to Fit Taylor Rule

---

Require: Files containing quarterly data for inflation and output gap. File containing daily data for effective federal funds rate (EFFR).

1: Read the files containing data for inflation and output gap.
2: Convert the data columns to floating-point data type and join the two tables.
3: Read the file containing EFFR data and convert the EFFR data column to floating-point data type.
4: Convert daily EFFR to quarterly EFFR by taking an average of daily EFFR values falling in a quarter.
5: Define the endogenous variable as $i_t - \pi_t$. This quantity is known as the real interest rate, obtained after subtracting inflation from interest rate (EFFR).
6: Fit the OLS model using the redefined endogenous variable and exogenous variables ($\pi_t - \pi^*$) and output gap ($y_t - \bar{y}_t$) after adding an intercept and print summary statistics.
7: Fit the Markov dynamic regime switching model using three regimes, regime switching variance, coefficients, and intercept. Print summary statistics.
8: Plot the output from fitted models and compare against actual EFFR rate. Because OLS and Markov dynamic regime switching models predict real interest rate, add inflation to obtain the predicted (fitted) for EFFR.

The final argument is the training-testing split. In this exercise, we use the entire dataset for training and set this argument to 1.

2. The code reads the effective federal funds file and converts the data column to floating-point type. It calls the method **convertToQuarterly** to convert EFFR data to quarterly frequency. This method extracts the year and quarter for each date and groups the data by year and quarter, taking the average of EFFR values falling in a year-quarter group.

3. The inflation file is read next, and the inflation target is subtracted from inflation to obtain inflation above the target. As explained earlier, an inflation target of 2% is used. Inflation data is already present at quarterly frequency.

4. The inflation dataframe is joined with a dataframe containing EFFR data on quarterly dates.

5. The output gap is read, and after changing column names, it is joined with the dataframe from the last step.

6. EFFR is specified as the endogenous variable, with excess inflation above the target and the output gap serving as exogenous variables.

7. The OLS model is fitted to predict EFFR using the abovementioned exogenous variables and an intercept.

8. The Markov regime switching model is fit to the data. The **MarkovRegression** class from the **statsmodels** library is used for the purpose. This API does not require prepending a column of ones to exogenous variables for inclusion of intercept. It includes an intercept if argument **trend=c** is provided to the constructor of class **MarkovRegression**. **c** represents the constant trend. In addition to endogenous and exogenous variables, the constructor of the **MarkovRegression** class is provided with arguments indicating the number of regimes **k_regimes**, argument **switching_trend** indicating if intercept coef-

ficients show change in different regimes, argument **switching_exog** indicating if coefficients of exogenous variables should change in different regimes, and argument **switching_variance** that indicates if error term covariance changes in different regimes. These last three arguments are supplied with a value of **True**.

9. The Markov model is fitted and summary statistics printed.
10. Marginal probabilities indicating the probability of occurrence of a regime are plotted using the **filtered_marginal_probabilities** API of the **MarkovRegression** class.
11. Predicted EFFR values on the training dataset are plotted against actual EFFR values and compared against the predictions of the OLS model.

The inflation used in fitting the Taylor rule is the seasonally adjusted GDP implicit price deflator published by the US Bureau of Economic Analysis (BEA). This measure of inflation is different from PCE (personal consumption expenditure) in that it includes volatile food and energy prices. PCE is the Federal Reserve's preferred inflation gauge. However, for fitting the Taylor rule, one must account for a representative basket of goods and services that include price volatile goods.

Time-Varying Transition Probabilities

We can employ time-varying transition probabilities (TVTP) in the Markov dynamic regime switching model for fitting the Taylor rule to see its impact on the closeness of fit. In order to use TVTP, we need to provide exogenous variables used in the definition of transition probabilities in Equation 5-9. Let us use the same set of exogenous variables used in the Taylor rule. The **statsmodels** library expects the addition of a row of ones to the matrix containing exogenous variables for transition probabilities in order to include a constant term. Accordingly, we add this column of ones. The code snippet in Listing 5-6 derives from the TaylorRule class used in the code from the last section and modifies the model to use TVTP.

Listing 5-6. Fitting Taylor Rule Using OLS and Markov Dynamic Regime Switching Model with Time-Varying Transition Probabilities

```
import numpy as np
import pandas as pd
import statsmodels.api as sm
import logging
import os
import matplotlib.pyplot as plt
from src.TaylorRule import TaylorRule

logging.basicConfig(level=logging.DEBUG)


class TaylorRuleTVTP(TaylorRule):
    def fitRegimeSwitch(self):
        y, X = self.trainData()
        ntrain = int(self.df.shape[0] * self.trainTestSplit)
        date = self.df.loc[0:ntrain, "DATE"].values
```

```python
np.random.seed(1024)
XWithConst = sm.add_constant(X, has_constant="add")
self.markovModel = sm.tsa.MarkovRegression(endog=y, k_regimes=3, trend='c',
    exog=X,
                                           exog_tvtp=XWithConst,
                                           switching_trend=True,
                                           switching_exog=True,
                                           switching_variance=True)
self.markovModel = self.markovModel.fit()

self.logger.info(self.markovModel.summary())
summaryfile = os.path.join(self.dirname, self.__class__.__name__ + "
    _regimeSwitch.txt")
with open(summaryfile, 'w') as fh:
    fh.write(self.markovModel.summary().as_text())

fig, axes = plt.subplots(3, figsize=(10, 7))

ax = axes[0]
ax.plot(date, self.markovModel.filtered_marginal_probabilities[:, 0])
ax.set(title="Probability of regime 1 (Low EFFR)")
ax.grid()

ax = axes[1]
ax.plot(date, self.markovModel.filtered_marginal_probabilities[:, 1])
ax.set(title="Probability of regime 2 (Medium EFFR)")
ax.grid()

ax = axes[2]
ax.plot(date, self.markovModel.filtered_marginal_probabilities[:, 2])
ax.set(title="Probability of regime 3 (High EFFR)")
ax.grid()

fig.tight_layout()
plt.savefig(os.path.join(self.dirname, f"regime_prob_{self.__class__.__name__
    }.jpeg"),
            dpi=500)
plt.show()

fig, axes = plt.subplots(3, figsize=(10, 7))

ax = axes[0]
ax.plot(date, y + self.df.loc[0: ntrain, "pi"].values)
ax.set(title="EFFR")
ax.grid()

ax = axes[1]
ax.plot(date, X[:, 0])
ax.set(title="Inflation Above Target (2%)")
ax.grid()

ax = axes[2]
ax.plot(date, X[:, 1])
```

```
67      ax. set ( title ="Output Gap")
68      ax. grid ()
69
70      fig . tight_layout ()
71      plt . savefig (os. path . join ( self .dirname, f"regime_vars_{ self .__class__.__name__
            }.jpeg"),
72                   dpi=500)
73      plt .show()
74
75
76  if __name__ == "__main__":
77      dirname = r"C:\prog\cygwin\home\samit_000\latex\ book_stats \code\data"
78      trule  = TaylorRuleTVTP(dirname, "DFF", "A191RI1Q225SBEA", "fredgraph_OutputGap"
            , trainTestSplit=1.0)
79      trule . fit ()
```

Using this change, we observe a small improvement in the closeness of fit, as evidenced by the reduction of AIC to -1054.69 from -1046.658. The fitted curve is shown in Figure 5-7 and is only marginally better than Figure 5-6.

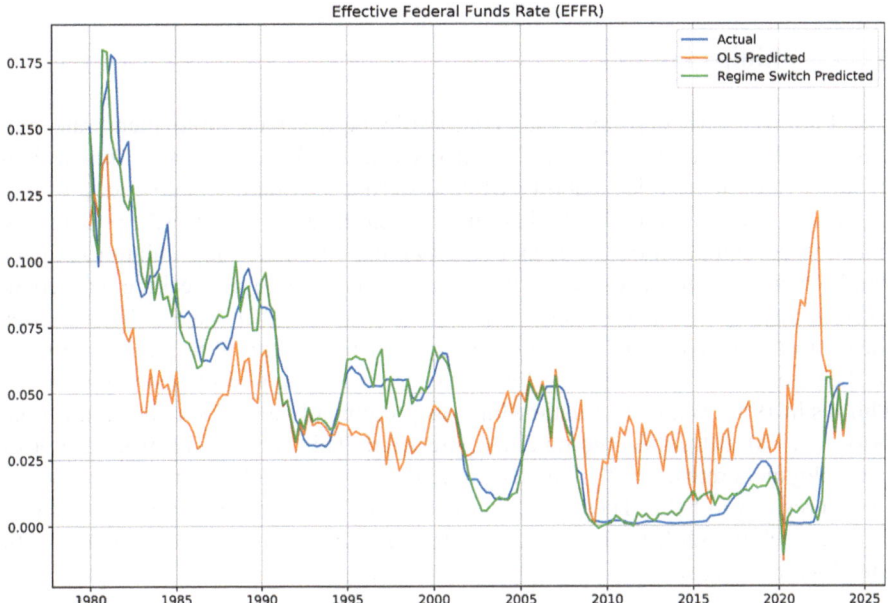

Figure 5-7. Comparison of Taylor Rule Fitted Using OLS and Dynamic Regime Switching Model with TVTP Against EFFR Set by the Federal Reserve

Code Explanation

A code walk-through is presented below:

1. As before, the code begins with instantiating an object of class **Taylor-RuleTVTP**, providing as arguments the directory name containing data files, followed by the name of files containing EFFR rates, inflation rate, and output gap. The final argument specifies the train-test split as 1.
2. The code proceeds by reading the files, processing them, and creating the endogenous and exogenous variables. This part of the code is identical to the code presented earlier without time-varying transition probabilities.
3. The point of departure from the code in the last section comes where class **MarkovRegression** from library **statsmodels** is instantiated. The constructor of this class is provided with an argument **exog_tvtp** specifying the exogenous variables to use in the expression for time-varying transition probabilities shown in Equation 5-9. This matrix also needs a column of ones for the inclusion of intercept term in the equation for transition probability in Equation 5-9. Accordingly, we add a column of ones using the API **add_constant**.
4. All the remaining code remains the same as the one presented in the last section.

5.6.2   Phillips Curve

The Phillips curve describes the relationship between unemployment rate and inflation. This relationship is important because the Federal Reserve has a dual mandate to promote price stability while fostering maximum employment. It may seem reasonable to think that higher employment leads to higher price pressure by boosting consumption of goods and services, thereby increasing inflation. Higher employment may also cause wage pressures because employers seeking workers in a shrinking pool of available workers are forced to offer higher wages. This increases the cost of production of goods and services for firms, who in turn raise the price of their goods and services. The Phillips curve was proposed by economist A. W. Phillips in 1958. While the relationship was prominently manifest in economic data prior to 1970, it has become muddled afterward. This can be seen from a plot of unemployment rate and inflation observed in the United States after 1950, as shown in Figures 5-8 to 5-11. These plots show unemployment vs. inflation on a scatter plot for each month from 1959 to 2024. To aid discernment, the year range is split into 20-year brackets:

1. From 1959 to 1979, higher unemployment was often associated with low inflation, though there were many outliers where high inflation and high unemployment coexisted (Figure 5-8). Higher unemployment depressed consumption, causing inflation to fall during this period.
2. From 1979 to 1999, higher inflation co-occurred with high unemployment as the Federal Reserve began raising interest rates aggressively to tame runaway inflation, even in the face of rising unemployment (Figure 5-9).

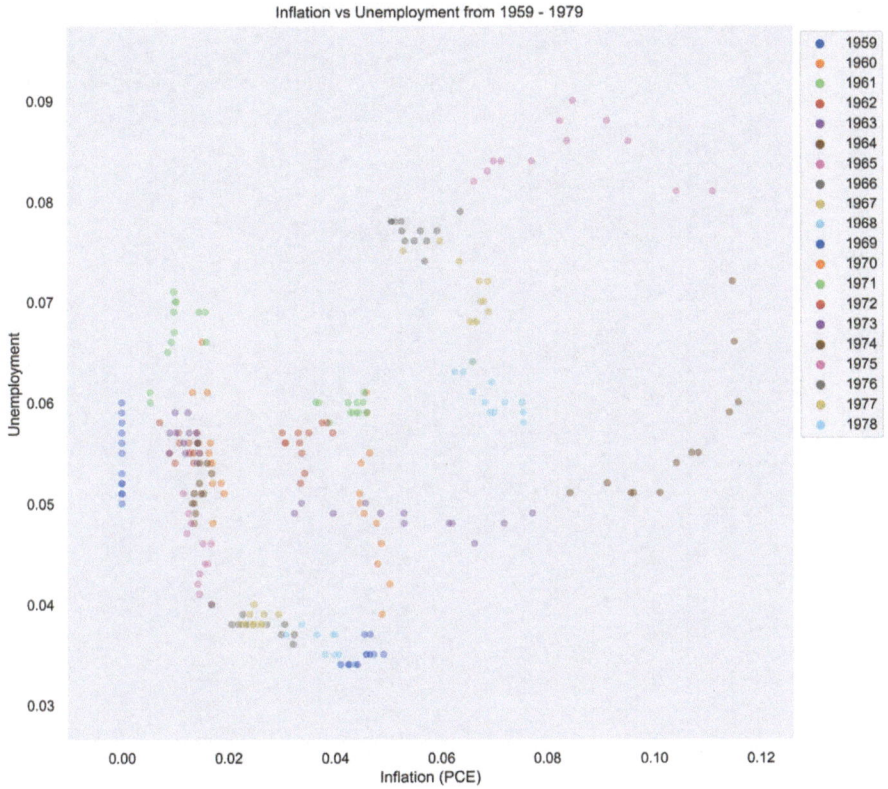

Figure 5-8. Unemployment vs. Inflation from 1959 to 1979

3. From 1999 to 2019, the relationship between inflation and unemployment had flattened. Inflation remaining low in the midst of both high and low unemployment (Figure 5-10).
4. From 2019 to 2024, inflation began rising, but unemployment barely budged from its low value (Figure 5-11).

The equation describing the augmented Phillips curve is shown in Equation 5-24.

$$\pi(t) = E_t\left[\pi(t+1)\right] + \beta\left(u(t) - u^n(t)\right) + \epsilon(t)$$
$$\beta < 0 \tag{5-24}$$

In Equation 5-24, $\pi(t)$ is the inflation, and $E_t\left[\pi(t+1)\right]$ is the expected value of the subsequent-period inflation calculated at time t, or forward-looking inflation. Inflation is measured using the PCE (personal consumption expenditure) index which excludes volatile food and energy prices from the CPI index. PCE is the Federal Reserve Board's preferred gauge to measure inflation in US economy.

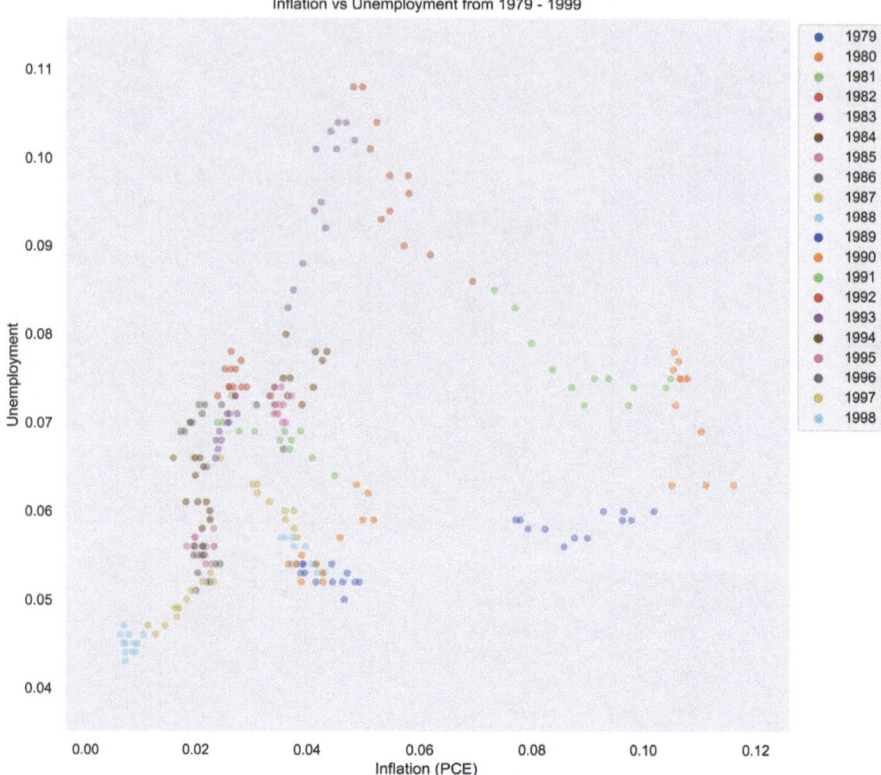

Figure 5-9. Unemployment vs. Inflation from 1979 to 1999

u is the unemployment rate, and u^n is the non-accelerating inflation rate of unemployment, or NAIRU. This is the unemployment rate that would prevail if inflation is at its expected level. Or stated another way, it is the equilibrium unemployment level from all natural sources that will not contribute to inflation. $\epsilon(t)$ denotes the idiosyncratic error.

Equation 5-24 has been generalized to include lagged inflation in order to account for the fact that certain producers of goods may lack the foresight to predict expected inflation and may use last period's inflation as a substitute. The modified equation is called the new Keynesian hybrid Phillips curve and is shown in Equation 5-25.

$$\pi(t) = \beta_1 E_t \left[\pi(t+1) \right] + \beta_2 \pi(t-1) + \beta_3 \left(u(t) - u^n(t) \right) + \epsilon(t)$$

$$\beta_3 < 0$$

(5-25)

As before, we are going to use OLS and Markov dynamic regime switching models to fit the new Keynesian hybrid Phillips curve to the data. In practice, economists use GMM (generalized method of moments) or DSGE (dynamic

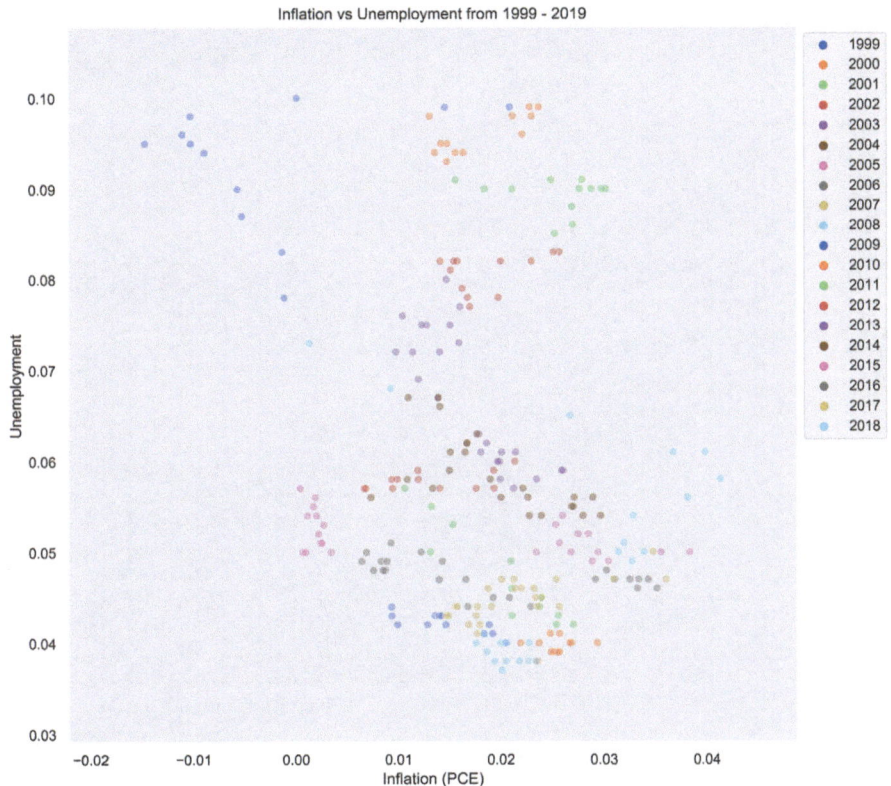

Figure 5-10. Unemployment vs. Inflation from 1999 to 2019

stochastic general equilibrium model) to fit this equation to econometric data. GMM will be covered in a subsequent chapter in greater detail. Let us use the difference between yields on ten-year nominal US government bonds and ten-year TIPS as a substitute for expected inflation, $E\left[\pi\left(t+1\right)\right]$. This rate is available from the FRED economic series database made available by the Federal Reserve Bank of St. Louis under the name "10-year breakeven inflation rate." This rate is forward-looking because it is derived from yields of government bonds which change according to the investor's expectation of forward-looking inflation. Data for NAIRU, $u^{n}(t)$, is also available from the FRED database as non-cyclical rate of unemployment, or NROU [21].

Inflation data (PCE index) and unemployment rate are available at a monthly frequency. Inflation expectation data is available daily, while NAIRU is available at a quarterly frequency. We select monthly data frequency and assign a NAIRU equal to that quarter's value for the three months falling in a quarter. Inflation expectation data is averaged monthly using all days falling in a month to obtain a monthly value.

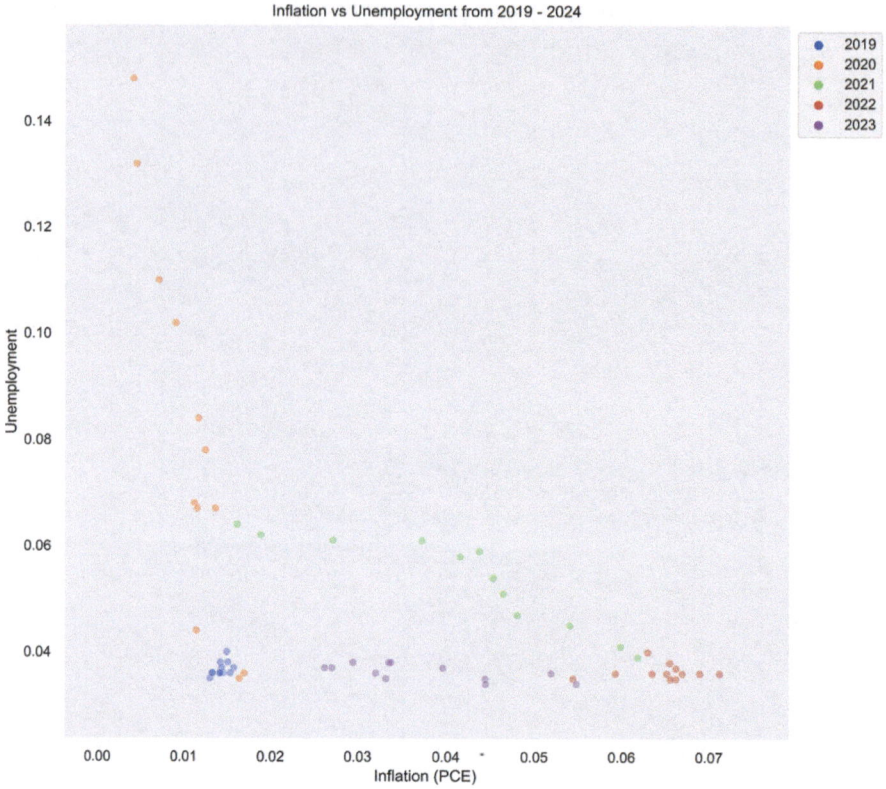

Figure 5-11.   Unemployment vs. Inflation from 2019 to 2024

Summary statistics of fitting the new Keynesian Phillips curve using the OLS model are shown in Listing 5-7. There is no intercept in Equation 5-25, as seen from the statistics. The "uncentered" qualifier used with R^2 statistic reflects this fact. Adjusted R^2 of 0.986 indicates a good model fit to the data. β_2 of 0.9358 indicates that the prior period's inflation plays an important role in determining the next period's inflation. β_3 has a negative value which aligns with traditional economic theory that predicts unemployment above the natural level will reduce inflation. However, the dependence is weak, as seen from the low value of β_3 of -0.0153 and from the fact that this coefficient is not statistically significant at 90% significance level. It is, however, significant at 80% level, as seen from the P-value for a one-sided test.

Listing 5-7. Fitting New Keynesian Hybrid Phillips Curve Using OLS Model

```
OLS Regression Results
==============================================================================
Dep. Variable :                     y    R—squared (uncentered):        0.986
Model:                            OLS    Adj. R—squared (uncentered):0.986
Method:                 Least Squares    F— statistic :                 5896.
Date:              Sun, 14 Jul 2024     Prob (F— statistic ):          1.90e−234
Time:                      00:59:37     Log—Likelihood:                1112.4
No. Observations:               257     AIC:                           −2219.
Df Residuals :                  254     BIC:                           −2208.
Df Model:                         3
Covariance Type:           nonrobust
==============================================================================
         coef    std err        t       P>|t |      [0.025      0.975]
------------------------------------------------------------------------------
x1     0.0866    0.023      3.719       0.000       0.041       0.132
x2     0.9358    0.017     55.424       0.000       0.903       0.969
x3    −0.0153    0.011     −1.419       0.157      −0.036       0.006
==============================================================================
Omnibus:                       31.650    Durbin—Watson:              1.027
Prob(Omnibus):                  0.000    Jarque—Bera (JB):         179.327
Skew:                          −0.153    Prob(JB):                 1.15e−39
Kurtosis :                      7.081    Cond. No.                    4.92
==============================================================================

Notes:
[1] R2 is computed without centering (uncentered) since the model does not contain a
    constant .
[2] Standard Errors assume that the covariance matrix of the errors is correctly
    specified .
```

Statistics for fitting the new Keynesian hybrid Phillips curve using the Markov dynamic regime switching model are shown in Listing 5-8. The regime switching model uses three regimes with no intercept and regime switching coefficients and variance of error term. Regime occurrence probabilities are shown in Figure 5-13 and value of exogenous variables is illustrated in Figure 5-12. From Figure 5-13, it is apparent that regime 3 has the highest probability of manifestation over the period. Regime 1 materialized in 2015, while regime 2 had a high probability of occurrence for a short period during the Great Financial Crisis of 2008–2009. From the plot of exogenous variables shown in Figure 5-12, we can ascribe economic meaning to the definition of regimes. Regime 1 seems to occur when inflation expectations are well anchored at 2% target articulated by the Federal Reserve and lagged inflation falls to a value close to 0% with unemployment close to its natural level. Under these circumstances, a decline in lagged inflation gives rise to expectations for reversion to the target value of 2%, as seen from the negative value for β_2 of -0.1456 in the

Listing 5-8. Fitting New Keynesian Hybrid Phillips Curve Using Regime Switching Model

```
Markov Switching Model Results
================================================================
Dep. Variable :              y    No. Observations :  257
Model:            MarkovRegression   Log Likelihood     1141.929
Date:            Sun, 14 Jul 2024   AIC                -2247.857
Time:                   14:21:20    BIC                -2183.974
Sample:                        0    HQIC               -2222.166

Covariance Type:            approx
Regime 0 parameters
================================================================
        coef    std err      z    P>|z|    [0.025    0.975]
----------------------------------------------------------------
x1     0.1474  0.021      7.010   0.000    0.106     0.189
x2    -0.1456  0.011    -13.405   0.000   -0.167    -0.124
x3    -0.0803  0.042     -1.931   0.053   -0.162     0.001
sigma2 3.371e-07 nan    nan       nan      nan       nan
Regime 1 parameters
================================================================
        coef    std err      z    P>|z|    [0.025    0.975]
----------------------------------------------------------------
x1     0.3390  0.031     10.939   0.000    0.278     0.400
x2     0.3657  0.024     15.502   0.000    0.319     0.412
x3    -0.1323  0.013    -10.400   0.000   -0.157    -0.107
sigma2 3.714e-07 nan nan          nan      nan       nan
Regime 2 parameters
================================================================
        coef    std err      z    P>|z|    [0.025    0.975]
----------------------------------------------------------------
x1     0.0885  0.022      3.947   0.000    0.045     0.132
x2     0.9402  0.016     58.624   0.000    0.909     0.972
x3    -0.0120  0.010     -1.196   0.232   -0.032     0.008
sigma2 8.419e-06 7.72e-07 10.903  0.000    6.91e-06  9.93e-06
Regime transition parameters
================================================================
         coef    std err      z    P>|z|    [0.025    0.975]
----------------------------------------------------------------
p[0->0] 0.8860   0.016    57.034   0.000    0.856     0.916
p[1->0] 0.1458   0.048     3.058   0.002    0.052     0.239
p[2->0] 6.844e-12 nan     nan      nan      nan       nan
p[0->1] 4.094e-72 nan     nan      nan      nan       nan
p[1->1] 0.5603   0.016    35.995   0.000    0.530     0.591
p[2->1] 0.0137   0.008     1.747   0.081   -0.002     0.029
================================================================

Warnings:
[1] Covariance matrix calculated using numerical (complex-step) differentiation .
[2] Covariance matrix is singular or near-singular, with condition number 1.14e+18.
    Standard errors may be unstable .
```

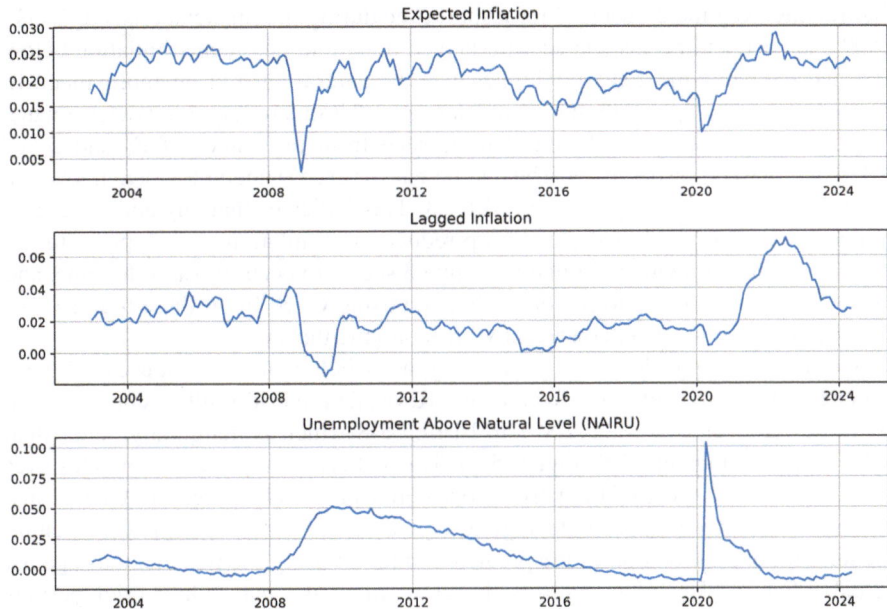

Figure 5-12.   Exogenous Variable Plots

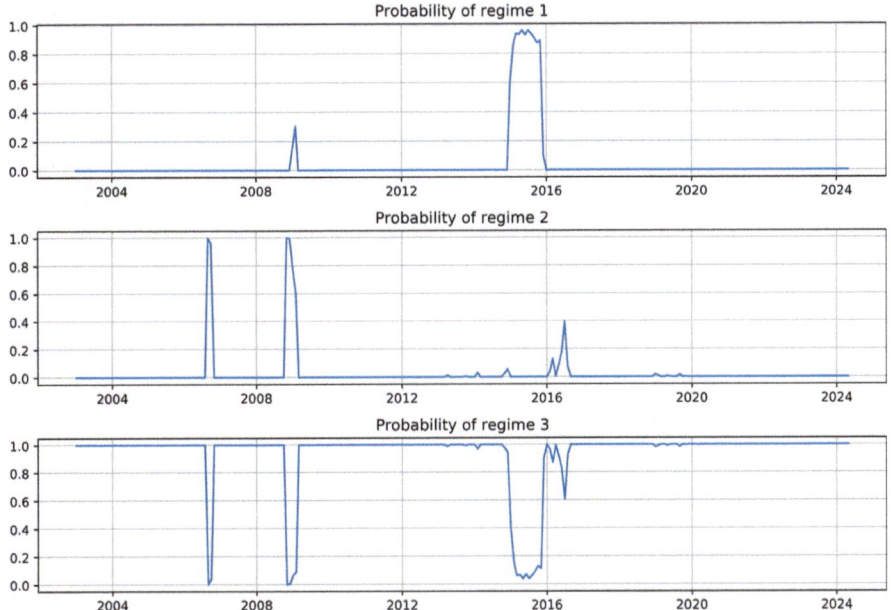

Figure 5-13.   Regime Probabilities for Fitting Phillips Curve Using Dynamic Regime Switching Model

first regime. Regime 2 occurred for a short time during the financial crisis of 2008–2009 and seems to be associated with lagged deflation, i.e., lagged inflation below 0%, and unemployment sharply higher than the natural level of unemployment. In this regime, both expected inflation and lagged inflation are important determinants for the current period's inflation. This is seen from the values of β_1 and β_2 of 0.339 and 0.3657, respectively. The coefficient for excess unemployment above the natural level is highly negative, at -0.1323. This indicates that any additional rise in unemployment would trigger a sharp reduction in inflation, as is typical during severe economic downturns. Finally, regime 3 seems to occur most of the time, and its coefficients are close to the ones obtained using OLS. In all three regimes, we observe β_3 to be negative, consistent with economic theory.

We observe that the Markov dynamic regime switching model produces a slightly better fit compared with OLS for the new Keynesian hybrid Phillips curve as seen from the AIC value of -2247.857 in Listing 5-8 which is marginally better than -2219 obtained using OLS (Figure 5-7). This is because regime 3 dominates over other regimes for most of the period considered in model fitting, and OLS is just as versatile in modeling single regime equations as dynamic regime switching model. Figure 5-14 compares the predicted inflation using the two methods against actual inflation. The plot confirms the close fit to the data. Only brief periods around 2008–2009 and 2015–2016 show the Markov dynamic regime switching

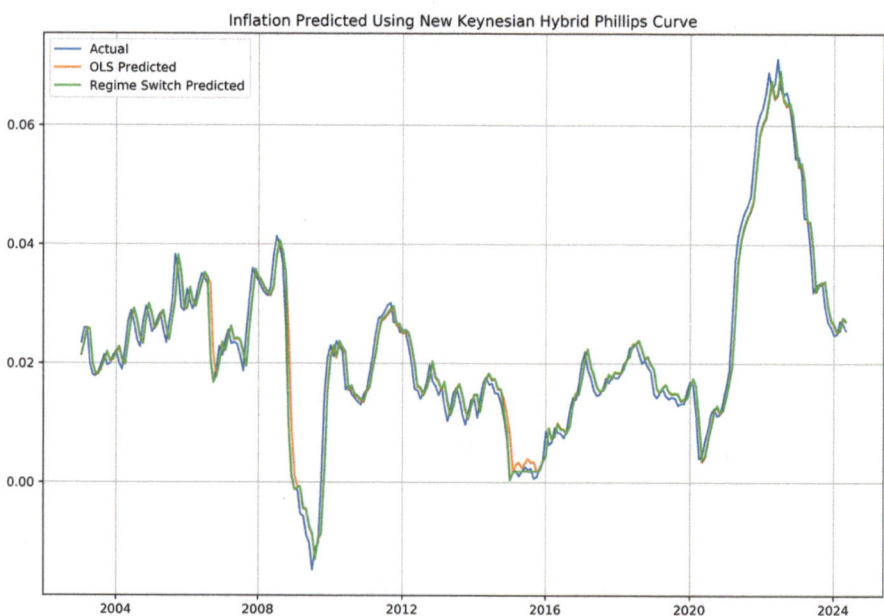

Figure 5-14. Comparison of New Keynesian Hybrid Phillips Curve Fitted Using OLS and Dynamic Regime Switching Model

model performing better than OLS. As described earlier, these are the periods when regimes 2 and 3, respectively, are likely to occur (Figure 5-13).

The code for fitting the Phillips curve to data is shown in Listing 5-9 and explained in pseudo-code 7.

Listing 5-9. Fitting New Keynesian Hybrid Phillips Curve Using OLS and Markov Dynamic Regime Switching Models

```python
import numpy as np
import pandas as pd
import statsmodels.api as sm
import logging
import os
import matplotlib.pyplot as plt


logging.basicConfig(level=logging.DEBUG)



class PhillipsCurve:
    def __init__(self, dirname, inflationFile, expectedInflFile, unemploymentFile,
        naturalUnempFile,
                        trainTestSplit=0.9):
        self.dirname = dirname
        self.trainTestSplit = trainTestSplit
        self.logger = logging.getLogger(self.__class__.__name__)

        inflation = pd.read_csv(os.path.join(dirname, f"{inflationFile}.csv"),
            parse_dates=["DATE"])
        self.convertColToFloat(inflation, inflationFile)
        pceValues = inflation.loc[:, inflationFile].values
        inflationVal = pceValues[12:] / pceValues[0:-12] - 1
        inflation.loc[:, "pi"] = 0
        inflation.loc[12:, "pi"] = inflationVal
        inflation.loc[:, "lagged_pi"] = 0
        inflation.loc[13:, "lagged_pi"] = inflationVal[0:-1]
        self.endog = "pi"

        expInflation = pd.read_csv(os.path.join(dirname, f"{expectedInflFile}.csv"),
            parse_dates=["DATE"])
        expInflation.rename(columns={expectedInflFile: "E_pi"}, inplace=True)
        self.convertColToFloat(expInflation, "E_pi", divideBy=100)
        expInflation = self.convertToMonthly(expInflation, ["E_pi"])
        df = pd.merge(inflation, expInflation, on=["DATE"], how="inner")

        unemp = pd.read_csv(os.path.join(dirname, f"{unemploymentFile}.csv"),
            parse_dates=["DATE"])
        unemp.rename(columns={unemploymentFile: "unemp"}, inplace=True)
        self.convertColToFloat(unemp, "unemp", divideBy=100)
        df = pd.merge(df, unemp, on=["DATE"], how="inner")

        nairu = pd.read_csv(os.path.join(dirname, f"{naturalUnempFile}.csv"),
            parse_dates=["DATE"])
        nairu.rename(columns={naturalUnempFile: "nairu"}, inplace=True)
```

```
41          self.convertColToFloat(nairu, "nairu", divideBy=100)
42          self.addYearAndQuarterColumn(df)
43          self.addYearAndQuarterColumn(nairu)
44          nairu.drop(columns=["DATE"], inplace=True)
45          df = pd.merge(df, nairu, on=["year", "quarter"], how="inner")
46          df.loc[:, "u_m_un"] = df.loc[:, "unemp"] - df.loc[:, "nairu"]
47
48          self.exog = ["E_pi", "lagged_pi", "u_m_un"]
49          self.df = df
50
51      def convertColToFloat(self, df, col, divideBy=1.0):
52          if (df.loc[:, col] == ".").sum() > 0:
53              df.drop(np.where(df.loc[:, col] == ".")[0], inplace=True)
54          df.loc[:, col] = df.loc[:, col].astype(np.float64) / divideBy
55          df.reset_index(drop=True, inplace=True)
56
57      def convertToMonthly(self, df, cols):
58          df.loc[:, 'year'] = df.DATE.dt.year
59          df.loc[:, 'month'] = df.DATE.dt.month.values
60          ypart = df[['year', 'month'] + cols]
61          ypart = ypart.groupby(['year', 'month']).mean().reset_index(drop=False)
62          ydate = df[["DATE", "year", "month"]].groupby(["year", "month"]).first().
                  reset_index(drop=False)
63          ydate.loc[:, "DATE"] = ydate.DATE + pd.offsets.MonthEnd(0) + pd.offsets.
                  MonthBegin(-1)
64          df = pd.merge(ydate, ypart, on=["year", "month"], how="inner")
65          df.drop(columns=["year", "month"], inplace=True)
66          return df
67
68      def addYearAndQuarterColumn(self, df):
69          df.loc[:, 'quarter'] = ((df.DATE.dt.month.values - 1) // 3)
70          df.loc[:, 'year'] = df.DATE.dt.year
71
72      def trainData(self):
73          ntrain = int(self.df.shape[0] * self.trainTestSplit)
74          y = self.df.loc[0:ntrain, self.endog].values
75          X = self.df.loc[:ntrain, self.exog].values
76          return y, X
77
78      def fitOLS(self):
79          y, X = self.trainData()
80          self.olsModel = sm.OLS(y, X)
81          self.olsModel = self.olsModel.fit()
82          self.logger.info(self.olsModel.summary())
83          summaryfile = os.path.join(self.dirname, self.__class__.__name__ + "_ols.txt")
84          with open(summaryfile, 'w') as fh:
85              fh.write(self.olsModel.summary().as_text())
86
87      def plotTrainingFit(self):
88          y, X = self.trainData()
89          fig, ax = plt.subplots(1, figsize=(10, 7))
90          ntrain = int(self.df.shape[0] * self.trainTestSplit)
91          date = self.df.loc[0:ntrain, "DATE"].values
```

```
92      ax. plot (date, y, label="Actual")
93      yPred = self .olsModel. fittedvalues
94      ax. plot (date, yPred, label="OLS Predicted")
95      yPred = self .markovModel. fittedvalues
96      ax. plot (date, yPred, label="Regime Switch Predicted")
97      ax. legend ()
98      ax. set ( title =" Inflation  Predicted  Using New Keynesian Hybrid  Phillips  Curve")
99      ax. grid ()
100     fig . tight_layout ()
101     plt . savefig (os. path . join ( self .dirname, f" infl_nkpc_{ self . __class__ . __name__}.
            jpeg"),
102                     dpi=500)
103     plt .show()
104
105 def fitRegimeSwitch( self ):
106     y, X = self . trainData ()
107     ntrain  = int ( self .df .shape[0] ∗ self . trainTestSplit )
108     date = self .df .loc [0: ntrain , "DATE"].values
109     np.random.seed(1024)
110     self .markovModel = sm.tsa.MarkovRegression(endog=y, k_regimes=3, trend='n',
            exog=X,
111                                     switching_trend =True,
112                                     switching_exog=True,
113                                     switching_variance =True)
114     self .markovModel = self.markovModel.fit ()
115
116     self . logger . info ( self .markovModel.summary())
117     summaryfile = os. path . join ( self .dirname, self . __class__ . __name__ + "
            _regimeSwitch.txt")
118     with open(summaryfile, 'w') as fh:
119         fh . write ( self .markovModel.summary().as_text())
120
121     fig , axes = plt . subplots (3, figsize =(10, 7))
122
123     ax = axes [0]
124     ax. plot (date, self .markovModel. filtered_marginal_probabilities [:, 0])
125     ax. set ( title =" Probability  of regime 1")
126     ax. grid ()
127
128     ax = axes [1]
129     ax. plot (date, self .markovModel. filtered_marginal_probabilities [:, 1])
130     ax. set ( title =" Probability  of regime 2")
131     ax. grid ()
132
133     ax = axes [2]
134     ax. plot (date, self .markovModel. filtered_marginal_probabilities [:, 2])
135     ax. set ( title =" Probability  of regime 3")
136     ax. grid ()
137
138     fig . tight_layout ()
139     plt . savefig (os. path . join ( self .dirname, f" regime_prob_{ self . __class__ . __name__
            }.jpeg"),
140                     dpi=500)
```

```
141     plt .show()
142
143     fig , axes = plt . subplots (3,  figsize =(10, 7))
144
145     ax = axes [0]
146     ax. plot (date ,  X[:, 0])
147     ax. set ( title ="Expected  Inflation ")
148     ax. grid ()
149
150     ax = axes [1]
151     ax. plot (date ,  X[:, 1])
152     ax. set ( title ="Lagged  Inflation ")
153     ax. grid ()
154
155     ax = axes [2]
156     ax. plot (date ,  X[:, 2])
157     ax. set ( title ="Unemployment Above Natural Level (NAIRU)")
158     ax. grid ()
159
160     fig . tight_layout ()
161     plt . savefig (os. path . join ( self .dirname, f"regime_vars_{ self .__class__.__name__
                }.jpeg"),
162                 dpi=500)
163     plt .show()
164
165  def  fit ( self ):
166      self .fitOLS()
167      self .fitRegimeSwitch ()
168      self . plotTrainingFit ()
169
170
171  if __name__ == "__main__":
172      dirname = r"C:\prog\cygwin\home\samit_000\latex\ book_stats \code\data"
173      nkpc_pc = PhillipsCurve (dirname,  "PCEPI", "T10YIE", "UNRATE", "NROU",
                trainTestSplit=1.0)
174      nkpc_pc. fit ()
```

Code Explanation

A code walk-through is presented below:

1. The code begins with instantiating an object of class **PhillipsCurve**. The constructor of this class is provided with the directory name containing data files, files containing the PCE price index, ten-year break-even inflation rate, unemployment rate, and natural rate of unemployment, also known as non-cyclical rate of unemployment. Because we are interested in fitting the Phillips curve to data rather than in making predictions, we use the entire dataset for model fitting (training) by specifying training-testing data split to be 1. Data for ten-year break-even inflation rate is used to determine expected inflation rate. All data files are in comma-separated format.

Algorithm 7 Using OLS and Markov Dynamic Regime Switching Model to Fit New Keynesian Hybrid Phillips Curve

---

Require: Files containing monthly data for the PCE index and unemployment level, daily data for break-even inflation rate and quarterly data for natural unemployment level.

1: Read the files containing data for the PCE index, unemployment and natural unemployment level.
2: Convert the data columns to floating-point data type.
3: Calculate monthly inflation as $\pi(t) = \frac{PCE(t)}{PCE(t-1)} - 1$.
4: Lag the calculated inflation by one month to obtain lagged inflation, $\pi(t-1)$.
5: Read the file containing break-even inflation rate (expected annual inflation). This value is obtained by subtracting the yield on par-value ten-year bond and ten-year TIPS.
6: Compute an average of daily expected inflation values to arrive at a monthly value.
7: Read the file containing natural rate of unemployment (NAIRU).
8: Assign the same quarterly value of natural unemployment rate to the three months falling in a quarter.
9: Join all the dataframes to obtain all exogenous variables needed for fitting the model in monthly frequency.
10: Fit the OLS model without using an intercept.
11: Fit the Markov dynamic regime switching model using three regimes, regime switching variance and coefficients but no intercept. Print summary statistics.
12: Plot the output from fitted models and compare against actual inflation rate.

---

2. The file containing the PCE price index is read and processed first, with the column containing the PCE index converted to floating-point decimal. Inflation rate is derived from this index by calculating the annual percentage change in index as $\pi(t) = \frac{PCE(t)}{PCE(t-12)} - 1$.

3. Lagged inflation is calculated by shifting the data for inflation calculated in the last step backward by one month. This implies $\pi_{\text{lagged}}(t) = \pi(t-1)$.

4. The file containing ten-year break-even inflation rate is read next. After converting the column containing the break-even rate to a floating-point decimal, it is converted to a monthly frequency from a daily frequency by calculating an average of all observations falling in a month. This processing occurs inside the method **convertToMonthly**.

5. The dataframe containing expected inflation is merged with the one containing inflation and lagged inflation on the date column. The resulting dataframe has data at monthly frequency.

6. The file containing unemployment rate is read, and after conversion to floating-point decimal, it is merged with the last dataframe. Unemployment data already has monthly frequency, and no further processing is required.

7. The data file containing natural rate of unemployment (or non-cyclical rate of unemployment) is read, and the data is converted to a floating-point decimal format. This data is at a quarterly frequency.

8. Columns containing year and quarter corresponding to each date are added to the dataframe containing non-cyclical rate of unemployment and the previous dataframe using the method **addYearAndQuarterColumn**.

9. The two dataframes are merged on year and quarter columns. The first dataframe had monthly frequency, while the dataframe containing non-cyclical rate of unemployment has quarterly frequency. The resulting dataframe has monthly frequency, with all months falling in a quarter having the same value of non-cyclical rate of unemployment for that quarter.

10. An additional column with the name **u_m_un** is computed as the difference between unemployment rate and non-cyclical rate of unemployment.

11. With all exogenous variables and endogenous variable computed, the code proceeds to fit the Phillips curve using OLS inside the method **fitOLS**. Summary statistics from model fitting are printed.

12. The Markov dynamic regime switching model is used to fit the Phillips curve inside the method **fitRegimeSwitch**. The **MarkovRegression** class from the **statsmodels** library is used for the purpose. This API does not require prepending a column of ones to exogenous variables for inclusion of intercept. It includes an intercept if argument **trend=c** is provided to the constructor of class **MarkovRegression**. **c** represents the constant trend. In addition to endogenous and exogenous variables, the constructor of the **MarkovRegression** class is provided with arguments indicating the number of regimes **k_regimes**, argument **switching_trend** indicating if intercept coefficients show change in different regimes, argument **switching_exog** indicating if coefficients of exogenous variables should change in different regimes, and argument **switching_variance** that indicates if error term covariance changes in different regimes. These last three arguments are supplied with a value of **True**.

13. The Markov model is fitted and summary statistics printed.

14. Marginal probabilities indicating the probability of occurrence of a regime are plotted using the **filtered_marginal_probabilities** API of the **MarkovRegression** class.

15. Predicted inflation rate on the training dataset is plotted against actual inflation rate and compared against the predictions of the OLS model.

Bayesian Methods

6

Bayesian statistics and frequentist statistics are the two pillars of quantitative probabilistic modeling. The frequentist approach is perhaps the easier of the two to grasp intuitively. Founded on the count-based modeling approach, it is premised on the hypothesis that the frequency of an event has a constant value that remains the same. The true parameter value is unique but unknown, and the experimental observations attempt to ferret out an estimate of this true parameter value.

Bayesian statistics, on the other hand, does not assume the existence of a true parameter value. Instead, parameters are hypothesized to belong to an unknown probability distribution. Observations are used to specify the probability distribution. Therefore, Bayesian methods produce a probability distribution of parameters, while frequentist methods produce a value. Another cornerstone feature of Bayesian methods is their use of the Bayes rule to modify the probability distribution of the parameter using a posterior distribution, as shown in Equation 6-1. The posterior distribution can be conceptualized as a distribution that utilizes the observed data to constrain the family of probability densities, while the prior distribution is an assumed distribution of parameters. The choice of prior is generally problem or domain dependent and can be supplemented with observations gleaned from the data. In Equation 6-1, set A denotes the observed data, and set B denotes model parameters. Model parameters can come from a discrete or continuous probability space. The cornerstone assumption of Bayes methods is that parameters B are realizations from a probability distribution space. In Equation 6-1, B_i spans the probability space of parameters B. If the space is discrete, the denominator in Equation 6-1 will involve a sum. If it is continuous, it will involve an integral.

© Samit Ahlawat 2025
S. Ahlawat, *Statistical Quantitative Methods in Finance*,
https://doi.org/10.1007/979-8-8688-0962-0_6

$P(B|A)$ denotes the posterior probability of parameters B given the observed data A.

$$P(B|A) = \frac{P(A|B)P(B)}{P(A)}$$

$$= \frac{P(A|B)P(B)}{\sum_{i=1}^{N} P(A|B_i)P(B_i)} \text{ for B} \in \text{ discrete space} \qquad (6\text{-}1)$$

$$= \frac{P(A|B)P(B)}{\int_i P(A|B_i)P(B_i)di} \text{ for B} \in \text{ continuous space}$$

Let us look at the mathematical formulation of Bayesian statistics to gain a better understanding of its fundamentals. Let us denote the set of observations by $(\mathbf{X}, y)$. We are attempting to fit a model $y = f(\mathbf{X}, \boldsymbol{\beta})$ to the observations where $\mathbf{X}$ is the vector of exogenous or explanatory variables, y is a vector of endogenous variables that the model is trying to predict, and $\boldsymbol{\beta}$ denotes a vector of model parameters. In Bayesian statistics, $\boldsymbol{\beta}$ is assumed to belong to a class of parametric probability distributions, and the model attempts to specify the hyper-parameters associated with the probability distribution. As a result, a distribution of model parameters, $\boldsymbol{\beta}$, is predicted. This is denoted as $P(\boldsymbol{\beta}|\mathbf{X}, \mathbf{y})$, also called the posterior density. $\mathbf{X}$ is assumed to be an exogenous variable whose probability is not being modeled; we are only concerned about the probability distribution of the dependent variable $\mathbf{y}$, the prior distribution of parameters $\boldsymbol{\beta}$, and the posterior distribution of parameters after observing the data.

A prior distribution of parameters, $\boldsymbol{\beta}$, is specified by the user as $P(\boldsymbol{\beta})$. This distribution is chosen to conform to prior estimates about the parameter distribution, i.e., a distribution that is not influenced by observed data but is guided by intuition. The choice of prior can be supplemented by insights gleaned from data. For example, a linear model that has normal errors may be expected to have normally distributed parameters. A priori estimates are intuitive guesses for a probability distribution. Model fitting using Bayesian statistics attempts to tweak this probability distribution to account for the observed data. This observation-aware distribution that refines the prior probability distribution is called the posterior distribution. It can be derived using the Bayes rule, as seen in Equation 6-2.

$$P(\boldsymbol{\beta}|\mathbf{X}, \mathbf{y}, \boldsymbol{\theta}) = \frac{P(\mathbf{y}|\boldsymbol{\beta}, \mathbf{X}, \boldsymbol{\theta})P(\boldsymbol{\beta})}{\int P(\mathbf{y}|\boldsymbol{\beta}, \mathbf{X}, \boldsymbol{\theta})P(\boldsymbol{\beta})d\beta} \qquad (6\text{-}2)$$

In frequentist methods, we typically use the principle of maximum likelihood to predict parameter values. Maximum likelihood begins with a functional form of the probability distribution using the data and parameter values, $P(y|\mathbf{X}, \boldsymbol{\beta})$. This expression is maximized with respect to $\boldsymbol{\beta}$.

Bayesian methods begin with the following two assumptions:

1. Prior distribution of parameters, $P(\beta)$.
2. Probability distribution of observing the data given parameter values and any other independent (exogenous) data $\mathbf{X}$, $P(y|\mathbf{X}, \beta)$. This expression is also used by the maximum likelihood method.

The point of departure between Bayesian methods and maximum likelihood–based frequentist methods is the formulation of unconditional probability density, $P(y|\mathbf{X})$. Maximum likelihood predicts this value as $\max_\beta P(y|\mathbf{X}, \beta)$, while Bayesian methods use posterior density, $P(\beta|y, \mathbf{X})$, to derive the unconditional probability density of data, y, as shown in Equation 6-3.

$$
\begin{aligned}
P(y|\mathbf{X}) &= \int P(y|\mathbf{X}, \beta) P(\beta|\mathbf{X}) d\beta \\
&= \int P(y|\mathbf{X}, \beta) P(\beta) d\beta \\
&= \int \text{model-likelihood} \times \text{prior} \times d\beta
\end{aligned}
\tag{6-3}
$$

For prediction, one can find a value of y that maximizes the unconditional probability density shown in Equation 6-3, i.e., $\operatorname{argmax}_\beta P(y|\mathbf{X})$. The posterior distribution of β takes the data distribution observed in training into account for making a prediction. After obtaining a distribution of y, one can obtain the most probable value or expected value, as desired in the problem. This is shown in Equation 6-4, where posterior density is used in the integral to obtain marginal density of the data. We can then pick the expected value of y as an inference using the marginal probability density.

$$
\begin{aligned}
P(y|\mathbf{X}) &= \int_{-\infty}^{\infty} P(y|\theta, \mathbf{X}) \times posterior \times d\theta \\
&= \int_{-\infty}^{\infty} P(y|\theta, \mathbf{X}) P(\theta|y, \mathbf{X}) d\theta \\
y_{inference} &= E[y] \\
&= \int_{-\infty}^{\infty} y P(y|\mathbf{X}) dy
\end{aligned}
\tag{6-4}
$$

In frequentist methods, confidence intervals for parameter estimates must be derived by assuming a distribution. In Bayesian methods by contrast, the probability distribution of parameter estimates is obtained naturally as a posterior distribution. Corresponding intervals are known as credible intervals in Bayesian statistics.

Hypothesis testing is another application where frequentist and Bayesian methods adopt markedly different approaches. In order to test a hypothesis using the frequentist method, one must first formulate a null hypothesis based on the model-likelihood function $f(y|\mathbf{X}, \boldsymbol{\beta})$. After this, a statistic is calculated under the assumption that the data is drawn randomly from the underlying distribution. A null hypothesis can then be accepted or rejected with a specified level of confidence depending upon the p-value. Bayesian methods, on the other hand, are endowed with a hypothesis-testing ability naturally because they formulate a posterior distribution of parameters. For example, one can compute the probability of hypothesis $\boldsymbol{\beta} < 0.2$ directly using the posterior distribution $P(\boldsymbol{\beta}|y, \mathbf{X})$ and then accept or reject the hypothesis based upon the computed probability and level of confidence, α. This is shown in Equation 6-5.

$$P(\boldsymbol{\beta} < p) = \int_{-\infty}^{p} P(\boldsymbol{\beta}|y, \mathbf{X}) P(\boldsymbol{\beta}) d\beta$$

$$P(\boldsymbol{\beta} < p) > \alpha \text{ accept the hypothesis} \tag{6-5}$$

$$< \alpha \text{ reject the hypothesis}$$

$$\alpha = \text{ confidence level}$$

As the number of observations goes to infinity, parameter values predicted by the frequentist approach converge with the parameter values that maximize the posterior density of parameters using the Bayesian approach. This will be verified in the next section using an example.

6.1   Application

Let us briefly look at a statistical problem to drive home the difference between frequentist and Bayesian approaches with respect to model fitting, inference, and hypothesis testing.

6.1.1   Model Fitting

We want to predict whether the five-day return for Bank of America's common stock (symbol BAC) is positive or negative. In previous chapters on linear regression, GLM, and Markov regime switching models, we studied different models for handling this problem. Those were all frequentist approach–based models. In this section, let us model the binary endogenous variable (if the five-day return for BAC is positive or negative) as the outcome of a coin toss with probability θ of giving a positive return (showing heads) and probability $1 - \theta$ of giving a non-positive (negative or zero) return. This model does not use any economic or other exogenous (explanatory) variables for simplicity. The probability of observing

k positive returns in a set of N observations can be written using a binomial distribution as shown in Equation 6-6.

$$P(y|\theta) = \binom{N}{k} \theta^k (1 - \theta)^{N-k} \tag{6-6}$$

Equation 6-6 is the likelihood of data; writing the log-likelihood and maximizing it with respect to model parameter θ gives the value $\frac{k}{N}$, as shown in Equation 6-7.

$$\log P(y|\theta) = \log \binom{N}{k} + k \log \theta + (N - k) \log (1 - \theta)$$

$$\frac{\partial \log P(y|\theta)}{\partial \theta} = 0 = \frac{k}{\theta} - \frac{N - k}{1 - \theta} \tag{6-7}$$

$$\frac{\theta}{1 - \theta} = \frac{k}{N - k}$$

$$\theta = \frac{k}{N}$$

Now let us turn to fitting the Bayesian model. First, we select a prior for the probability of observing a positive five-day return for Bank of America's common stock. Let us use a beta distribution shown in Equation 6-8. We keep the same probability function for likelihood of data $P(y|\theta)$ as was used for the frequentist approach in Equation 6-6.

$$P(\theta) = \frac{\theta^{\alpha-1} (1 - \theta)^{\beta-1}}{\int_0^1 \theta^{\alpha-1} (1 - \theta)^{\beta-1} \, d\theta} \tag{6-8}$$

Using the prior and likelihood, we can write the unconditional probability of data as shown in Equation 6-9.

$$P(y, \theta) = P(y|\theta)P(\theta)$$

$$= \binom{N}{k} \theta^k (1 - \theta)^{N-k} \frac{\theta^{\alpha-1} (1 - \theta)^{\beta-1}}{\int_0^1 \theta^{\alpha-1} (1 - \theta)^{\beta-1} \, d\theta} \tag{6-9}$$

$$= \binom{N}{k} \frac{\theta^{k+\alpha-1} (1 - \theta)^{N-k+\beta-1}}{\int_0^1 \theta^{\alpha-1} (1 - \theta)^{\beta-1} \, d\theta}$$

The posterior probability of parameter θ can be written using the Bayes rule, as shown in Equation 6-10.

$$P(\theta|y) = \frac{P(y|\theta)P(\theta)}{P(y)}$$

$$= \frac{\binom{N}{k}\theta^k(1-\theta)^{N-k}\frac{\theta^{\alpha-1}(1-\theta)^{\beta-1}}{\int_0^1 \theta^{\alpha-1}(1-\theta)^{\beta-1}d\theta}}{\int_0^1 P(y|\theta)P(\theta)d\theta} \tag{6-10}$$

Let us look at the numerator of the final equation in Equation 6-10. It can be recast into Equation 6-11. In this equation, we have used the notation for the beta function, $B(\alpha, \beta)$, which is defined as $\int_0^1 \theta^{\alpha-1}(1-\theta)^{\beta-1}d\theta$.

$$\text{Numerator} = \binom{N}{k}\frac{\theta^{k+\alpha-1}(1-\theta)^{N-k+\beta-1}}{\int_0^1 \theta^{\alpha-1}(1-\theta)^{\beta-1}d\theta}$$

$$= \binom{N}{k}\frac{\theta^{k+\alpha-1}(1-\theta)^{N-k+\beta-1}}{B(\alpha, \beta)} \tag{6-11}$$

The denominator of Equation 6-10 can be simplified, as shown in Equation 6-12.

$$\int_0^1 P(y|\theta)P(\theta)d\theta = \int_0^1 \binom{N}{k}\theta^k(1-\theta)^{N-k}\frac{\theta^{\alpha-1}(1-\theta)^{\beta-1}}{\int_0^1 \theta^{\alpha-1}(1-\theta)^{\beta-1}d\theta}d\theta$$

$$= \frac{\binom{N}{k}}{B(\alpha, \beta)}\int_0^1 \theta^{k+\alpha-1}(1-\theta)^{N-k+\beta-1}$$

$$= \binom{N}{k}\frac{B(k+\alpha, N-k+\beta)}{B(\alpha, \beta)} \tag{6-12}$$

Finally, putting together the expression for the numerator (Equation 6-11) and denominator (Equation 6-12), we get the expression for the posterior probability of parameter θ, as shown in Equation 6-13.

$$P(\theta|y) = \frac{\binom{N}{k}\frac{\theta^{k+\alpha-1}(1-\theta)^{N-k+\beta-1}}{B(\alpha,\beta)}}{\binom{N}{k}\frac{B(k+\alpha,N-k+\beta)}{B(\alpha,\beta)}}$$

$$= \frac{\theta^{k+\alpha-1}(1-\theta)^{N-k+\beta-1}}{B(k+\alpha, N-k+\beta)} \tag{6-13}$$

$$\sim \text{Binomial}(k+\alpha, N-k+\beta)$$

Expressions for the posterior parameter distribution in Equation 6-13 and the unconditional probability distribution of data in Equation 6-9 constitute the fitted Bayesian model. Using these two equations, we can make predictions and undertake hypothesis testing, as shown in the next two sections.

One important feature to note about Bayesian modeling is that the parameters that maximize the joint likelihood function of data and parameters $P(y, \theta)$ also maximize the posterior probability $P(\theta|y)$. For the specific case of the binomial distribution, this can be seen from Equations 6-9 and 6-13. Both expressions are proportional to $\theta^{k+\alpha-1}(1-\theta)^{N-k+\beta-1}$. For the general case, we can likewise establish that the joint probability density and the posterior probability share the same parameter value θ as the maximizer. This is shown in Equation 6-14.

$$P(y, \theta) = P(\theta|y)P(y)$$
$$\therefore \underset{\theta}{\operatorname{argmax}}\, P(y, \theta) = \underset{\theta}{\operatorname{argmax}}\, P(\theta|y)P(y)$$
$$= P(y)\underset{\theta}{\operatorname{argmax}}\, P(\theta|y) \qquad (6\text{-}14)$$
$$= \underset{\theta}{\operatorname{argmax}}\, P(\theta|y) \because P(y) > 0$$

6.1.2 Inference

Let us use the fitted binomial model to predict if the five-day return of BAC stock is positive. The frequentist approach–based binomial model will assign a constant probability θ of observing a positive return. We can refit the model to data once a new observation is available. In this setup, θ will be updated after each observation. To make a prediction, we toss a coin with probability θ of showing heads (success) and predict the five-day return to be positive if the toss reveals a head and predict a non-positive return for a tail. Let us train the model using data from January 2000 to January 2022 and make predictions for a 2.5-year period from January 2022 to July 2024. The accuracy of model predictions is 48.62%.

Turning to the Bayesian model, let us set the hyper-parameters of the selected prior for θ (binomial distribution) as $\alpha = \beta = 0.5$. As with the frequentist approach, we train the model from January 2000 to January 2022 and make predictions for a 2.5-year period from January 2022 to July 2024. After each prediction, once the final data is available, we update the posterior distribution. In order to make a prediction if the five-day return on BAC stock is positive or not, we first maximize the posterior density to find the optimum value of θ and then toss a coin with θ probability of heads (success) to predict if the return is positive. A head translates to a positive return. Let us first find the value of θ that maximizes the probability of data. This value of θ also maximizes the posterior probability of θ as seen from Equation 6-14.

In general, one will need to resort to numerical methods for maximization of posterior probability in Equation 6-19. However, for the present problem of coin toss, we can derive an analytical expression for θ that maximizes posterior density, as shown in Equation 6-15. For the general case, posterior probability may be intractable for analytic maximization. One may need to resort to numerical differentiation to obtain the optimum parameter value.

$$P(\theta|y) = \frac{\theta^{k+\alpha-1}(1-\theta)^{N-k+\beta-1}}{B(k+\alpha, N-k+\beta)}$$

$$= \text{Constant} \times \theta^{k+\alpha-1}(1-\theta)^{N-k+\beta-1}$$

$$\log P(\theta|y) = \log C + (k+\alpha-1)\log\theta + (N-k+\beta-1)\log(1-\theta)$$

$$\frac{\partial \log P(\theta|y)}{\partial\theta} = \frac{k+\alpha-1}{\theta} - \frac{N-k+\beta-1}{1-\theta} = 0$$

$$\frac{k+\alpha-1}{\theta} = \frac{N-k+\beta-1}{1-\theta}$$

$$\theta = \frac{k+\alpha-1}{N+\alpha+\beta-2}$$

$$(6\text{-}15)$$

From Equation 6-15, we can observe that as the number of observations goes to infinity, i.e., $N \to \infty$, $k \to \infty$, the optimum value of θ converges to $\frac{k}{N}$, as shown in Equation 6-16. As we can recall from Equation 6-7, this is the parameter value from the frequentist approach.

$$N \to \infty, k \to \infty, \theta_{\text{bayes}}^{\text{optimum}} \to \frac{k}{N} \qquad (6\text{-}16)$$

For making an inference, we perform the following steps shown in pseudo-code 8.

Algorithm 8 Inference Using Bayesian Model

Require: Hyper-parameters defining the prior probability: α, β, and input binary data. In the problem under consideration, binary data is a flag indicating if the five-day return on BAC stock is positive.

1: **for** i = 1, 2, $\cdots$, testing data rows **do**
2:     Compute the posterior density using Equation 6-13.
3:     Maximize it and obtain the value of θ. For this problem, maximization can be done analytically, as shown in Equation 6-15.
4:     Compute the probability of observing a positive return with the value of θ obtained from maximizing the posterior density, $P(y = 1|\theta)$. Use Equation 6-6.
5:     Update the posterior when actual observation becomes available.
6: **end for**
7: Compare the predictions against ground truth values.

6.1.3 Hypothesis Testing

For hypothesis testing, the frequentist approach assumes an underlying distribution and considers the null hypothesis to be the proposition that the hypothesized value is obtained by chance or a random draw from the assumed distribution. It computes a t-statistic of the calculated parameter value and associated p-value. If the p-value is too small, we can reject the null hypothesis using a prespecified confidence interval and state that we do not have enough confidence (at the specified level) in the underlying distribution and calculation of θ. Failure to reject the null hypothesis is considered an implicit confirmation that the underlying distribution and calculated parameter value are correct because we were not able to disprove it.

Returning to our example, let us try to test the hypothesis that $\theta = 0.5$ for predicting five-day returns of BAC stock. The null hypothesis is that five-day returns follow a binomial distribution with $\theta = 0.5$. The calculated value of $\theta = 0.527656$ on 2024-07-08 (July 8, 2024). Using this, the p-value is 1.0, and we cannot reject the null hypothesis at any significance level. This is taken as evidence that the model and parameter value is correct.

For the Bayesian approach, hypothesis testing involves integrating the posterior density over the relevant region of θ to obtain the probability of a parameter value exceeding a limit. For example, if we wish to compute the confidence that θ falls between 0.49 and 0.51, we integrate the posterior probability in Equation 6-13 between 0.49 and 0.51 and accept or reject the hypothesis based on a confidence level. This is shown in Equation 6-17.

$$P(0.49 \leq \theta \leq 0.51) = \int_{0.49}^{0.51} P(\theta|y)d\theta$$

(6-17)

Accept if $P(0.49 \leq \theta \leq 0.51) \geq$ confidence level

The code for model fitting, drawing inference, and testing hypothesis for the example discussed is presented in Listing 6-1.

Listing 6-1. Model Fitting, Inference Drawing, and Hypothesis Testing in Frequentist and Bayesian Approaches

```
import numpy as np
import pandas as pd
import logging
import os
import matplotlib.pyplot as plt
from abc import ABC, abstractmethod
import scipy.stats as ss

logging.basicConfig( level =logging.DEBUG)


class BinomialModel(ABC):
    PRICE_COL = "Close"
```

```
PERIOD = 5

def __init__( self , dirname, security , trainTestSplit =0.9, seed=10):
    self . logger = logging.getLogger( self .__class__.__name__)
    df = pd.read_csv(os.path . join (dirname, f"{ security }.csv"), parse_dates =["Date"
        ])
    df = self . calculateReturns (df)
    self .df = df
    self .dirname = dirname
    self . security  = security
    self .trainingRows = int (df.shape[0] ∗ trainTestSplit )
    np.random.seed(seed)

def calculateReturns ( self , df):
    price  = df.loc [:,  self .PRICE_COL].values
    returnCol = price [ self .PERIOD:] / price[:− self .PERIOD] − 1
    df.loc [:,  "return"] = 0
    df.loc [0: df.shape[0]  − self .PERIOD − 1, "return"] = returnCol
    df = df.loc [0: df.shape[0]  − self .PERIOD − 1, :]. reset_index (drop=True)
    return  df

@abstractmethod
def  fit ( self , endIndex=None):
    raise  NotImplementedError(f"Sub class { self .__class__.__name__} needs to
        implement")

@abstractmethod
def  predict ( self , index):
    raise  NotImplementedError(f"Sub class { self .__class__.__name__} needs to
        implement")

@abstractmethod
def  testHypothesis ( self , theta =None, nobservation=None, nsuccess=None):
    raise  NotImplementedError(f"Sub class { self .__class__.__name__} needs to
        implement")

def  test ( self ):
    actual  = np.zeros( self .df.shape[0]  − self .trainingRows, dtype=np.int8 )
    predicted  = np.zeros( actual .shape [0], dtype=np.int8 )
    thetaArr = np.zeros( actual .shape [0], dtype=np.float32 )
    for  i in range( self .trainingRows, self .df.shape[0], 1):
        predicted [i  − self .trainingRows] = self . predict (i)
        actual [i  − self .trainingRows] = np.where( self .df.loc [i, "return"] > 0, 1,
            0)
        thetaArr [i  − self .trainingRows] = self . theta
        self . fit (i)

    accuracy  = (sum(actual  == ( predicted )) / actual .shape[0]) ∗ 100
    self . logger .info ("Overall  accuracy  of %s: %.2f", self .__class__.__name__,
        accuracy)
    return  thetaArr

def compareResults( self , freq , bayes):
```

```
61        dates = self.df.loc[self.trainingRows :, "Date"].values
62        fig, ax = plt.subplots(nrows=2, figsize =(10, 7))
63        ax[0].plot(dates, freq, label="Frequentist")
64        ax[0].plot(dates, bayes, label="Bayesian")
65        ax[0].set(title ="Frequentist and Bayesian Values of Parameter Theta")
66        ax[0].set_ylabel("Theta")
67        ax[0].set_xlabel("Date")
68        ax[0].legend()
69        ax[0].grid()
70
71        diffs = freq − bayes
72        ax[1].hist(diffs, bins=20)
73        ax[1].set(title ="Histogram of Difference Between Frequentist and Bayesian
              Predictions")
74        ax[1].grid()
75        fig.tight_layout()
76        plt.savefig(os.path.join(self.dirname, f"diff_{self.__class__.__name__}.jpeg"),
77                    dpi=500)
78        plt.show()
79
80
81   class Frequentist(BinomialModel):
82       def fit(self, endIndex=None):
83           if endIndex is None:
84               endIndex = self.trainingRows
85           returns = self.df.loc[0:endIndex−1, "return"].values
86           self.theta = np.sum(returns > 0) / returns.shape[0]
87           self.logger.info("class: %s, Date: %s, theta = %f", self.__class__.__name__,
                  str(self.df.loc[endIndex, "Date"]), self.theta)
88
89       def predict(self, index):
90           return np.random.binomial(1, self.theta, 1)
91
92       def testHypothesis(self, theta=None, nobservation=None, nsuccess=None):
93           if theta is None:
94               theta = self.theta
95           if nobservation is None:
96               nobservation = self.df.shape[0]
97           if nsuccess is None:
98               nsuccess = sum(self.df.loc[:, "return"].values > 0)
99           result = ss.binom_test(nsuccess, nobservation, theta)
100          self.logger.info("P−value: %f", result)
101
102
103  class Bayesian(BinomialModel):
104      def __init__(self, dirname, security, trainTestSplit =0.9, seed=10, alpha=0.5, beta
             =0.5):
105          super().__init__(dirname, security, trainTestSplit, seed)
106          self.alpha = alpha
107          self.beta = beta
108
109      def fit(self, endIndex=None):
110          if endIndex is None:
```

```
111            endIndex = self .trainingRows
112            returns  = self .df. loc [0: endIndex−1, "return" ]. values
113            N = returns .shape [0]
114            k = np.sum(returns > 0)
115            self . theta  = (k + self .alpha − 1) / (N + self .alpha + self .beta − 2)
116            self .logger .info ("class : %s, Date: %s, theta = %f", self .__class__.__name__,
                     str ( self .df. loc [endIndex, "Date" ]),  self . theta )

118      def  predict ( self ,  index ):
119            return  np.random.binomial(1,  self . theta ,  1)

121      def  testHypothesis ( self ,  theta =None, nobservation=None, nsuccess=None):
122            if  theta  is  None:
123                 theta  = self . theta
124            if  nobservation  is  None:
125                 nobservation  = self . df .shape [0]
126            if  nsuccess  is  None:
127                 nsuccess = sum(self .df. loc [:,  "return" ]. values  > 0)
128            result  = ss .binom_test(nsuccess,  nobservation ,  theta )
129            self .logger .info ("P−value: %f",  result .pvalue)


132  if  __name__ == "__main__":
133       dirname = r"C:\prog\cygwin\home\samit_000\latex\ book_stats \code\data"
134       freqModel = Frequentist (dirname,  "BAC")
135       freqModel. fit ()
136       freq  = freqModel. test ()
137       freqModel. testHypothesis ()

139       bayesianModel = Bayesian(dirname,  "BAC", alpha=0.5,  beta =0.5)
140       bayesianModel. fit ()
141       bayes = bayesianModel. test ()

143       freqModel.compareResults(freq ,  bayes)
```

Before doing a code walk-through, let us take a look at the results of fitting the two models. As shown earlier, as the number of observations increases, the predicted parameter values from both frequentist and Bayesian approaches converge. This can be seen in the plot of results in Figure 6-1. Predictions of θ are virtually identical, and the two plots lie on top of each other. There is a small difference between them, as observed in the histogram of differences between their predictions. We observe that the Bayesian estimate of θ is slightly higher, though the difference is insignificantly small.

Code Explanation
Let us walk through the code in Listing 6-1.

1. The code uses a base class that contains the boilerplate code shared between frequentist and Bayesian binomial models. Using class derivation enhances code modularity, reduces duplication, and improves readability and code maintenance. The base class is **BinomialModel**. It is an abstract base class.

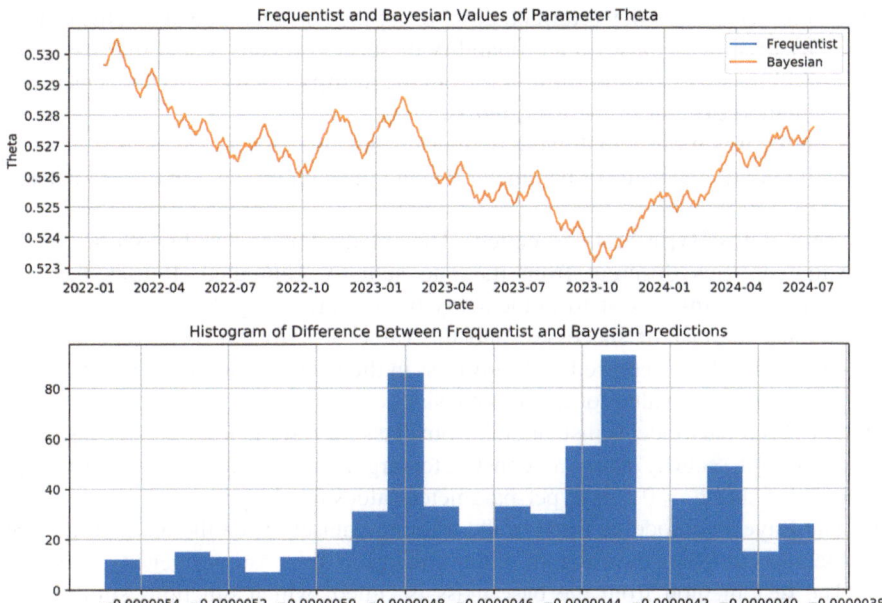

Figure 6-1. Comparison of Frequentist and Bayesian Predictions for Five-Day Return on BAC Common Stock

In Python, one creates an abstract base class by deriving from the **ABC** class in Python library **abc**.

2. Abstract base classes cannot be instantiated. They define interfaces, i.e., they provide declaration of certain methods that a concrete subclass must implement. Interface methods are declared using the decorator **abstractmethod**.

3. Abstract base class **BinomialModel** provides the following interface declarations: **fit**, **predict**, and **testHypothesis**. It provides concrete implementation for its constructor that reads the CSV file containing security price, calculates returns, and computes the rows to use for model training. The class also provides a method for testing the model, invoking the **predict** method which must be implemented by a concrete subclass.

4. The code begins by instantiating a concrete subclass **Frequentist** that implements the abstract interface **BinomialModel**. The constructor of class **Frequentist** is provided with the directory name containing data files and the name of security (BAC).

5. The file is read and five-day returns computed inside the base class constructor.

6. The frequentist model is fit to training data. This method calculates θ.

7. It calls method **test** on the frequentist model. This method is implemented in base class **BinomialModel**. The method steps through each day in the test dataset. For each day, it does the following:

- Predict the five-day return on day i using the current value of θ. This is done by tossing a coin with a probability of success (heads) equal to θ using the API **numpy.random.binomial** from the **numpy** library. This happens inside method **predict** of the **Frequentist** class.
- It records if the actual five-day return is positive or not.
- It updates the value of parameter θ using the data for day i in the test dataset.

8. Method **testHypothesis** is called. This method tests the hypothesis if the value of θ is different from the one actually computed. It uses the API **scipy.stats.binom_test** from the **scipy** library, providing the number of observations, number of successes, and the value of θ to test as arguments. This method has been replaced by **binomtest** in the latest version of **scipy**. It returns the computed p-value for the hypothesis.
9. The Bayesian model is instantiated from class **Bayesian**, providing the directory name and security name as constructor arguments, as before, and providing $\alpha = 0.5$ and $\beta = 0.5$ as hyper-parameter values for the prior.
10. The Bayesian model is fitted to the training dataset inside the method **fit**. This method computes the value of θ that maximizes the posterior density of θ.
11. The **test** method from the base class is called, which functions similarly as described before, except that the **predict** method from the subclass is called.
12. Hypothesis testing is performed for the **Bayesian** class.

6.2   Bayesian Linear Regression

Linear regression involves fitting a linear model to a set of observations, y, as shown in Equation 6-18. Unlike ordinary linear regression based on frequentist methods, Bayesian linear regression assumes model parameters to follow a probability distribution specified by an initial prior, $P(\boldsymbol{\beta})$.

$$y = \mathbf{X}\beta + \epsilon$$

$$\epsilon \sim N(0, \sigma^2)$$

$$P(\boldsymbol{\beta}) = N(\boldsymbol{\mu_0}, \boldsymbol{\Sigma_0}) \text{ prior}$$

$$P(y|\mathbf{X}, \beta) = N(\mathbf{X}\beta, \sigma^2)$$

(6-18)

Let us derive the posterior probability density of $\boldsymbol{\beta}$ using Equations 6-1 and 6-18. Substituting the prior for $\boldsymbol{\beta}$ from Equation 6-18 into Equation 6-1, one gets the expression shown in Equation 6-19. We have used the fact that $\mathbf{X}$ is an independent or exogenous variable. Also, the denominator in Equation 6-1 can be thought of as a normalization factor, and we only look at the numerator.

$$P(\boldsymbol{\beta}|y, \mathbf{X}) = \frac{P(y|\mathbf{X}, \boldsymbol{\beta})P(\boldsymbol{\beta})}{P(y|\mathbf{X})}$$

$$\propto P(y|\mathbf{X}, \boldsymbol{\beta})P(\boldsymbol{\beta})$$

$$\propto \exp\left[-\frac{1}{2\sigma^2}\left((y - \mathbf{X}\boldsymbol{\beta})'(y - \mathbf{X}\boldsymbol{\beta}) + (\boldsymbol{\beta} - \mu_0)'\left(\sigma^{-2}\mathbf{\Sigma}\right)^{-1}(\boldsymbol{\beta} - \mu_0)\right)\right]$$

$$\propto \exp\left[-\frac{1}{2\sigma^2}\left(\boldsymbol{\beta}'\left(\mathbf{X}'\mathbf{X} + \left(\sigma^{-2}\mathbf{\Sigma}\right)^{-1}\right)\boldsymbol{\beta} - 2\left(\mu_0'\left(\sigma^{-2}\mathbf{\Sigma}\right)^{-1} + \mathbf{X}'y\right)\boldsymbol{\beta} + \cdots\right)\right]$$

$$\propto \exp\left[-\frac{1}{2\sigma^2}(\boldsymbol{\beta} - \chi)'\mathbf{\Lambda}^{-1}(\boldsymbol{\beta} - \chi)\right]$$

where $\chi = \left(\mathbf{X}'\mathbf{X} + \left(\sigma^{-2}\mathbf{\Sigma}\right)^{-1}\right)^{-1}\left(\mu_0'\left(\sigma^{-2}\mathbf{\Sigma}\right)^{-1} + \mathbf{X}'y\right)$

and $\mathbf{\Lambda} = \left(\mathbf{X}'\mathbf{X} + \left(\sigma^{-2}\mathbf{\Sigma}\right)^{-1}\right)$

$$= N(\chi, \mathbf{\Lambda})$$

(6-19)

As can be seen from Equation 6-19, a Gaussian prior density for $\boldsymbol{\beta}$ yields a Gaussian posterior density. If we assume the prior density of $\boldsymbol{\beta}$ to be a normal distribution centered at 0 with a variance of σ_0^2, i.e., $\mu_0 = 0$ and $\mathbf{\Sigma} = \sigma_0^2\mathbf{I}$, the above expression can be simplified as shown in Equation 6-20.

$$P(\boldsymbol{\beta}|y, \mathbf{X}) \sim N\left(\left(\mathbf{X}'\mathbf{X} + \sigma^{-2}\sigma_0^2\mathbf{I}\right)^{-1}\mathbf{X}'y, \left(\mathbf{X}'\mathbf{X} + \sigma^{-2}\sigma_0^2\mathbf{I}\right)^{-1}\right) \qquad (6\text{-}20)$$

From Equation 6-20, we observe that the expression for posterior parameter density is similar to the parameter density derived from the OLS solution. OLS predicts the true parameter value to be $(\mathbf{X}'\mathbf{X})\mathbf{X}y$, as shown in Equation 2-12. The variance of parameter estimates is due to the variance of error term, ϵ, in OLS and is shown in Equation 2-10. If we assume error term ϵ is normally distributed (**note** this assumption is not required by OLS), we can write the probability distribution of parameter $\boldsymbol{\beta}$ as shown in Equation 6-21.

$$\boldsymbol{\beta}_{OLS} = N\left((\mathbf{X}'\mathbf{X})^{-1}\mathbf{X}'y, (\mathbf{X}'\mathbf{X})^{-1}\sigma_\epsilon^2\mathbf{I}\right) \qquad (6\text{-}21)$$

If we set the variance of the posterior distribution of $\boldsymbol{\beta}$ to 0 by setting $\sigma_0 = 0$ in Equation 6-20, we obtain the distribution implied by OLS in Equation 6-21 (after adjusting for a constant σ_ϵ). If there is no uncertainty in our prior belief about the value of parameter β, Bayesian estimates reduce to frequentist estimates.

Using normally distributed priors in Bayesian linear regression with the Gaussian data distribution is equivalent to adding L2 regularization to OLS as discussed in Section 2.7. To see why this is true, let us write the objective function for OLS with L2 regularization, as shown in Equation 6-22. Differentiating with respect to parameter β, we get the value of β which corresponds to the mean of the posterior distribution of β using Bayesian regression, as seen from Equation 6-20.

$$\max_{\beta} \frac{1}{2} (\mathbf{y} - \mathbf{X}\boldsymbol{\beta})' (\mathbf{y} - \mathbf{X}\boldsymbol{\beta}) + \frac{\lambda}{2} \boldsymbol{\beta}' \boldsymbol{\beta}$$

Differentiate with respect to β (6-22)

$$- \mathbf{X}' (\mathbf{y} - \mathbf{X}\boldsymbol{\beta}) + \lambda \boldsymbol{\beta} = 0$$

$$\boldsymbol{\beta} = (\mathbf{X}'\mathbf{X} + \lambda \mathbf{I})^{-1} \mathbf{X}'\mathbf{y}$$

For other prior distributions of β, a closed-form solution of the posterior distribution may not be possible, and one must use numerical methods.

For inference, we are provided a set of exogenous variables as part of the test dataset, and we want to predict the output for each test data point. The Bayesian model furnishes a probability distribution $P(y|\mathbf{X_{test}})$, as shown in Equation 6-23.

$$P(y|\mathbf{X_{test}}) = \int P(y|\mathbf{X_{test}}, \boldsymbol{\beta}) P(\boldsymbol{\beta}|y_{train}, \mathbf{X_{train}}) d\boldsymbol{\beta}$$ (6-23)

Integrating the parameter β, we get a distribution of y. In Equation 6-23, $P(\boldsymbol{\beta}|y_{train}, \mathbf{X_{train}})$ is the posterior distribution calculated using the training dataset. In order to make an inference, we can maximize this probability with respect to y and predict the value of y that maximizes the probability density. For example, for Bayesian linear regression, the predicted value of y that maximizes the probability $P(y|\mathbf{X_{test}})$ can be shown to be equal to $\mathbf{X_{test}}\hat{\boldsymbol{\beta}}$ where $\hat{\boldsymbol{\beta}}$ is the expected value of β obtained from its posterior density in Equation 6-19, i.e., $\hat{\boldsymbol{\beta}}$ = $\left(\mathbf{X}'_{train}\mathbf{X_{train}} + \left(\sigma^{-2}\boldsymbol{\Sigma}\right)^{-1} \right)^{-1} \left(\mu'_0 \left(\sigma^{-2}\boldsymbol{\Sigma}\right)^{-1} + \mathbf{X}'_{train}y_{train} \right)$

Let us look at an example to compare the predictions of frequentist-based OLS with Bayesian linear regression.

6.2.1   Application

The Fama-French five-factor model is an enhancement to the capital-asset pricing model and its progeny. The model uses additional risk factors in addition to the market return to explain a security's return. The model was proposed by Eugene Fama and Kenneth French [22]. Prior to this model, Fama and French (1993) had proposed a three-factor model [23]. The models were motivated by research that found systematic influence of stock size and momentum on future returns of stocks after accounting for market beta.

According to the Fama-French five-factor model, expected excess future returns on a stock can be explained using a linear regression equation shown in Equation 6-24. In this equation, $R(t)$ refers to the expected security return on day t. $R_{rf}(t)$ is the risk-free rate with the annual yield on a one-month US treasury bill providing the value. SMB refers to "small minus big," or the difference in return between a diversified portfolio of small market capitalization stocks (small cap) and large market capitalization (large cap) stocks. HML refers to "high minus low," or the difference in return between a diversified portfolio of high book value to market value stocks and low book value to market value stocks. Book-to-market value ratio is a measure of leverage. Book value is the liquidation value of a company's assets. Market value is the present market value of a company's common stock and bonds. Typically, companies that are well established and have steady business tend to have higher book-to-market value of equity. RMW refers to "robust minus weak," or the difference in return between a diversified portfolio of stocks with robust profitability and those with weak profitability. Profitability is measured using earnings per share. Finally, CMW refers to "conservative minus weak," or the difference in return between a diversified portfolio of stocks with conservative (steady) and weak investments into business. Portfolios are constructed by sorting stocks on specific criteria followed by grouping them into deciles. For further details on portfolio construction, please refer to the research paper [22].

$$R(t) - R_{rf}(t) = \alpha + \beta_1 \left(R_{mkt}(t) - R_{rf}(t) \right) + \beta_2 \text{SMB}(t) +$$

$$\beta_3 \text{HML}(t) + \beta_4 \text{RMW}(t) + \beta_5 \text{CMA}(t) + \epsilon(t)$$

$$R(t) = \text{Expected stock return at time t}$$

$$R_{mkt}(t) = \text{Expected market return at time t}$$

$$R_{rf}(t) = \text{Risk-free rate of return at time t}$$

$$\text{SMB}(t) = \text{Excess return on small-cap over large-cap stocks at time t}$$

$$\text{RMW}(t) = \text{Excess return on robust over weak profitability stocks}$$

$$\text{CMA}(t) = \text{Excess return of conservative over aggressive investment stocks}$$

(6-24)

Let us also fit a ridge regression version of Equation 6-24. As shown in the last section, this corresponds to Bayesian linear regression with isotropic prior, i.e., a prior with variance as $\sigma^2 \mathbf{I}$. Following model fitting, let us use the two models to predict returns for the test dataset and compare the predictions.

Values of risk factors $\left(R_{mkt}(t) - R_{rf}(t) \right)$, SMB$(t)$, HML$(t)$, RMW$(t)$, and CMA$(t)$ have been made available by Fama and French on the website [24].

   The code for fitting the Fama-French five-factor model using OLS and Bayesian linear regression models for one-month returns on Bank of America (BAC) common stock is presented in Listing 6-2. As seen from the output, the Bayesian approach gives a slightly better fit for the test dataset with a root-mean-square (RMS) error of 0.086662, as compared with 0.086962 from the OLS model.

Listing 6-2. Fitting Fama-French Five-Factor Model to BAC Stock One-Month Returns Using OLS and Bayesian Approaches

```
import numpy as np
import pandas as pd
from abc import ABC, abstractmethod
import logging
import os
import statsmodels.api as sm

logging.basicConfig(level=logging.DEBUG)


class FF5FactorBase(ABC):
    PRICE_COL = "Close"
    PERIOD = 21

    def __init__(self, dirname, ff5FactorFile, security, trainTestSplit=0.9):
        self.logger = logging.getLogger(self.__class__.__name__)
        self.dirname = dirname
        self.security = security
        dfFF5 = pd.read_csv(os.path.join(dirname, ff5FactorFile), parse_dates=["Date"])
        dfSecurity = pd.read_csv(os.path.join(dirname, f"{security}.csv"), parse_dates
            =["Date"])
        self.endog, self.exog = None, None
        self.df = self.processColumns(dfFF5, dfSecurity)
        self.nTrainRows = int(self.df.shape[0] * trainTestSplit)
        self.model = None

    def processColumns(self, dfFF5, dfSecurity):
        cols = ["Mkt-RF", "SMB", "HML", "RMW", "CMA", "RF"]
        for col in cols:
            dfFF5.loc[:, col] = dfFF5.loc[:, col].astype(np.float32)
        dfSecurity.loc[:, self.PRICE_COL] = dfSecurity.loc[:, self.PRICE_COL].astype(
            np.float32)
        price = dfSecurity.loc[:, self.PRICE_COL].values
        return1Mo = price[self.PERIOD:] / price[0:-self.PERIOD] - 1
        dfSecurity.loc[:, "1MoReturn"] = 0
        dfSecurity.loc[self.PERIOD:, "1MoReturn"] = return1Mo
        dfMerged = pd.merge(dfSecurity, dfFF5, on=["Date"], how="inner")
        self.endog = "1MoRetMinusRF"
        dfMerged.loc[:, self.endog] = dfMerged.loc[:, "1MoReturn"] - dfMerged.loc[:, "
            RF"]

        self.exog = ["Mkt-RF", "SMB", "HML", "RMW", "CMA"]
        dfMerged = dfMerged.loc[self.PERIOD:, ["Date", self.endog] + self.exog].
            reset_index(drop=True)
```

```
41              return  dfMerged
42
43       @abstractmethod
44       def  fit ( self ):
45              raise  NotImplementedError(f"Sub class  { self . __class__ . __name__} needs to
                     implement")
46
47       def  test ( self ):
48              groundTruth = self . df . loc [ self .nTrainRows:,  self .endog]. values
49              exog = self . df . loc [ self .nTrainRows:,  self .exog]. values
50              exog = sm.add_constant(exog,  has_constant="add")
51              predicted  = self .model. predict (exog)
52              diff  = predicted  − groundTruth
53              mse = np. sqrt (np.dot ( diff ,  diff ) /  diff . shape [0])
54              self . logger . info ("MSE error on  test  dataset :  %f", mse)
55
56
57  class  OLSModel(FF5FactorBase):
58       def  fit ( self ):
59              endog = self . df . loc [0: self .nTrainRows, self .endog]. values
60              exog = self . df . loc [0: self .nTrainRows, self .exog]. values
61              exog = sm.add_constant(exog,  has_constant="add")
62              self . model = sm.OLS(endog, exog). fit ()
63              self . logger . info ( self .model.summary())
64              summaryfile = os. path . join ( self .dirname,  self . __class__ . __name__ + ".txt")
65              with  open(summaryfile,  'w') as  fh:
66                   fh . write ( self .model.summary().as_text ())
67
68
69  class  BayesianLinearRegModel(FF5FactorBase):
70       def  fit ( self ):
71              endog = self . df . loc [0: self .nTrainRows, self .endog]. values
72              exog = self . df . loc [0: self .nTrainRows, self .exog]. values
73              exog = sm.add_constant(exog,  has_constant="add")
74              self . model = sm.OLS(endog, exog). fit_regularized  (alpha=1.0,  L1_wt=0.0)
75
76
77  if  __name__ == "__main__":
78       dirname = r"C:\prog\cygwin\home\samit_000\latex \ book_stats \code\data "
79       ff5file  = " ff5Factors .csv"
80       security  = "BAC"
81       ols  = OLSModel(dirname, ff5file ,  security )
82       ols . fit ()
83       ols . test ()
84
85       bayesian = BayesianLinearRegModel(dirname, ff5file ,  security )
86       bayesian . fit ()
87       bayesian . test ()
```

Code Explanation

Let us walk through the code in Listing 6-2.

1. Let us quickly recapitulate the code structure before stepping through the logic in the order of execution. Abstract base class **FF5FactorBase** contains the boilerplate code for reading the Fama-French five-factor model file and security file and testing the model on the test dataset.

2. Abstract base class **FF5FactorBase** derives from the **ABC** class of the **abc** library. This class has abstract method **fit** which needs to be implemented inside a concrete subclass. Additionally, the **model** attribute of this class is set to **None** and needs to be instantiated by concrete subclasses.

3. **OLSModel** and **BayesianLinearRegModel** are the two concrete subclasses that derive from abstract base class **FF5FactorBase**. These two classes provide an implementation of abstract method **fit**.

4. Base class **FF5FactorBase** defines a member **PERIOD** which determines the number of days used for calculating return. We are calculating a one-month return, so this value is set to 21 trading days.

5. Stepping through the code in the order of execution, the **OLSModel** class is instantiated first, with three arguments provided to its constructor—the directory name containing data files, the name of the file containing historical values of Fama-French five factors, and the name of the security.

6. The constructor of class **OLSModel** first calls the constructor of the base class. Inside the base class, input files are read as dataframes using the pandas library.

7. The one-month return of the BAC common stock is computed as $r(t) = \frac{P(t)}{P(t-21)} - 1$. It should be noted that these are contemporaneous returns.

8. It joins the dataframes containing Fama-French factors and security returns.

9. The code computes the difference between the one-month security return and the risk-free rate of return and stores it in column **1MoRetMinusRF** of the dataframe.

10. The OLS model is fitted on the training dataset after adding an intercept to the regression. It uses the **fit** method of the **OLS** class from the **statsmodels** library.

11. The test dataset is used for predicting the security's one-month returns and root-mean-square error against actual returns computed and reported.

12. A similar process is followed for the **BayesianLinearRegModel** class. The only difference from the **OLSModel** class is the API used for fitting the model. Here, we fit a regularized model with L2 regularization (ridge regression). To achieve this, the **fit_regularized(alpha=1.0, L1_wt=0.0)** API is invoked. **Alpha** of 1.0 indicates the weight for the regularization term, and **L1_wt** indicates the proportion of L1 or norm-1 regularization to add. This is also called lasso regularization—here, we use none. Entire regularization comes from L2 penalty on model weights.

13. The root-mean-square error on the test dataset is computed and printed inside the **fit** method defined in the base class.

6.2.2 Conjugate Priors

In Bayesian statistics, we have encountered two related probability distributions for model parameters $\boldsymbol{\theta}$: prior distribution $P(\boldsymbol{\theta})$ and posterior distribution $P(\boldsymbol{\theta}|y, \mathbf{X})$. When the two probability functions belong to the same class of probability density functions, we say that the prior and the probability density function $P(y|\boldsymbol{\theta}, \mathbf{X})$ are conjugate priors.

We have encountered two conjugate priors so far. We observed that a Gaussian prior density and a Gaussian probability distribution gave a Gaussian posterior density. Similarly, a prior with a beta distribution along with binomial probability density gave a binomial posterior density. Therefore, beta and binomial distributions are conjugate priors.

A gamma prior distribution along with Poisson probability density yields a gamma posterior density. Therefore, gamma and Poisson distributions are conjugate priors. This is shown in Equation 6-25. It is useful to know about conjugate priors because it affords analytic tractability in writing out the posterior density function. Knowing conjugate priors, we can select appropriate prior distributions for enhancing analytic tractability.

$$
\begin{aligned}
P(\theta|y, \mathbf{X}) &\propto P(y|\theta, \mathbf{X})P(\theta) \\
&= \text{Poisson}\,(y, \theta)\,\text{Gamma}\,(\theta; \alpha, \beta) \\
&= \frac{e^{-\theta}\theta^y}{y!}\frac{\beta^\alpha}{\Gamma(\alpha)}\theta^{(\alpha-1)}e^{-\beta\theta} \\
&= \frac{\beta^\alpha}{\Gamma(\alpha)}\theta^{(\alpha-1+y)}e^{-(\beta+1)\theta} \\
&= \frac{\beta^\alpha}{\Gamma(\alpha)}\frac{\Gamma(\alpha+y)}{(\beta+1)^{(\alpha+y)}}\text{Gamma}\,(\theta; \alpha+y, \beta+1) \\
&= \text{Gamma distribution}
\end{aligned}
\tag{6-25}
$$

A table showing commonly encountered conjugate priors is shown in Table 6-1.

Let us look at another conjugate prior pair: a categorical distribution for the likelihood function and a Dirichlet distribution for prior. The categorical distribution is a generalization of the Bernoulli distribution because it has K distinct categories in contradistinction with the Bernoulli distribution that has two categories. The categorical distribution has $K-1$ free parameters: $p_1, p_2, \cdots, p_K$ with the condition $\sum_{i=1}^{K} p_i = 1$, which reduces the number of free parameters by one. Its probability density is shown in Equation 6-26. $I(x_i = i)$ denotes the indicator function that takes the value of one if x observation belongs to category i.

$$
\text{PDF}_{\text{categorical}} = \prod_{i=1}^{K} p_i^{I(x=i)}
\tag{6-26}
$$

Table 6-1. Conjugate priors

Probability Density	Model Parameters	Prior	Prior Hyper-parameters
Binomial	θ	Beta	α, β
Negative binomial	θ	Beta	α, β
Multinomial	$p_1, p_2, \cdots, k_k$	Dirichlet	$\alpha_1, \alpha_2, \cdots, \alpha_k$
Gaussian	σ_0	Gaussian	μ, σ
Gaussian	σ_0	Inverse gamma	α, β
Gaussian	σ_0	Gamma	α, β
Multivariate normal	$\boldsymbol{\mu}, \boldsymbol{\Sigma}$	Inverse Wishart	$v, \boldsymbol{\psi}$
Multivariate normal	$\boldsymbol{\mu}, \boldsymbol{\Sigma}$	Multivariate normal	$\boldsymbol{\mu_0}, \boldsymbol{\Sigma_0}$
Multivariate normal	$\boldsymbol{\mu}, \boldsymbol{\Sigma}$	Wishart	$v, \mathbf{V}$
Uniform	θ	Pareto	x_{max}, k
Exponential	λ	Gamma	α, β
Poisson	λ: rate	Gamma	α, β
Categorical	$p_1, \cdots, p_K$	Dirichlet	$p_1, \cdots, p_K, \alpha_1, \cdots, \alpha_K$
Gamma	α_p, β_p	Gamma	α, β
Inverse gamma	α_p, β_p	Gamma	α, β

The Dirichlet distribution is a multivariate generalization of the beta distribution. It has a density function shown in Equation 6-27, with parameters $\alpha_1, \alpha_2, \cdots, \alpha_K$.

$$\text{PDF}_{\text{Dirichlet}} = \frac{1}{B(\boldsymbol{\alpha})} \prod_{i=1}^{K} p_i^{\alpha-1}$$

$$p_i \in [0, 1]$$

$$\sum_{i=1}^{K} p_i = 1 \tag{6-27}$$

$$B(\boldsymbol{\alpha}) = \frac{\prod_{i=1}^{K} \Gamma(\alpha_i)}{\Gamma\left(\sum_{i=1}^{K} \alpha_i\right)}$$

As can be seen from the expressions of probability density functions, the product of the categorical and Dirichlet distributions' probability density function will yield the PDF of another Dirichlet distribution with parameters $\alpha_1 + I(x = 1), \cdots, \alpha_K + I(x = K)$.

6.3   Bayesian Kernel Regression

Bayesian kernel regression applies the Bayesian approach to parameter estimation on top of kernel regression developed in an earlier chapter. The posterior probability of parameters is the linchpin of Bayesian regression. As is customary in Bayesian statistics, we use the Bayes rule to compute the posterior distribution. The selection of prior probability is application specific. Likelihood (or probability of data conditional on parameter values) uses kernel regression.

For the specific case of linear kernel regression with Gaussian noise and Gaussian prior, we observed that regression was equivalent to ridge regression with L2 (norm 2) regularization. The same is true for kernel regression. Using the loss function from Equation 4-3 and appending an L2 loss term, we can write the equivalent loss function for Bayesian regression for this specific case.

$$L(\boldsymbol{\beta}) = \frac{1}{2N} \sum_{i=1}^{N} \left(y_i - \sum_{j=1}^{k} \beta_j \psi_j(\mathbf{X_i}) \right)^2 + \frac{1}{K} \sum_{j=1}^{K} \beta_j^2 \qquad (6\text{-}28)$$

6.4 Bayesian Generalized Linear Regression

Bayesian generalized linear regression uses the likelihood function specific to the case of the generalized linear model and adds an assumed prior distribution for the parameters to formulate a posterior distribution of parameters following the Bayes rule. Because the likelihood from the generalized linear model is usually not from a class of conjugate priors with the assumed prior probability, we have to resort to numerical approximation or numerical simulation. We will consider each of these methods with a concrete example.

Let us consider a binomial generalized linear model, also known as logistic regression. As shown in Equation 3-17, the likelihood function is $\binom{N}{k} p^k (1-p)^{N-k}$. Setting the number of trials to 1, i.e., $N = 1$, and the probability of success to θ, the likelihood for the entire dataset $(y_i, \mathbf{X_i})$ where y_i is 1 or 0 can be written as shown in Equation 6-29.

$$P(y|X, \theta) = \prod_{i=1}^{N} \theta^{y_i} (1-\theta)^{1-y_i} \qquad (6\text{-}29)$$

Let us use a Gaussian prior for θ centered at θ_0 with variance σ_0^2. Using the Bayes rule, the posterior density for θ is written in Equation 6-30.

$$P(\theta|y, \mathbf{X}) = \frac{P(y|\theta, \mathbf{X}) P(\theta)}{\int_{-\infty}^{\infty} P(y|\theta, \mathbf{X}) P(\theta) d\theta}$$

$$\propto P(y|\theta, \mathbf{X}) P(\theta)$$

$$= \prod_{i=1}^{N} \theta^{y_i} (1-\theta)^{1-y_i} \frac{1}{\sqrt{2\pi\sigma_0^2}} \exp\left(-\frac{(\theta - \theta_0)^2}{2\sigma_0^2} \right) \qquad (6\text{-}30)$$

$$\propto \prod_{i=1}^{N} \theta^{y_i} (1-\theta)^{1-y_i} \exp\left(-\frac{(\theta - \theta_0)^2}{2\sigma_0^2} \right)$$

In order to make an inference, we need to calculate the expected value of y using the posterior density of θ to obtain the marginal distribution of y, as shown in Equation 6-31.

$$y_{inference} = E[y]$$

First compute posterior density $P(\theta|y, \mathbf{X})$

Then compute marginal density $P(y|\mathbf{X})$

$$P(y|\mathbf{X}) = \int \text{likelihood} \times \text{posterior} \times d\theta \qquad (6\text{-}31)$$

$$= \int P(y|\theta, \mathbf{X}) P(\theta|y, \mathbf{X}) d\theta$$

$$E[y] = \int y P(y|\mathbf{X}) dy$$

Since the posterior function in Equation 6-30 does not have a closed-form analytical solution, the integral in Equation 6-31 for computing marginal density needs to be evaluated numerically or approximated. Let us look at these two methods in the following two sections.

6.4.1   Laplace Approximation

The Laplace approximation is a method for calculating an approximate analytical expression for a posterior distribution in the neighborhood of its maxima using a Taylor expansion. Let θ_m be the parameter value that maximizes the posterior density. The method is best understood by looking at the following steps involved in computing an approximation to posterior density.

1. Write the expression of posterior density, as shown, for example, in Equation 6-30.
2. Find the parameter θ_m that maximizes the posterior density. Use numerical methods if necessary.
3. Let θ_m denote the parameter value that maximizes the posterior density. This means that $\frac{dP(\theta|y,X)}{d\theta}\big|_{\theta=\theta_m} = 0$.
4. Use a Taylor expansion in the neighborhood of θ_m to write an expression for posterior density, as shown in Equation 6-32.

$$P(\theta|y, X) = P(\theta_m|y, X) + \frac{1}{2}(\theta - \theta_m)^T H(\theta - \theta_m) \qquad (6\text{-}32)$$

5. Use this expression in calculating the marginal density of y in Equation 6-31.

An apparent shortcoming of the method is that it only linearizes the posterior density in the neighborhood of its maxima. Away from the maximum, the value may be quite different from the linearized approximation. Due to this, numerical methods such as Markov Chain Monte Carlo are usually preferred in practice. We will look at this method in the next section.

6.4.2   Markov Chain Monte Carlo Method

The Markov Chain Monte Carlo (MCMC) method is the workhorse of Bayesian statistics. In practice, most expressions of posterior density and marginal likelihood are not amenable to analytical calculations. This necessitates using an approach leveraging numerical computation. As seen from Equation 6-31, one needs to compute the integral of posterior density or the product of prior probability and model likelihood in order to make an inference. This integral also occurs in the denominator of the posterior density as a normalization factor. Hence, the method chosen must not rely on the computation of the normalization factor of posterior density because that is the problem being tackled by the method. It needs to work by calculating the posterior density up to a normalization constant. To state succinctly, the method needs to evaluate the expression shown in Equation 6-33.

$$\text{Evaluate } I = \int_{-\infty}^{\infty} P(y|\theta, \mathbf{X}) P(\theta) d\theta \tag{6-33}$$

For one-dimensional problems where the parameter vector θ is a scalar, we could discretize the grid and approximate the integral as **Riemann sums,** as shown in Equation 6-34. However, this approach suffers from the curse of dimensionality. For multidimensional parameter vector θ, the number of grid points for discretization increases exponentially, by a power of the parameter vector's dimensionality. Additionally, for applications with an infinite-size domain, it is not straightforward to determine where to stop discretization.

Discretize $\theta \in (-\infty, \infty)$ as $\theta_1, \theta_2, \cdots, \theta_N$ over its one-dimensional domain

Determine where to cut off discretization

$$I \approx \frac{1}{N-1} \sum_{i=1}^{N-1} P(y|\theta = \frac{\theta_i + \theta_{i+1}}{2}, \mathbf{X}) P(\theta = \frac{\theta_i + \theta_{i+1}}{2}) (\theta_{i+1} - \theta_i)$$

$$\tag{6-34}$$

In order to tackle the twin challenge of deciding where to stop discretization and the curse of dimensionality, one can use **importance sampling**. Let us look at this approach briefly because the Markov Chain Monte Carlo method is motivated by importance sampling.

Importance Sampling

Let us assume we have a probability density $Q(\theta)$ that has the same support as the posterior density $P(\theta|y, \mathbf{X})$. This means that the two probability functions have the same domain on θ. If we can sample from probability density $q(\theta)$, the integral in Equation 6-33 can be computed using the following method:

1. Sample $\theta_1, \theta_2, \cdots, \theta_N$ from $Q(\theta)$.
2. Evaluate the posterior density at the sampled points, i.e., evaluate the following functions $\frac{P(y|\theta_1, \mathbf{X})P(\theta_1)}{Q(\theta_1)}, \frac{P(y|\theta_2, \mathbf{X})P(\theta_2)}{Q(\theta_2)}, \ldots, \frac{P(y|\theta_N, \mathbf{X})P(\theta_N)}{Q(\theta_N)}$.
3. Approximate the integral as the average of the above computed values, as shown in Equation 6-35.

$$I \approx \frac{1}{N} \sum_{i=1}^{N} \frac{P(y|\theta_i, \mathbf{X})P(\theta_i)}{Q(\theta_i)} \tag{6-35}$$

Weight $\frac{1}{Q(\theta_N)}$ is multiplied with each posterior probability value to account for the fact that we have sampled from distribution $Q(\theta)$.

The MCMC algorithm generates a set of correlated points $\theta_1, \theta_2, \cdots, \theta_N$ such that the points begin to approach a stationary density. If we choose the stationary density to be the posterior density we wish to evaluate, we can use the samples after a cutoff index such as i for the evaluation of the integral in Equation 6-33. This algorithm of selecting correlated points that begin to approach the desired probability density after a cutoff value is also known as the **Metropolis-Hastings algorithm**.

Metropolis-Hastings Algorithm

The Metropolis-Hastings algorithm is depicted in pseudo-code 9. In order to state it in a generic fashion, let us say we wish to compute the integral in Equation 6-36. Our original problem of computing the normalization factor of posterior density can be recast to this form by setting $h(y|\theta) \equiv P(y|\theta, \mathbf{X})$ and $Q(\theta, y, \mathbf{X}) \equiv P(\theta)$. It is difficult to sample directly from $P(\theta|y, \mathbf{X})$. $Q(\theta, y, \mathbf{X})$ does not need to be a normalized probability distribution. It is sufficient for it to be proportional to a probability distribution. For example, in order to evaluate the expected value of y in order to make an inference in Bayesian statistics using Equation 6-23, we set $h(y|\theta) \equiv y$ and $Q(\theta, y, \mathbf{X}) \equiv P(y|\theta, \mathbf{X})P(\theta)$. In this formulation, $Q(\theta, y, \mathbf{X})$ is proportional to the posterior distribution $P(\theta|y, \mathbf{X})$. The MCMC algorithm will give us a sample of draws from the posterior density of θ using the aforementioned value of $Q(\theta, y, \mathbf{X})$ because it is proportional to the posterior density.

$$I = \int_{-\infty}^{\infty} h(y|\theta)Q(\theta, y, \mathbf{X})d\theta \tag{6-36}$$

There are two prerequisites that must be met to ensure that the generated chain of parameter values converges to the target probability distribution. The proposal

---

Algorithm 9 Metropolis-Hastings Algorithm

Require: A **proposal density function** $g(\theta_{t+1}|\theta_t)$ and initial state θ_0. We should be able to sample from proposal density function $g(\theta_{t+1}|\theta_t)$.

1: Set $t = 0$.
2: **for** $t = 1, 2, \cdots,$ **do**
3:     Sample a random θ_t from the proposal density function, $\theta_t \sim g(\theta|\theta_{t-1})$.
4:     Calculate the **acceptance probability** $A(\theta, \theta_{t-1})$ using Equation 6-37.

$$A(\theta, \theta_{t-1}) = \min\left(\frac{g(\theta_{t-1}|\theta)}{g(\theta|\theta_{t-1})} \frac{Q(\theta, y, \mathbf{X})}{Q(\theta_{t-1}, y, \mathbf{X})}, 1 \right) \tag{6-37}$$

5:     Sample a random number u from a uniform distribution, i.e., $u \sim \mathcal{U}[0, 1]$.
6:     If $u \le A(\theta, \theta_{t-1})$, accept θ as the parameter value, i.e., $\theta_t \leftarrow \theta$.
7:     If $u > A(\theta, \theta_{t-1})$, reject θ, i.e., $\theta_t \leftarrow \theta_{t-1}$.
8: **end for**
9: After a certain number of time steps K, $\theta_k \sim Q(\theta, y, \mathbf{X})$ for all $k \ge K$. This period is known as the **burn-in** period.
10: Calculate the integral in Equation 6-36 as $\frac{1}{N-K} \sum_{i=K}^{N} h(y|\theta_i)$.

---

density function must obey a condition known as **detailed balance**, and the target probability density $Q(\theta, y, \mathbf{X})$ must be aperiodic. Once these conditions are met, we can establish that the values of the parameter, θ, produced by the Metropolis-Hastings algorithm converge to the target probability distribution. Let us look at the two conditions below:

1. **Detailed balance condition for the proposal distribution and the target probability distribution**: This condition states that every transition $\theta_t \rightarrow \theta_{t+1}$ is reversible, i.e., the probability of going from θ_t to θ_{t+1} is equal to the probability of transitioning from θ_{t+1} to θ_t. This is shown in Equation 6-38.

$$P(\theta_t)P(\theta_{t+1}|\theta_t) = P(\theta_{t+1})P(\theta_t|\theta_{t+1}) \tag{6-38}$$

2. **Aperiodicity of the target probability distribution**: The target probability distribution should be aperiodic. This means that the target distribution should be unique. It must not have closed cycles. For example, if a system has four states, two of which lead exclusively to one another, the system is periodic.

Once the aforementioned conditions are satisfied, the Metropolis-Hastings algorithm will produce a chain of values that will converge to the unique stationary probability distribution. This assertion is proven in Equation 6-39.

$$P(\theta_t)P(\theta_{t+1}|\theta_t) = P(\theta_{t+1})P(\theta_t|\theta_{t+1}) \text{ detailed balance}$$

$$\implies \frac{P(\theta_{t+1}|\theta_t)}{P(\theta_t|\theta_{t+1})} = \frac{P(\theta_{t+1})}{P(\theta_t)}$$

Let $A(\theta_t \to \theta_{t+1})$ denote the acceptance probability

Also $g(\theta_{t+1}|\theta_t)$ denotes the proposal density function

$$P(\theta_{t+1}|\theta_t) = P(\theta_t)A(\theta_t \to \theta_{t+1})$$

$$\therefore \frac{P(\theta_{t+1}|\theta_t)}{P(\theta_t|\theta_{t+1})} = \frac{P(\theta_t)}{P(\theta_{t+1})} \frac{A(\theta_t \to \theta_{t+1})}{A(\theta_{t+1} \to \theta_t)}$$

$$= \frac{P(\theta_{t+1})}{P(\theta_t)} \tag{6-39}$$

$$\implies \frac{A(\theta_t \to \theta_{t+1})}{A(\theta_{t+1} \to \theta_t)} = \frac{P(\theta_{t+1})}{P(\theta_t)} \frac{P(\theta_{t+1})}{P(\theta_t)}$$

Select an acceptance ratio that satistfies the above condition

$$A(\theta_t \to \theta_{t+1}) \equiv A(\theta_{t+1}, \theta_t)$$

$$= \min\left(\frac{g(\theta_t|\theta_{t+1})}{g(\theta_{t+1}|\theta_t)} \frac{P(\theta_{t+1})}{P(\theta_t)}, 1\right)$$

$$= \min\left(\frac{g(\theta_t|\theta_{t+1})}{g(\theta_{t+1}|\theta_t)} \frac{Q(\theta_{t+1}, y, \mathbf{X})}{Q(\theta_t, y, \mathbf{X})}, 1\right)$$

In the last step of Equation 6-39, if $\frac{g(\theta_t|\theta_{t+1})}{g(\theta_{t+1}|\theta_t)} \frac{P(\theta_{t+1})}{P(\theta_t)} \le 1$, then its reciprocal is > 1. This implies that if $A(\theta_{t+1}, \theta_t) \le 1$, then $A(\theta_t, \theta_{t+1}) = 1$. Therefore, the condition $\frac{A(\theta_t \to \theta_{t+1})}{A(\theta_{t+1} \to \theta_t)} \equiv \frac{A(\theta_{t+1}, \theta_t)}{A(\theta_t, \theta_{t+1})} = \frac{P(\theta_{t+1})}{P(\theta_t)} \frac{P(\theta_{t+1})}{P(\theta_t)}$ is satisfied. The same is true when $\frac{g(\theta_t|\theta_{t+1})}{g(\theta_{t+1}|\theta_t)} \frac{P(\theta_{t+1})}{P(\theta_t)} > 1$.

Equation 6-39 shows that with the choice of acceptance ratio $A(\theta_{t+1}, \theta_t) = \min\left(\frac{g(\theta_t|\theta_{t+1})}{g(\theta_{t+1}|\theta_t)} \frac{P(\theta_{t+1})}{P(\theta_t)}, 1\right)$, the Markov Chain converges to the stationary distribution $P(\theta)$ which we chose to be $Q(\theta_t, y, \mathbf{X})$. This gives us a method of sampling from the density $Q(\theta_t, y, \mathbf{X})$. Once we have samples from the density, we can evaluate the integral in Equation 6-36 as shown in Equation 6-40.

$$I = \int_{-\infty}^{\infty} h(y|\theta)Q(\theta, y, \mathbf{X})d\theta$$

$$= \frac{1}{N-K} \sum_{i=K}^{N} h(y|\theta_i) \tag{6-40}$$

Similarly, in order to evaluate the normalization factor of the posterior density function in Equation 6-33, we set $h(y|\theta) \equiv P(y|\theta, \mathbf{X})$ and $Q(\theta, y, \mathbf{X}) = P(\theta)$

and draw samples from that distribution using the MCMC algorithm. Once the MCMC chain converges to its stationary distribution, we will obtain samples from $P(\theta)$ and can compute the normalization factor of posterior density as shown in Equation 6-41. This assumes that it is hard to sample from the prior distribution $P(\theta)$. If it is easy to sample from that distribution, we can simply draw samples from the prior directly and use a vanilla Monte Carlo algorithm to approximate the integral in normalization factor, as shown in Equation 6-42.

Compute the normalization factor for posterior density $P(\theta|y, \mathbf{X})$

If it is not easy to sample from prior density $P(\theta)$, use the MCMC algorithm

Set $Q(\theta, y, \mathbf{X}) = P(\theta)$ and obtain samples from $P(\theta)$ using MCMC

$$I = \frac{1}{N - K} \sum_{i=K}^{N} P(y|\theta_i, \mathbf{X}) \text{ with } \theta_i \sim P(\theta) \text{ using MCMC}$$

K is the burn-in period

(6-41)

Compute the normalization factor for posterior density $P(\theta|y, \mathbf{X})$

If it is easy to sample from prior density $P(\theta)$,

use the simple Monte Carlo algorithm

(6-42)

$$I = \frac{1}{N} \sum_{i=1}^{N} P(y|\theta_i, \mathbf{X}) \text{ with } \theta_i \sim P(\theta) \text{ using direct samples from } P(\theta)$$

This is shown succinctly in pseudo-code 10.

Algorithm 10 Compute Normalization Factor for Posterior Density

---

1: **Objective:** Compute $I(y) = \int_{-\infty}^{\infty} P(y|\theta, \mathbf{X})P(\theta)d\theta$
2: **if** it is easy to sample from prior density $P(\theta)$ **then**
3:     Sample $\theta_i \sim P(\theta)$
4:     Compute $I(y) = \frac{1}{N} \sum_{i=1}^{N} P(y|\theta_i, \mathbf{X})$
5: **else**
6:     **if** it is easy to sample from another distribution $H(\theta)$ having the same domain as $P(\theta)$ **then**
7:         Use **importance sampling**.
8:         Sample $\theta_i \sim H(\theta)$
9:         Compute $I(y) = \frac{1}{N} \sum_{i=1}^{N} P(y|\theta_i, \mathbf{X}) \frac{P(\theta_i)}{H(\theta_i)}$
10:     **else**
11:         Use the **Metropolis-Hastings algorithm**.
12:         Set $Q(\theta, y, \mathbf{X}) = P(\theta)$
13:         Obtain samples from $P(\theta)$ using the MCMC algorithm.
14:         Compute $I(y) = \frac{1}{N-K} \sum_{i=K}^{N} P(y|\theta_i, \mathbf{X})$ where K is the burn-in period.
15:     **end if**
16: **end if**

The MCMC algorithm is frequently used to compute expected values of the endogenous variable or its functions using posterior density of parameters. Because it is difficult to sample directly from the posterior density, we set $h(y|\theta)$ to be the function of interest and $Q(\theta, y, \mathbf{X}) \equiv P(y|\theta, \mathbf{X})P(\theta)$ in order to draw samples from the posterior and approximate the integral using Equation 6-40. The algorithm is summarized in pseudo-code 11.

Algorithm 11 Compute Expected Value of Exogenous Variable in Bayesian Model

1: **Objective**: Compute $E[y] = \int y P(y|\mathbf{X})dy$ for a Bayesian model.
2: $E[y] = \int y P(y|\mathbf{X})dy = \int_y y \int_{\theta=-\infty}^{\infty} P(y|\theta, \mathbf{X})P(\theta)d\theta dy$
3: $I(\tilde{y}) = \int_{-\infty}^{\infty} P(y|\theta, \mathbf{X})P(\theta)d\theta$ is the problem of computing the normalization factor for posterior density from pseudo-code 10.
4: Use pseudo-code 10 to write $I(\tilde{y})$. This will be a probability density for y. It will generally be hard to sample y from this distribution.
5: Use the **Metropolis-Hastings algorithm** to sample from $I(\tilde{y})$.
6: Set $Q(\theta, \cdots) = I(\tilde{y})$ with parameter $\theta \equiv y$. We want to draw samples of y from this distribution.
7: Set $h(y|\theta) \equiv y$.
8: Obtain samples y_i from $I(\tilde{y})$ using the MCMC algorithm in pseudo-code 9.
9: Compute $I = \frac{1}{N-K} \sum_{i=K}^{N} y_i$ where K is the burn-in period.

Gibbs Sampling

Gibbs sampling is a simplification of the Metropolis-Hastings algorithm obtained by using a set of conditional probability distributions. This algorithm is applied for sampling from the joint probability distribution of parameters which is hard to sample from directly, but the conditional probability distribution of each parameter is easy to sample from.

Let us suppose the parameter space is N-dimensional, with $\vec{\theta} \equiv (\theta_1, \theta_2, \cdots, \theta_N)$. Symbol $\vec{\theta}$ denotes that θ is a vector with components $(\theta_1, \theta_2, \cdots, \theta_N)$. We want to sample from $P(\theta_1, \theta_2, \cdots, \theta_N)$, but it is hard to sample from this distribution directly. This could be due to the fact that $P(\theta_1, \theta_2, \cdots, \theta_N)$ has a complex analytical expression. In order to apply Gibbs sampling, we must know how to sample from each of the N conditional probability distributions below:

- $P(\theta_1|\theta_2, \cdots, \theta_N)$
- $P(\theta_2|\theta_1, \theta_3, \cdots, \theta_N)$
- $P(\theta_i|\theta_1, \theta_2, \cdots, \theta_{i-1}, \theta_{i+1}, \cdots, \theta_N)$
- $P(\theta_N|\theta_1, \theta_2, \cdots, \theta_{N-1})$

The algorithm is summarized in pseudo-code 12. One of its attractive features is ease of implementation—at each step, we can update the parameters in place and proceed to the next step.

Algorithm 12 Sample from Joint Probability Distribution Using Gibbs Sampling

---

Require: N conditional probability distributions $P(\theta_i | \theta_1, \theta_2, \cdots, \theta_{i-1}, \theta_{i+1}, \cdots, \theta_N)$ for $i \in$ $[1, 2, \cdots, N]$. Initial values of parameters $\vec{\theta}(0) = (\theta_1(0), \theta_2(0), \cdots, \theta_N(0))$

1: **Objective:** Sample $\vec{\theta} = (\theta_1, \theta_2, \cdots, \theta_N)$ from $P(\theta_1, \theta_2, \cdots, \theta_N)$.
2: **for** t = 1, 2, $\cdots$, **do**
3:      **for** i = 1, 2, $\cdots$, N **do**
4:          Sample θ_i from distribution $P(\theta_i | \theta_1(t), \theta_2(t), \cdots, \theta_{i-1}(t), \theta_{i+1}(t-1), \cdots, \theta_N(t-1))$.
5:          Set $\theta_i(t) \leftarrow \theta_i$.
6:      **end for**
7:      Store $\vec{\theta}(t) = (\theta_1(t), \theta_2(t), \cdots, \theta_N(t))$
8: **end for**
9: After a burn-in period t^*, use the sample values of $\theta(t \geq t^*)$ as samples from the joint probability distribution $P(\theta_1, \theta_2, \cdots, \theta_N)$.

In order to understand why the algorithm works, let us look at the acceptance probability $A(\theta(t+1), \theta(t))$ of a sample in the Gibbs sampling algorithm in pseudo-code 12. This is shown in Equation 6-43. We observe that acceptance probability reduces to one, which means all proposals for updated parameters are accepted.

$$A\left(\vec{\theta}(t+1), \vec{\theta}(t)\right) \equiv A\left(\vec{\theta}(t) \rightarrow \vec{\theta}(t+1)\right)$$

$$= \min\left(\frac{g(\vec{\theta}(t) | \vec{\theta}(t+1))}{g(\vec{\theta}(t+1) | \vec{\theta}(t))} \frac{P(\vec{\theta}(t+1))}{P(\vec{\theta}(t))}, 1\right)$$

$$= \min\left(\frac{P(\theta_1(t) | \theta_2(t+1), \theta_3(t+1), \cdots) P(\theta_2(t) | \theta_1(t), \theta_3(t+1), \cdots) \cdots}{P(\theta_0(t+1) | \theta_1(t), \theta_2(t), \cdots) P(\theta_2(t+1) | \theta_1(t+1), \theta_3(t), \cdots) \cdots}\right.$$

$$\left. \frac{P(\vec{\theta}(t+1))}{P(\vec{\theta}(t))}, 1\right)$$

$$= \min\left(\frac{P(\vec{\theta}(t))}{P(\vec{\theta}(t+1))} \frac{P(\vec{\theta}(t+1))}{P(\vec{\theta}(t))}, 1\right)$$

$$= 1$$

$$(6\text{-}43)$$

The algorithm sketched in pseudo-code 12 is known as sequential scan Gibbs sampling because we update the components of $\vec{\theta}$ sequentially, stepping from index 1 to N. If we select a random component i and update it to produce the next parameter value, the resulting Gibbs sampler is known as the **random scan Gibbs sampler**.

6.4.3 Application

Market returns are frequently modeled using a Gaussian distribution centered around a historical return with historical volatility as the variance. However, deviations from a normal distribution are widely known and acknowledged among financial analysts and researchers. For example, Cont (2001), Fama (1965), and Kon (1984) document deviations of security returns from a normal distribution. A few of those deviations are listed below:

1. **Fat tails**: Returns falling in the tail of distribution occur more frequently than that predicted by a normal distribution.
2. **Taller peak around mean**: Distribution around the mean is higher, i.e., more frequent, than that predicted by a normal distribution.

In order to model these known deviations, researchers have often used the Student-T distribution as a substitute for the normal distribution. For example, Blattberg and Gonedes (1974) investigate using the Student-T distribution to model asset returns.

Volatility of returns are not constant either. Heteroskedasticity in asset returns is a widely documented phenomenon. For example, Engle (1982) uses an ARCH model to represent the dynamics of volatility. Bollerslev (1986) and Bollerslev (1987) proposed an enhancement to the ARCH model by including an auto-regressive term for volatility and christened the model as GARCH—an acronym that stands for Generalized Autoregressive Conditional Heteroskedasticity. GARCH models have motivated a long list of successors, such as EGARCH (Nelson, 1991).

Let us use the Metropolis-Hastings algorithm to sample security returns from the joint distribution of returns conditional on volatility (Student-T distribution) and the normal distribution of volatility derived using the GARCH(1,1) model. The joint probability distribution cannot be sampled directly due to its intractable analytic form.

1. Returns, conditional on mean and variance, are distributed according to a Student-T distribution. The governing equation defines the likelihood of data and is shown in Equation 6-44.

$$
L(x|\mu, v) = \frac{\Gamma\left(\frac{v+1}{2}\right)}{\sqrt{\pi v}\,\Gamma\left(\frac{v}{2}\right)}\left(1 + \frac{(x-\mu)^2}{v}\right)^{-\frac{v+1}{2}}
$$

$$
\text{Variance}(x) = \frac{v}{v-2} \text{ for } v > 2 \tag{6-44}
$$

$$
x \in (-\infty, \infty)
$$

2. Mean return is the same as average daily return observed over the last five days (or one week).

3. Volatility of returns follows a GARCH(1,1) process, as shown in Equation 6-45.

$$\text{Variance}(x(t)|x(t-1)) = \sigma(t)^2 = \beta_0 + \beta_1 x(t-1)^2 + \beta_2 \sigma(t-1)^2$$

$$x(t) = \sigma(t)\epsilon(t)$$

$$\epsilon(t) \sim N(0, 1)$$

$$(6\text{-}45)$$

Let us use end-of-day prices for S&P 500 from 2000 to 2021 for fitting the GARCH(1,1) model in Equation 6-45 to volatility of daily returns. Following this, we use the fitted GARCH(1,1) to predict the next day's volatility and set $\frac{\nu}{\nu-2}$ equal to this value to compute the value of parameter ν in the Student-T distribution in Equation 6-44. μ is computed using the average of the last five days of returns. We sample from the joint distribution of returns and volatility using the Metropolis-Hastings algorithm, compute the volatility of sampled returns, and compare the bootstrap estimate of volatility against GARCH(1,1) predicted volatility and actual volatility observed over the last five days. The algorithm is sketched in pseudo-code 13.

Algorithm 13 Calculating Daily Volatility of Security Returns Sampled Using Metropolis-Hastings Algorithm

---

Require: End-of-day prices for S&P 500 from 2000 to 2024.
1: **Objective**: Sample S&P 500 returns from the joint distribution of returns and volatility. Returns are assumed to belong to a Student-T distribution, and the evolution of volatility is assumed to be governed by the GARCH(1,1) model.
2: Calculate daily returns, five-day volatility, and lagged volatility.
3: Fit the GARCH(1,1) model to the training data.
4: Use the Metropolis-Hastings algorithm to sample from the joint distribution of returns and volatility. The joint distribution is shown in Equation 6-46.

$$P(x, \sigma^2) = \frac{\Gamma\left(\frac{\nu+1}{2}\right)}{\sqrt{\pi\nu}\Gamma\left(\frac{\nu}{2}\right)} \left(1 + \frac{(x-\mu)^2}{\nu}\right)^{-\frac{\nu+1}{2}}$$

$$\times \frac{1}{\sqrt{2\pi}} \exp\left(-\frac{\sigma(t)^2 - \beta_0 - \beta_1 x(t-1)^2 \, \beta_2 \sigma(t-1)^2}{2}\right) \qquad (6\text{-}46)$$

$$\text{where } \frac{\nu}{\nu-2} = \sigma^2$$

5: For testing, predict the next period's volatility using the GARCH(1,1) model.
6: Use the Metropolis-Hastings sampler that has reached steady state to sample from the joint distribution to sample returns. Compute the empirical volatility.
7: Plot the three volatilities.

The code for fitting the model and sample from joint probability density is shown in Listing 6-3.

Listing 6-3. Computing Empirical Volatility of S&P 500 Returns Using Metropolis-Hastings Algorithm

```python
import numpy as np
import pandas as pd
import os
import logging
from abc import ABC, abstractmethod
import statsmodels.api as sm
from statsmodels.base.model import GenericLikelihoodModel
from scipy import stats
import matplotlib.pyplot as plt

logging.basicConfig(level=logging.DEBUG)


class MetropolisHastings(ABC):
    def __init__(self, burnIn=1000):
        self.logger = logging.getLogger(self.__class__.__name__)
        self.burnIn = burnIn

    @abstractmethod
    def sampleFromProposalDensity(self, state0):
        raise NotImplementedError("Base class needs to implement")

    @abstractmethod
    def proposalDensity(self, state0, state1):
        raise NotImplementedError("Base class needs to implement")

    @abstractmethod
    def targetProb(self, state, params):
        raise NotImplementedError("Base class needs to implement")

    def sample(self, N, initial, params, burnIn=None):
        if burnIn is None:
            burnIn = self.burnIn
        samples = np.zeros(N, dtype=np.float64)
        state0 = initial
        i = 0
        while i < burnIn + N:
            state = self.sampleFromProposalDensity(state0)
            fac1 = self.proposalDensity(state, state0) / self.proposalDensity(state0,
                state)
            fac2 = self.targetProb(state, params) / self.targetProb(state0, params)
            acceptanceProb = min(fac1 * fac2, 1)
            u = np.random.random(1)
            if u <= acceptanceProb:
                state0 = state
                if i >= burnIn:
                    samples[i - burnIn] = state0
            i += 1

        return samples
```

```python
class Garch11Model(GenericLikelihoodModel):
    def __init__(self, endog, exog):
        super().__init__(endog=endog, exog=exog)
        self.endog = endog
        self.exog = sm.add_constant(exog, has_constant="add")
        assert self.exog.shape[1] == 3
        self.parameters = np.random.random(3)

    def loglikeobs(self, params):
        pred = np.einsum("ij,j->i", self.exog, params)
        return np.sum(stats.norm.logpdf(pred, self.endog, 1))

    def fit(self, **kwargs):
        return super().fit(self.parameters, method="bfgs")

    def predict(self, exog):
        exog = sm.add_constant(exog, has_constant="add")
        return np.einsum("ij,j->i", exog, self.parameters)


class SP500ReturnPosterior(MetropolisHastings):
    PRICE_COL = "Close"
    PERIOD = 5

    def __init__(self, dirname, security, trainTestRatio=0.9):
        super().__init__()
        self.logger = logging.getLogger(self.__class__.__name__)
        self.dirname = dirname
        self.df = pd.read_csv(os.path.join(dirname, f"{security}.csv"), parse_dates=[
            "Date"])
        self.trainTestRatio = trainTestRatio
        self.ntraining = int(self.df.shape[0] * trainTestRatio)
        self.garchModel = None
        self.calculateEndogExogVars()
        self.volatForProb = None

    def calculateEndogExogVars(self):
        price = self.df.loc[:, self.PRICE_COL].values
        returns = price[1:] / price[0:-1] - 1
        self.df.loc[:, "returns"] = 0
        self.df.loc[1:, "returns"] = returns
        self.df.loc[:, "returns_square"] = self.df.loc[:, "returns"] ** 2
        self.df.loc[:, "volat"] = 0
        self.df.loc[:, "lagged_volat"] = 0
        sumsq = np.sum(returns[0:self.PERIOD] ** 2)
        for i in range(self.PERIOD, self.df.shape[0]-1, 1):
            self.df.loc[i, "volat"] = sumsq / self.PERIOD
            self.df.loc[i+1, "lagged_volat"] = self.df.loc[i, "volat"]
            sumsq += returns[i] * returns[i] - returns[i - self.PERIOD] * returns[i -
                self.PERIOD]

    def fitGarch(self):
```

```
102      endog = self.df.loc[self.PERIOD+1:self.ntraining, "volat"].values
103      exog = self.df.loc[self.PERIOD+1:self.ntraining, ["returns_square", "
             lagged_volat"]].values
104
105      self.garchModel = Garch11Model(endog=endog, exog=exog)
106      res = self.garchModel.fit()
107      self.logger.info(res.summary())
108      self.garchModel.parameters = res.params
109      self.volatForProb = res.params[0] / (1 − res.params[2])
110
111  def fitMHSampler(self):
112      state = self.df.loc[self.ntraining, "returns"]
113      mu = np.mean(self.df.loc[self.ntraining − self.PERIOD:self.ntraining, "returns"
             ].values)
114      volat = self.df.loc[self.ntraining, "volat"]
115      params = (mu, volat)
116      self.sample(1, state, params)
117
118  def fit(self):
119      self.fitGarch()
120      self.fitMHSampler()
121
122  def sampleFromProposalDensity(self, state0):
123      return np.random.normal(size=1, loc=state0, scale=self.volatForProb)
124
125  def proposalDensity(self, state0, state1):
126      return stats.norm.pdf(state0 − state1, 0, 1)
127
128  def targetProb(self, state, params):
129      mu, volat = params
130      nu = 2 * volat / (1 − volat)
131      return (1 + (state − mu)**2/nu) ** (−(nu+1)/2) * stats.norm.pdf(state, mu, 1)
132
133  def test(self):
134      exog = self.df.loc[self.ntraining:, ["returns_square", "lagged_volat"]].values
135      actual = self.df.loc[self.ntraining:, "volat"].values
136      predictedVol = self.garchModel.predict(exog)
137      sampledVol = np.zeros(self.df.shape[0]−1−self.ntraining, dtype=np.float64)
138      x = self.df.loc[self.ntraining:, "Date"].values
139
140      for i in range(self.ntraining, self.df.shape[0]−1, 1):
141          vol = predictedVol[i−self.ntraining]
142          self.volatForProb = self.df.loc[i, "lagged_volat"]
143          mu = np.mean(self.df.loc[i−self.PERIOD:i, "returns"].values)
144          initial = self.df.loc[i, "returns"]
145          params = (mu, vol)
146          returns = self.sample(20, initial, params, burnIn=0)
147          sampledVol[i−self.ntraining] = np.std(returns)
148
149      plt.figure(figsize=(10, 10))
150      plt.plot(x[0:−1], sampledVol, label="Sampled")
151      plt.plot(x[0:−1], predictedVol[0:−1], label="GARCH(1,1)")
152      plt.plot(x[0:−1], actual[0:−1], label="Empirical")
```

```
153        plt . grid ()
154        plt . legend ()
155        plt . xlabel ("Date")
156        plt . ylabel ("Daily    Volatility ")
157        plt . savefig (os . path . join ( self .dirname,  "mcmc_variance.jpeg"),
158                        dpi=500)
159        plt . show()
160
161
162  if   __name__ == "__main__":
163      dirname = r"C:\prog\cygwin\home\samit_000\latex\ book_stats \code\data"
164      posterior  = SP500ReturnPosterior(dirname, "SPY")
165      np.random.seed(32)
166      posterior . fit ()
167      posterior . test ()
```

The plot produced by the code in Listing 6-3 is shown in Figure 6-2.

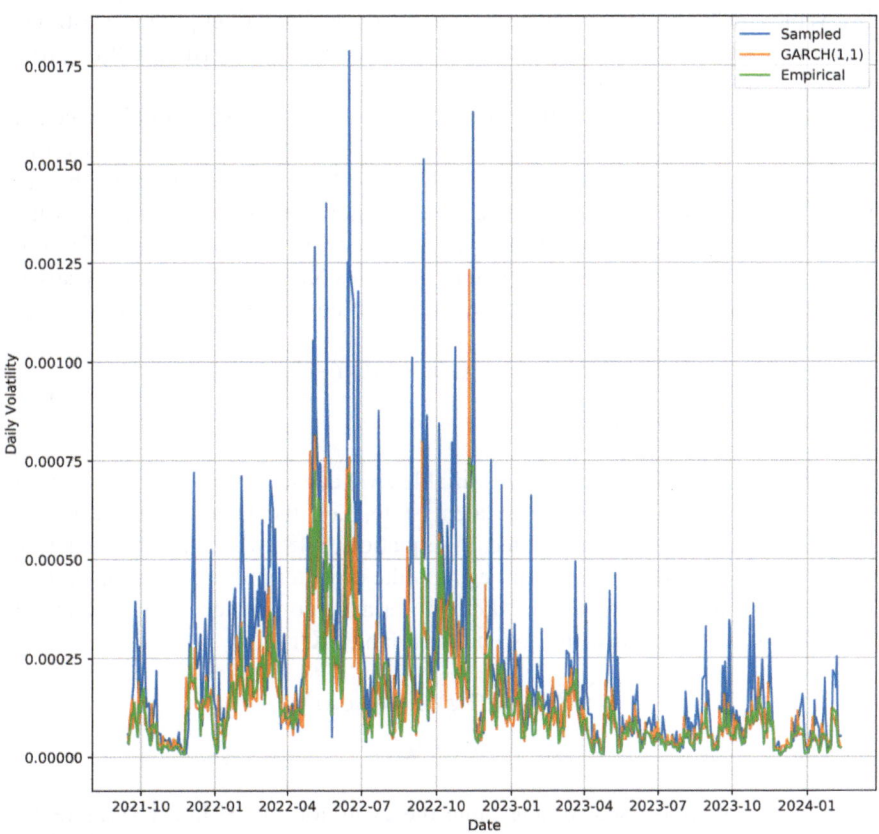

Figure 6-2. Comparison of Volatility of S&P 500 Returns Obtained Using MCMC Algorithm, GARCH(1, 1) Model, and Empirically Observed Volatility

As seen from Figure 6-2, bootstrap volatility from the Metropolis-Hastings algorithm spikes more than GARCH(1,1) predicted volatility during periods of market gyrations and tracks the GARCH(1,1) volatility closely during periods of market calm. The Student-T distribution has greater probability density in the tails compared with the normal distribution and enables the MCMC sampler to produce values with higher volatility than the GARCH(1,1) model.

6.5   Kalman Filter

The Kalman filter is a method for updating our beliefs about the value of a hidden or latent (unobservable) variable given noisy observations of another variable dependent on the latent variable. The filter has some similarities with the Bayesian posterior distribution, but is specific to the Gaussian distribution for noise terms. The term "filter" connotes removing or alleviating the impact of noise. If all errors are Gaussian and the model is linear in its evolution of latent state and the dependence of observables on latent state, it can be shown that the Kalman filter produces estimates of latent variables with minimum mean-square error. This assertion will be proven later.

Let us represent latent variables at time t by vector $\mathbf{X}(t)$. This could be the state of a dynamic system that we do not observe directly. Let us represent the governing equation for the evolution of hidden state by 6-47. $\mathbf{F}(t)$ is a known matrix that may evolve as a function of time. $\mathbf{U}(t)$ represents known exogenous variables whose measurement entails no error. Matrix $\mathbf{G}(t)$ is likewise known a priori. The error term $\mathbf{e}_{\mathbf{x}}(t)$ is uncorrelated with exogenous variables $\mathbf{X}(t-1)$. It follows a Gaussian distribution with zero mean and a known variance-covariance matrix, $\mathbf{Q}(t)$.

$$\mathbf{X}(t) = \mathbf{F}(t-1)\mathbf{X}(t-1) + \mathbf{G}(t-1)\mathbf{U}(t-1) + \mathbf{e}_{\mathbf{x}}(t)$$
$$\mathbf{e}_{\mathbf{x}}(t) \sim N\left(0, \mathbf{Q}(t)\right)$$

(6-47)

We record noisy observations of a variable $\mathbf{Z}(t)$ at each time step t. This variable is dependent on contemporaneous latent state, i.e., the latent state $\mathbf{X}(t)$ at the same time step. This is represented in Equation 6-48, where $\mathbf{H}(t)$ is a known, possibly time-dependent matrix and $\mathbf{e}_{\mathbf{z}}(t)$ is the error term that follows a Gaussian distribution with zero mean and a known variance-covariance matrix, $\mathbf{R}(t)$. The error term is uncorrelated with exogenous variables $\mathbf{X}(t)$.

$$\mathbf{Z}(t) = \mathbf{H}(t)\mathbf{X}(t) + \mathbf{e}_{\mathbf{z}}(t)$$
$$\mathbf{e}_{\mathbf{z}}(t) \sim N\left(0, \mathbf{R}(t)\right)$$

(6-48)

A Kalman filter is used to determine the evolution of the latent variable $\mathbf{X}(t)$ as observations $\mathbf{Z}(t)$ become available. It formulates an expression for the latent variable at time step t before and after the observations at time step t are known. This iterative procedure of moving from one time step to the next and updating our predictions of the latent variable once observations for $\mathbf{Z}(t)$ are available is shown in pseudo-code 14. Superscript T denotes matrix transpose.

Algorithm 14 Kalman-Filter Iterative Procedure for Predicting Latent Variable

Require: 1. Exogenous variable $\mathbf{U}(t)$ for all time steps.
  2. Matrix $\mathbf{F}(t)$ and $\mathbf{G}(t)$ governing the evolution of latent variable $\mathbf{X}(t)$ for all time steps.
  3. Matrix $\mathbf{H}(t)$ describing the dependence of observation $\mathbf{Z}(t)$ on the latent variable for all time steps.
  4. Variance-covariance matrices of error terms, $\mathbf{Q}(t)$ and $\mathbf{R}(t)$, as described in Equations 6-47 and 6-48. If the matrices are a function of time, these are required for all time steps.
  5. Initial value of latent state $\mathbf{X}(0)$.
  6. Initial variance-covariance matrix of latent variable $\mathbf{X}(0)$. Denote this as $\mathbf{P}(0)$.
1: Set t = 0.
2: Set $\mathbf{X}(0|0) = \mathbf{X}(0)$.
3: Set $\mathbf{P}(0|0) = \mathbf{P}(0)$
4: **for** t = 1, 2, $\cdots$, **do**
5:      Predict the value of latent state $\mathbf{X}(t|t-1)$ **ex-ante**, i.e., without knowing the value of observation $\mathbf{Z}(t)$ at time step t. This prediction is denoted as $\hat{\mathbf{X}}(t|t-1)$ because it uses observations before time t. Use Equation 6-49.

$$\hat{\mathbf{X}}(t|t-1) = \mathbf{F}(t-1)\hat{\mathbf{X}}(t-1|t-1) + \mathbf{G}(t-1)\mathbf{U}(t-1) \qquad (6\text{-}49)$$

6:      Predict the observation $\hat{\mathbf{Z}}(t|t-1)$ using Equation 6-50.

$$\hat{\mathbf{Z}}(t|t-1) = \mathbf{H}(t)\hat{\mathbf{X}}(t|t-1) \qquad (6\text{-}50)$$

7:      Calculate the variance-covariance matrix of the latent variable, $\mathbf{P}(t|t-1)$, using Equation 6-51.

$$\mathbf{P}(t|t-1) = \mathbf{F}(t-1)\mathbf{P}(t-1|t-1)\mathbf{F}(t-1)^T + \mathbf{Q}(t) \qquad (6\text{-}51)$$

8:      Calculate the variance-covariance matrix of observations, $\mathbf{S}(t|t-1)$, using Equation 6-52.

$$\mathbf{S}(t) = \mathbf{H}(t)\mathbf{P}(t|t-1)\mathbf{H}(t)^T + \mathbf{R}(t) \qquad (6\text{-}52)$$

Let us now look at the derivation of the expressions used in the Kalman filter in order to comprehend how it works.

The expression for the variance-covariance matrix of a latent variable, $\mathbf{P}(t|t-1)$, can be obtained by writing the expression of $E\left[\left(\hat{\mathbf{X}}(t|t-1) - \mathbf{X}(t)\right)^T \left(\hat{\mathbf{X}}(t|t-1) - \mathbf{X}(t)\right)\right]$, as shown in Equation 6-57.

9:      Calculate **Kalman gain** $\mathbf{K}(t)$ using Equation 6-53.

$$\mathbf{K}(t) = \mathbf{P}(t|t-1)\mathbf{H}(t)\mathbf{S}(t)^{-1} \tag{6-53}$$

10:     Update the latent state variance-covariance matrix $\mathbf{P}(t|t)$ using Equation 6-54.

$$\mathbf{P}(t|t) = \mathbf{P}(t|t-1) - \mathbf{K}(t)\mathbf{S}(t)\mathbf{K}(t)^{-1}$$
$$= (\mathbf{I} - \mathbf{K}(t)\mathbf{H}(t))\,\mathbf{P}(t|t-1) \tag{6-54}$$

11:     Observation becomes available as $\mathbf{Z}(t)$ for time step t.
12:     Calculate the observation error using Equation 6-55.

$$\mathbf{e_z}(t) = \mathbf{Z}(t) - \hat{\mathbf{Z}}(t|t-1) \tag{6-55}$$

13:     Update the prediction of latent state, $\mathbf{X}(t|t)$, using Equation 6-56. This prediction is known
        as **ex-post** prediction because it uses the observation $\mathbf{Z}(t)$.

$$\hat{\mathbf{X}}(t|t) = \hat{\mathbf{X}}(t|t-1) + \mathbf{K}(t)\mathbf{e_z}(t) \tag{6-56}$$

14: **end for**

$$\mathbf{P}(t|t-1) = E\left[\left(\hat{\mathbf{X}}(t|t-1) - \mathbf{X}(t)\right)^T \left(\hat{\mathbf{X}}(t|t-1) - \mathbf{X}(t)\right)\right]$$

$$\hat{\mathbf{X}}(t|t-1) = \mathbf{F}(t-1)\hat{\mathbf{X}}(t-1|t-1) + \mathbf{G}(t-1)\mathbf{U}(t-1)$$

$$\mathbf{X}(t) = \mathbf{F}(t-1)\mathbf{X}(t-1) + \mathbf{G}(t-1)\mathbf{U}(t-1) + \mathbf{e_x}(t)$$

Substituting the expressions, we get

$$\mathbf{P}(t|t-1) = \mathbf{F}(t-1)E\left[\left(\hat{\mathbf{X}}(t|t-1) - \mathbf{X}(t-1)\right)^T\right]\mathbf{F}(t-1)^T + \tag{6-57}$$

$$E\left[\mathbf{e_x}(t)^T\mathbf{e_x}(t)\right]$$

Because $\mathbf{e_x}$ is uncorrelated with latent variables

$$\mathbf{P}(t|t-1) = \mathbf{F}(t-1)\mathbf{P}(t-1|t-1)\mathbf{F}(t-1)^T + \mathbf{Q}(t)$$

Let us define the Kalman gain as the coefficient of the observation-error term used for updating our estimate of the latent variable when observation at time t is available. This is shown in Equation 6-58.

$$\hat{\mathbf{X}}(t|t) = \hat{\mathbf{X}}(t|t-1) + \mathbf{K}(t)\left(\mathbf{Z}(t) - \hat{\mathbf{Z}}(t|t-1)\right) \tag{6-58}$$

$\mathbf{K}(t)$ is called Kalman gain

Now let us write an expression for the variance-covariance matrix of the latent variable at time step t when observation for time step t is known, i.e., $\mathbf{P}(t|t)$. This is shown in Equation 6-59.

$$\mathbf{P}(t|t) = E\left[\left(\hat{\mathbf{X}}(t|t) - \mathbf{X}(t)\right)^T \left(\hat{\mathbf{X}}(t|t) - \mathbf{X}(t)\right)\right]$$

Use the definition of Kalman gain

$$\mathbf{P}(t|t) = E\left[\left(\hat{\mathbf{X}}(t|t-1) + \mathbf{K}(t)\left(\mathbf{Z}(t) - \hat{\mathbf{Z}}(t|t-1)\right) - \mathbf{X}(t)\right)^T\right.$$

$$\left.\left(\hat{\mathbf{X}}(t|t-1) + \mathbf{K}(t)\left(\mathbf{Z}(t) - \hat{\mathbf{Z}}(t|t-1)\right) - \mathbf{X}(t)\right)\right] \tag{6-59}$$

Substitute $\hat{\mathbf{Z}}(t|t-1) = \mathbf{H}(t)\hat{\mathbf{X}}(t|t-1)$

$$\mathbf{P}(t|t) = E\left[\left(\hat{\mathbf{X}}(t|t-1) + \mathbf{K}(t)\left(\mathbf{Z}(t) - \mathbf{H}(\hat{t})\mathbf{X}(t|t-1)\right) - \mathbf{X}(t)\right)^T\right.$$

$$\left.\left(\mathbf{H}(t)\hat{\mathbf{X}}(t|t-1) + \mathbf{K}(t)\left(\mathbf{Z}(t) - \hat{\mathbf{Z}}(t|t-1)\right) - \mathbf{X}(t)\right)\right]$$

We also know that $\mathbf{Z}(t) = \mathbf{H}(t)\mathbf{X}(t) + \mathbf{e_z}(t)$ from Equation 6-48. Substituting it in Equation 6-59, we get Equation 6-60.

$$\mathbf{P}(t|t) = E\left[\left(\hat{\mathbf{X}}(t|t-1) + \mathbf{K}(t)\left(\mathbf{H}(t)\mathbf{X}(t) + \mathbf{e_z}(t) - \mathbf{H}(t)\hat{\mathbf{X}}(t|t-1)\right) - \mathbf{X}(t)\right)^T\right.$$

$$\left.\left(\mathbf{H}(t)\hat{\mathbf{X}}(t|t-1) + \mathbf{K}(t)\left(\mathbf{H}(t)\mathbf{X}(t) + \mathbf{e_z}(t) - \hat{\mathbf{Z}}(t|t-1)\right) - \mathbf{X}(t)\right)\right]$$

$$= (\mathbf{I} - \mathbf{K}(t)\mathbf{H}(t))\, E\left[\left(\mathbf{X}(t) - \hat{\mathbf{X}}(t|t-1)\right)\left(\mathbf{X}(t) - \hat{\mathbf{X}}(t|t-1)\right)^T\right]$$

$$(\mathbf{I} - \mathbf{K}(t)\mathbf{H}(t))^T + \mathbf{K}(t) E\left[\mathbf{e_z}(t)\mathbf{e_z}(t)^T\right]\mathbf{K}(t)^T$$

$$= (\mathbf{I} - \mathbf{K}(t)\mathbf{H}(t))\,\mathbf{P}(t|t-1)\,(\mathbf{I} - \mathbf{K}(t)\mathbf{H}(t))^T + \mathbf{K}(t)\mathbf{R}(t)\mathbf{K}(t)^T \tag{6-60}$$

The Kalman filter seeks to minimize the mean square error of the posterior estimate for the latent variable. It minimizes the term $\left\|\mathbf{X}(t) - \hat{\mathbf{X}}(t|t)\right\|^2$. This is the trace of variance-covariance matrix $\mathbf{P}(t|t)$, whose expression was derived in

Equation 6-60. We can write the trace of variance-covariance matrix $\mathbf{P}(t|t)$ as shown in Equation 6-61. The last step uses the definition of $\mathbf{S}(t)$ shown in Equation 6-52.

$$\text{trace}(\mathbf{P}(t|t)) = \mathbf{P}(t|t-1) - \mathbf{K}(t)\mathbf{H}(t)\mathbf{P}(t|t-1)^T - \mathbf{P}(t|t-1)\mathbf{H}(t)^T\mathbf{K}(t)^T +$$

$$\mathbf{K}(t)\left(\mathbf{H}(t)\mathbf{P}(t|t-1)\mathbf{H}(t)^T + \mathbf{R}(t)\right)\mathbf{K}(t)^T$$

$$= \mathbf{P}(t|t-1) - \mathbf{K}(t)\mathbf{H}(t)\mathbf{P}(t|t-1)^T - \mathbf{P}(t|t-1)\mathbf{H}(t)^T\mathbf{K}(t)^T +$$

$$\mathbf{K}(t)\mathbf{S}(t)\mathbf{K}(t)^T$$

$$(6\text{-}61)$$

Finally, differentiating the expression for trace in Equation 6-61 with respect to $\mathbf{K}(t)$ gives us the value of the Kalman gain, as shown in Equation 6-62.

$$\frac{\partial\text{trace}(\mathbf{P}(t|t))}{\partial\mathbf{K}(t)} = 0$$

$$\implies -2\left(\mathbf{H}(t)\mathbf{P}(t|t-1)\right) + 2\mathbf{K}(t)\mathbf{S}(t) = 0$$

$$\implies \mathbf{K}(t) = \mathbf{P}(t|t-1)\mathbf{H}(t)^T\mathbf{S}(t)^{-1}$$

$$(6\text{-}62)$$

Tobit Regression

Tobit regression – also known as censored regression – refers to the regression methodology when observations are truncated at a threshold value. Truncation can be applied at a lower and, optionally, at an upper threshold. Applying ordinary linear regression methodology to truncated data can lead to biased parameter estimates that give poor in-sample and out-of-sample predictions. This occurs because of the non-linear shape of observable variables due to truncation. Censored regression handles this complexity by accounting for truncation of observations and modeling the non-truncated part of data separately.

In order to motivate the necessity of accounting for truncation, let us consider an unobservable variable y^* dependent on a vector of exogenous variables $\mathbf{X}$ according to the relation shown in Equation 7-1. $\boldsymbol{\beta}$ is a column vector of coefficients. For including an intercept, we can append a column containing one to matrix $\mathbf{X}$.

$$y^* = \mathbf{X}\boldsymbol{\beta} + \epsilon$$
$$\epsilon \sim N(0, \sigma^2)$$

$$(7\text{-}1)$$

y^* is an unobservable or latent variable; we only observe y that is related to y^* according to Equation 7-2.

$$y = \begin{cases} y^* \text{ if } y^* > 0 \\ 0 \text{ if } y^* \leq 0 \end{cases}$$

$$(7\text{-}2)$$

Let us simulate a few data points from the distribution and plot them, as shown in Figure 7-1. If we attempt to fit a simple linear regression model to this data, the estimated coefficient values will be biased and inconsistent. This can be seen from the output in Listing 7-1. The underlying (latent) process $y^* = 1 + 3x + \epsilon$, where $\epsilon \sim N(0, 1)$ is drawn from a normal distribution with mean 0 and variance 1. The process is shown in Equation 7-3. The observable variable is y and is related to y^*

© Samit Ahlawat 2025
S. Ahlawat, *Statistical Quantitative Methods in Finance*,
https://doi.org/10.1007/979-8-8688-0962-0_7

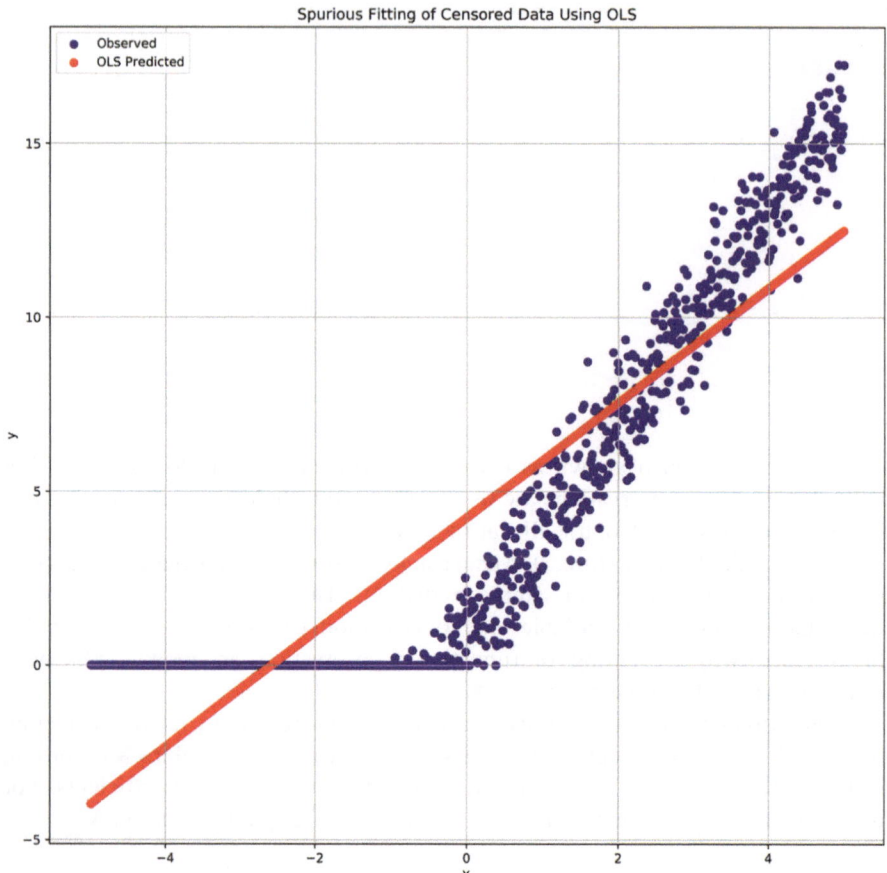

Figure 7-1. Scatter Plot of Left-Censored Data and OLS Model Fit

according to Equation 7-2. Discarding the points with $y = 0$ loses information. OLS obtains the coefficients of fitted line as $y_{\text{OLS fitted}} = 4.2572 + 1.6470x$, as seen in Listing 7-1.

$$y^* = 1 + 3x + \epsilon$$

$$\epsilon \sim N(0, 1)$$

$$y = \begin{cases} y^* \text{ if } y^* > 0 \\ 0 \text{ if } y^* \leq 0 \end{cases}$$

(7-3)

Listing 7-1. Left-Censored Data Fitted Using OLS Model

INFO:CensoredData:		OLS Regression Results	

```
===========================================================
Dep. Variable :    y              R−squared:        0.813
Model:             OLS            Adj. R−squared:   0.812
Method:            Least Squares  F− statistic :    4325.
Date:              Wed, 07 Aug 2024 Prob (F− statistic ) 0.00
Time:              22:28:34       Log−Likelihood:   −2244.8
No. Observations:1000             AIC:              4494.
Df Residuals :     998            BIC:              4503.
Df Model:          1
Covariance Type:              nonrobust
===========================================================
         coef    std err    t      P>|t|    [0.025      0.975]
- - - - - - - - - - - - - - - - - - - - - - - - - - - - - -
const    4.2572  0.072   58.885   0.000    4.115       4.399
x1       1.6470  0.025   65.765   0.000    1.598       1.696
===========================================================
Omnibus:            331.542     Durbin−Watson:      0.219
Prob(Omnibus):      0.000       Jarque−Bera (JB):   48.684
Skew:               −0.007      Prob(JB):           2.68e−11
Kurtosis :          1.919       Cond. No.           2.89
===========================================================

Notes:
[1] Standard Errors assume that the covariance matrix of the errors is correctly
    specified .
```

The code for performing the spurious OLS regression and plotting the results is shown in Listing 7-2.

Listing 7-2. Fitting Censored Data Using OLS Produces Biased and Inconsistent Estimates

```
import numpy as np
import statsmodels . api as sm
import logging
from matplotlib . pyplot import cm
import matplotlib . pyplot as plt
import os


logging . basicConfig ( level =logging . DEBUG)


class CensoredData(object):
    def __init__ ( self , dirname, coeff=3, variance=1, ll=0.0):
        self . logger = logging . getLogger( self . __class__ . __name__ )
        x = np.arange(−5, 5, 0.01)
        sdev = np. sqrt ( variance )
        epsilon = sdev ∗ np.random.standard_normal(x.shape [0])
        ystar = 1 + coeff∗x + epsilon
        y = np.where( ystar < ll , ll , ystar )
```

```
20        self .x = x
21        self .y = y
22        self .dirname = dirname
23
24     def fitOLS( self ):
25        x = sm.add_constant( self .x [:,  np.newaxis],  has_constant="add")
26        olsModel = sm.OLS(self.y,  x). fit ()
27        self . logger . info (olsModel.summary())
28        predicted  = olsModel. predict (x)
29
30        fig , axs  = plt . subplots (1,  1,  figsize =(10,  10))
31        colors  = cm.rainbow(np. linspace (0,  1,  2))
32        axs. scatter ( self .x,  self .y,  c=colors [0],  label ="Observed")
33        axs. scatter ( self .x,  predicted ,  c=colors [1],  label ="OLS Predicted")
34        axs. grid ()
35        axs. legend ()
36        axs. set_xlabel ("X")
37        axs. set_ylabel ("y")
38        axs. set ( title ="Spurious  Fitting  of Censored Data Using OLS")
39        fig . tight_layout ()
40        plt . savefig (os. path . join ( self .dirname, f" plot_{ self . __class__ . __name__ }.jpeg"),
41                        dpi=500)
42        plt .show()
43
44
45 if __name__ == "__main__":
46     dirname = r"C:\prog\cygwin\home\samit_000\latex\ book_stats \code\data"
47     censoredData = CensoredData(dirname)
48     censoredData. fitOLS ()
```

Tobit regression is used to fit censored data. In the example sketched earlier, the data was censored or cut off at a lower limit. This is called **left-censoring**. It can additionally have **right-censoring**, where we apply an upper limit and censor the values above that limit.

7.1   Problem Formulation

Let us formulate the problem of censored linear regression in a generic framework. Equation 7-1 describes the dynamics of a linear process. We can generalize the censoring equation to the one shown in Equation 7-4.

$$
y = \begin{cases} L_1 \text{ if } y^* < L_1 \\ y^* \text{ if } L_1 \leq y^* < L_2 \\ L_2 \text{ if } y^* \geq L_2 \end{cases} \tag{7-4}
$$

We can write the probability density of observing a left-censored value $y = L_1$ as shown in Equation 7-5.

$$
\begin{aligned}
P\left(y = L_1\right) &= P\left(y^* < L_1\right) \\
&= P\left(\mathbf{X}\beta + \epsilon < L_1\right) \\
&= P\left(\epsilon < L_1 - \mathbf{X}\beta\right) \\
&= \Phi\left(\frac{L_1 - \mathbf{X}\beta}{\sigma}\right) \quad \text{because } \epsilon \sim N(0, \sigma^2)
\end{aligned}
\tag{7-5}
$$

Φ denotes the CDF of standard normal distribution

$$
\Phi\left(z\right) = \frac{1}{\sqrt{2\pi}} \int_{-\infty}^{z} \exp\left(-\frac{x^2}{2}\right) dx
$$

Similarly, we can write the probability of observing a right-censored value $y = L_2$ as shown in Equation 7-6.

$$
\begin{aligned}
P\left(y = L_2\right) &= P\left(y^* > L_2\right) \\
&= P\left(\mathbf{X}\beta + \epsilon > L_2\right) \\
&= P\left(\epsilon > L_2 - \mathbf{X}\beta\right) \\
&= 1 - \Phi\left(\frac{L_2 - \mathbf{X}\beta}{\sigma}\right) \quad \text{because } \epsilon \sim N(0, \sigma^2) \\
&= \Phi\left(-\frac{L_2 - \mathbf{X}\beta}{\sigma}\right)
\end{aligned}
\tag{7-6}
$$

Φ denotes the CDF of standard normal distribution

For a non-censored value of y, probability density is shown in Equation 7-7. This corresponds to the usual probability distribution of a data point drawn from a normal distribution.

$$
\begin{aligned}
P\left(y = y^*\right) &= P\left(\mathbf{X}\beta + \epsilon = y^*\right) \text{ if } L_1 \leq y^* < L_2 \\
&= P\left(\epsilon = y^* - \mathbf{X}\beta\right) \\
&= P\left(\frac{\epsilon}{\sigma} = \frac{y^* - \mathbf{X}\beta}{\sigma}\right) \\
&= \frac{1}{\sigma}\phi\left(\frac{y^* - \mathbf{X}\beta}{\sigma}\right)
\end{aligned}
\tag{7-7}
$$

$$\phi(x) = \frac{1}{\sqrt{2\pi}} \exp\left(\frac{(y^* - \mathbf{X}\beta)^2}{2}\right)$$

ϕ denotes the PDF of standard normal distribution

Using Equations 7-5, 7-6, and 7-7, we can write the likelihood expression for the data, as shown in Equation 7-8.

$$L(\beta) = \prod_{i=1}^{N} \left(\Phi\left(\frac{L_1 - \mathbf{X_i}\beta}{\sigma}\right)\right)^{I(y_i = L_1)} \left(\frac{1}{\sigma}\phi\left(\frac{y_i - \mathbf{X_i}\beta}{\sigma}\right)\right)^{I(L_1 \le y_i < L_2)}$$

$$\left(\Phi\left(-\frac{L_2 - \mathbf{X_i}\beta}{\sigma}\right)\right)^{I(y_i \ge L_2)} \tag{7-8}$$

$$I(z) = \begin{cases} 0 \text{ if } z \text{ is false} \\ 1 \text{ if } z \text{ is true} \end{cases}$$

Using Equation 7-8, one can write the log-likelihood function by taking the logarithm. Differentiation with respect to model parameters β and setting it to zero to get the maxima of likelihood yields the value of parameters β using the method of maximum likelihood.

7.2   Marginal Effects

Marginal effects, also known as partial effects, measure the impact of change of an exogenous variable on an endogenous variable, $\frac{\partial y}{\partial X_i}$. In a vanilla linear regression model, the marginal effect of exogenous variable X_i is simply the coefficient β_i. In Tobit (censored) regression, that is no longer the case.

Before calculating marginal effects, we must estimate two values below that frequently crop up in Tobit regression.

1. $E\left[y|L_1 \le y^* < L_2\right]$: This is the expected value of the endogenous variable conditional on the condition $L_1 \le y^* < L_2$.
2. $E[y]$: This is the unconditional expected value of the endogenous variable.

Let us first calculate $E\left[y|L_1 \le y^* < L_2\right]$. This region corresponds to values where $y = y^*$, with a normal distribution of $y - \mathbf{X}\beta$. This expression is derived in Equations 7-9, 7-10, and 7-11.

$$E\left[y|L_1 \le y^* < L_2\right] = E\left[\mathbf{X}\beta + \epsilon|L_1 \le \mathbf{X}\beta + \epsilon < L_2\right]$$

$$= E\left[\mathbf{X}\beta + \epsilon|\frac{L_1 - \mathbf{X}\beta}{\sigma} \le \frac{\epsilon}{\sigma} < \frac{L_2 - \mathbf{X}\beta}{\sigma}\right] \tag{7-9}$$

$$P\left(\frac{L_1 - \mathbf{X}\beta}{\sigma} \leq \frac{\epsilon}{\sigma} < \frac{L_2 - \mathbf{X}\beta}{\sigma}\right) = \Phi\left(\frac{L_2 - \mathbf{X}\beta}{\sigma}\right) - \Phi\left(\frac{L_1 - \mathbf{X}\beta}{\sigma}\right)$$

$$\equiv F$$

(7-10)

$$E\left[y|L_1 \leq y^* < L_2\right] = \frac{1}{F}\int_{low}^{high}\left(\frac{\mathbf{X}\beta}{\sigma} + z\right)\phi(z)\,dz$$

$$\text{where } z = \frac{\epsilon}{\sigma}$$

$$low = \frac{L_1 - \mathbf{X}\beta}{\sigma} \text{ and high} = \frac{L_2 - \mathbf{X}\beta}{\sigma}$$

$$\therefore E\left[y|L_1 \leq y^* < L_2\right] = \frac{1}{F}\left(\mathbf{X}\beta F + \sigma\int_{low}^{high} z\phi(z)dz\right)$$

$$= \mathbf{X}\beta + \frac{\sigma}{F}\left(\phi\left(\frac{L_2 - \mathbf{X}\beta}{\sigma}\right) - \phi\left(\frac{L_1 - \mathbf{X}\beta}{\sigma}\right)\right)$$

$$\text{where } F \equiv \Phi\left(\frac{L_2 - \mathbf{X}\beta}{\sigma}\right) - \Phi\left(\frac{L_1 - \mathbf{X}\beta}{\sigma}\right)$$

(7-11)

Equation 7-11 shows why ignoring the censored values and fitting a linear model gives biased and inconsistent results. $E\left[y|L_1 \leq y^* < L_2\right] = \mathbf{X}\beta + \frac{\sigma}{F}\left(\phi\left(\frac{L_2-\mathbf{X}\beta}{\sigma}\right) - \phi\left(\frac{L_1-\mathbf{X}\beta}{\sigma}\right)\right)$, whereas a naive linear regression would give $E\left[y|L_1 \leq y^* < L_2\right] = \mathbf{X}\beta$.

For the case where we have just left-censoring of data, we can simplify the expression in Equation 7-11 to the one shown in Equation 7-12.

$$E\left[y|L_1 \leq y^*\right] = \mathbf{X}\beta + \sigma\frac{-\phi\left(\frac{L_1-\mathbf{X}\beta}{\sigma}\right)}{1 - \Phi\left(\frac{L_1-\mathbf{X}\beta}{\sigma}\right)}$$

$$= \mathbf{X}\beta - \sigma\frac{-\phi\left(-\frac{L_1-\mathbf{X}\beta}{\sigma}\right)}{\Phi\left(-\frac{L_1-\mathbf{X}\beta}{\sigma}\right)}$$

$$= \mathbf{X}\beta - \sigma\lambda\left(-\frac{L_1 - \mathbf{X}\beta}{\sigma}\right)$$

For left-censored data (7-12)

$$\lambda(z) = \frac{\phi(z)}{\Phi(z)}$$

$\lambda(z)$ is called inverse Mills-ratio

$$\phi(z) = \frac{1}{\sqrt{2\pi}} \exp\left(\frac{z^2}{2}\right)$$

$$\Phi(z) = \frac{1}{\sqrt{2\pi}} \int_{-\infty}^{z} \exp\left(-\frac{x^2}{2}\right) dx$$

In Equation 7-12, $\lambda\left(-\frac{L_1 - \mathbf{X}\beta}{\sigma}\right)$ is called **inverse Mills ratio** and is defined as the ratio of the probability density function (ϕ) to the cumulative probability density function (Φ) of a standard normal distribution.

The expression for the unconditional expected value of endogenous variable y can be derived by considering the censored areas and uncensored areas separately, as shown in Equation 7-13.

$$E[y] = E[y|y \in \text{left censored area}] P(y \in \text{left censored area}) +$$

$$E[y|y \in \text{un-censored area}] P(y \in \text{un-censored area}) +$$

$$E[y|y \in \text{right censored area}] P(y \in \text{right censored area}) +$$

where $P(y \in \text{left censored area}) = \Phi\left(\frac{L_1 - \mathbf{X}\beta}{\sigma}\right)$

$$P(y \in \text{un-censored area}) = \Phi\left(\frac{L_2 - \mathbf{X}\beta}{\sigma}\right) - \Phi\left(\frac{L_1 - \mathbf{X}\beta}{\sigma}\right)$$

$$P(y \in \text{right censored area}) = 1 - \Phi\left(\frac{L_2 - \mathbf{X}\beta}{\sigma}\right) = \Phi\left(-\frac{L_2 - \mathbf{X}\beta}{\sigma}\right)$$

$$E[y] = L_1 \Phi\left(\frac{L_1 - \mathbf{X}\beta}{\sigma}\right) + \mathbf{X}\beta F + \sigma\left(\phi\left(\frac{L_2 - \mathbf{X}\beta}{\sigma}\right) - \phi\left(\frac{L_1 - \mathbf{X}\beta}{\sigma}\right)\right) +$$

$$L_2 \Phi\left(-\frac{L_2 - \mathbf{X}\beta}{\sigma}\right)$$

where $F = \Phi\left(\frac{L_2 - \mathbf{X}\beta}{\sigma}\right) - \Phi\left(\frac{L_1 - \mathbf{X}\beta}{\sigma}\right)$

(7-13)

For the special case where we have only left-censoring of data and the left-censoring value for y is $L_1 = 0$, we can simplify Equation 7-13 to Equation 7-14.

$$E[y] = \Phi\left(\frac{\mathbf{X}\beta}{\sigma}\right) \mathbf{X}\beta - \sigma\phi\left(\frac{\mathbf{X}\beta}{\sigma}\right)$$

For the special case of left-censoring only and

(7-14)

left-censored value for y, $L_1 = 0$

In order to calculate the marginal effects, we can differentiate the expression for $E[y]$ in Equation 7-13 with respect to model coefficients β. For the special case of left-censoring only with a left-censored value of zero, we can use Equation 7-14.

7.3 Tobit-I Implementation

In this section, let us look at the code for fitting a Tobit-I model using a maximum likelihood method in order to illustrate the mathematical concepts presented earlier. Tobit models are not currently available in the **statsmodels** library, so let us write our own implementation. Due to the reusable nature of this implementation, we will implement it as a modular and reusable library with a unittest to test it.

Let us use the earlier example of fitting a Tobit model to the censored data in Equation 7-3. Summary statistics of fitting the model are shown in Listing 7-3.

As seen from the output, the Tobit-I model correctly deduces the parameters of the model as $y^* = 0.9546 + 2.9868x$, with the variance of error 0.9983. This is close to the original data generation process in Equation 7-3. The implementation uses the Broyden–Fletcher–Goldfarb–Shanno (BFGS) method to find the minimum of the negative log-likelihood function. Statistics from minimization are shown in Listing 7-4, confirming that the process finds a minimum. For numerical optimization applied to multi-modal functions, it is important to verify that iterations terminate successfully.

Listing 7-3. Left-Censored Data Fitted Using Tobit-1 Model

```
 1  INFO:TobitTest:                              Tobit1  Results
 2  ================================================================
 3  Dep. Variable :    y                Log–Likelihood: −645.41
 4  Model:             Tobit1           AIC:            1291.
 5  Method:            Maximum Likelihood BIC:          1291.
 6  Date:              Thu, 08 Aug 2024
 7  Time:              23:17:26
 8  No. Observations : 900
 9  Df Residuals :     899
10  Df Model:          0
11  ================================================================
12     coef    std err    z     P>|z|    [0.025    0.975]
13  ----------------------------------------------------------------
14  x1    2.9868  0.034   86.581  0.000    2.919     3.054
15  par0  0.9546  0.074   12.853  0.000    0.809     1.100
16  par1  0.9983  0.033   30.000  0.000    0.933     1.064
17  ================================================================
```

Listing 7-4. Minimization of Tobit-I Negative Log-Likelihood

```
Optimization  terminated  successfully .
Current  function  value :  0.717120
 Iterations  :  17
Function  evaluations :  20
Gradient  evaluations :  20
```

Finally, we can compare the root-mean-square errors between predicted and actual values on the test dataset to compare the fit produced by Tobit-I and OLS models. As seen from the output, RMSE for the Tobit-I model is 1.143183, while for OLS is 4.785384, confirming that the Tobit-I model produces a better fit.

The code for implementing and testing the Tobit-I model is presented in Listing 7-5.

Listing 7-5. Implementation of Tobit-1 Model

```python
from statsmodels.base.model import GenericLikelihoodModel
import numpy as np
import statsmodels.api as sm
from scipy import stats


class Tobit1 (GenericLikelihoodModel):
    def __init__( self , endog, exog, low=None, high=None, add_constant=True, method="
        bfgs"):
        super().__init__(endog=endog, exog=exog)
        assert (low is not None) or (high is not None), "both low and high cannot be
            None"
        if (low is not None) and (high is not None):
            assert low < high, "low must be  strictly  less than high"
        self.low = low
        self.high = high
        self.endog = endog
        self.exog = exog
        self.add_constant = add_constant
        if add_constant :
            self.exog = sm.add_constant(exog, has_constant="add", prepend=False)
        self.parameters = np.zeros( self.exog.shape[1] + 1, dtype=np.float64 )
        self.parameters[-1] = 1.
        self.method = method

    def loglikeobs ( self , params):
        error = self.endog - ( self.exog @ params[:-1])
        ll = 0
        condition = np.ones( self.endog.shape[0], dtype=bool)
        if self.low is not None:
            ll += np.sum(stats.norm.logcdf( error [ self.endog <= self.low], self.low,
                params[-1]))
            condition [ self.endog <= self.low] = False
        if self.high is not None:
```

```
32            ll  += np.sum( stats .norm.logcdf(−error [ self .endog > self .high ],  self .high,
                    params[−1]))
33            condition [ self .endog > self .high ] = False
34        ll  += np.sum( stats .norm.logpdf( error [ condition ],  0,  params[−1]))
35        return  ll
36
37    def  fit ( self , ∗∗kwargs):
38        if  "method" in  kwargs:
39            kwargs.pop("method")
40        return  super () . fit ( self . parameters , method=self.method, ∗∗kwargs)
41
42    def  predict ( self , params, exog, ∗args, ∗∗kwargs):
43        if  self . add_constant :
44            exog = sm.add_constant(exog, has_constant="add", prepend=False)
45        pred = np.einsum(" ij , j−>i", exog, params[0:−1])
46        if  self .low is not None:
47            pred = np.where(pred <= self .low,  self .low, pred)
48        if  self .high is not None:
49            pred = np.where(pred > self .high,  self .high, pred)
50        return  pred
51
52 # unittest  below
53
54 import  unittest
55 import  logging
56 import  statsmodels . api as sm
57
58 logging . basicConfig ( level =logging .DEBUG)
59
60
61 class  TobitTest ( unittest .TestCase):
62    def  setUp( self ) −> None:
63        self . logger = logging . getLogger( self . __class__ . __name__)
64        x = np.arange(−5, 5, 0.01)
65        np.random.seed(1024)
66        coeff = 3
67        variance = 1
68        ll = 0.0
69        sdev = np. sqrt ( variance )
70        epsilon = sdev ∗ np.random.standard_normal(x.shape [0])
71        ystar = 1 + coeff ∗ x + epsilon
72        y = np.where( ystar < ll,  ll,  ystar )
73        self .x = x
74        self .y = y
75        self .ll = ll
76        self . tobit  = None
77        self . trainTestRatio = 0.9
78
79    def  test_regression ( self ):
80        training = int ( self . trainTestRatio ∗ self .x.shape [0])
81        x = self .x[0: training , np.newaxis]
82        y = self .y[0: training ]
83        self . tobit = Tobit1 (y, x, low=self .ll, add_constant=True)
```

```
84
85    res  =  self . tobit . fit ()
86    self . logger . info ( res . summary())
87    self . assertIsNotNone ( res )
88
89    exogTest  =  self .x[ training :]
90    testPred  =  self . tobit . predict ( res .params,  exogTest)
91    actual  =  self .y[ training :]
92    diff  =  testPred  − actual
93    mse1 = np. sqrt (np.mean(diff * diff ))
94
95    #  fit  OLS model
96    x = sm.add_constant( self .x[0: training ,  np.newaxis],  has_constant="add")
97    olsModel = sm.OLS(self.y[0: training ], x). fit ()
98    testX  = sm.add_constant( self .x[ training :,  np.newaxis],  has_constant="add")
99    olsPred  = olsModel. predict ( testX )
100   self . logger . info (olsModel.summary())
101   diff  = olsPred − actual
102   mse2 = np. sqrt (np.mean(diff * diff ))
103   self . logger . info ("RMSE from tobit: %f, from OLS: %f", mse1, mse2)
104   self . assertLess (mse1, mse2)
```

Code Explanation

Let us do a code walk-through to understand the implementation of the Tobit-I
model as well as the unittest for verifying it:

1. Class **Tobit1** implements the Tobit-I model. It uses **GenericLikelihoodModel**
   from the **statsmodels** library as the base class. The base class defines methods
   for calculating the parameter values by minimizing the negative log-likelihood
   function. The base class requires the following two items from the subclass:

   • Negative log-likelihood function with the signature **loglikeobs(self, params)**.
   • Initial values for parameters. This is provided as an argument to the **loglikeobs**
     method above.

2. The constructor of derived class **Tobit1** accepts the following arguments:

   • An array of type **numpy.ndarray** containing endogenous or dependent
     variables. This is a one-dimensional array with length equal to the number
     of samples in the training dataset.
   • A two-dimensional array of type **numpy.ndarray** containing exogenous or
     independent variables. This array does not include a column containing one
     for the intercept.
   • Low threshold for censoring. This argument is optional. If provided, the
     output variable is left-censored at this threshold.

- High threshold for censoring. This argument is optional. If provided, the output variable is right-censored at this threshold.
- **add_constant** boolean flag. If true, an intercept term is added to the regression by appending a column of ones to the exogenous variable array.
- **method**: This is the name of numerical method used for minimizing the negative log-likelihood function.

The constructor initializes an array of parameters equal to the number of exogenous variables after including an additional one for the intercept, if requested, and an additional parameter for the variance σ^2 of error ϵ.

3. The class provides an implementation of method **loglikeobs** for computing the negative log-likelihood. As described earlier, this can be represented by the analytical expression in Equation 7-15.

$$
\begin{aligned}
-\log L(\beta) = \sum_{i=1}^{N} & -I(y_i = L_1) \log\left(\Phi\left(\frac{L_1 - \mathbf{X_i}\beta}{\sigma}\right)\right) \\
& -I(L_1 \leq y_i < L_2) \log\left(\frac{1}{\sigma}\phi\left(\frac{y_i - \mathbf{X_i}\beta}{\sigma}\right)\right) \\
& -I(y_i \geq L_2) \log\left(\Phi\left(-\frac{L_2 - \mathbf{X_i}\beta}{\sigma}\right)\right)
\end{aligned}
$$

(7-15)

$$
I(z) = \begin{cases} 0 \text{ if } z \text{ is false} \\ 1 \text{ if } z \text{ is true} \end{cases}
$$

4. The **fit** method calls the base class method for fitting the model, passing the initial value of parameters and numerical method used to minimize the negative log-likelihood function as arguments.
5. The class has a method **predict** for predicting values using the fitted model. **fit** must have been called before invocation of **predict**.
6. Now let us look at the unittest. The class **TobitTest** derives from base class **unittest.TestCase**. The base class provides boilerplate code for running a suite of tests and reporting results.
7. Method **setUp** sets the seed for the numpy library's random number generation. This is set for reproducibility of results. The method generates the data using the data generation process defined in Equation 7-3. It also sets the train-test ratio to 0.9, implying that 90% of data will be used for training.
8. Method **test_regression** runs the unittest. It creates an object of class **Tobit1** and fits the model using the training dataset. It also uses an OLS model to fit the training data. After fitting the two models, it uses the test dataset to evaluate their relative performance. It computes the root-mean-square error of predicted and

actual values and asserts that the RMSE value is smaller for the Tobit-I model as compared with the OLS model.

7.4   Heteroskedasticity in Tobit-I Model

In Equation 7-1 formulating the Tobit-I model, we assumed that error terms have a constant variance σ^2. In the presence of heteroskedasticity or changing variance, we must treat σ^2 as a variable in the expression for maximum likelihood and optimize it with respect to the variance parameter in addition to model coefficients. In cases where heteroskedasticity is present, we hypothesize a parametric model for variance, such as the GARCH model, and then maximize the likelihood expression with respect to parameters in the parametric variance model, in addition to model coefficients.

It should be noted that if heteroskedasticity is ignored when it is actually present, we will get biased and inconsistent coefficient estimates from the Tobit-I model. This is in contrast with linear regression using OLS where heteroskedasticity does not impact the consistency of model coefficient estimates. It only makes OLS inefficient. This occurs because the expression for $E[y]$ in the Tobit-I model (Equation 7-13) has a term that depends on σ, making model coefficient estimates biased and inconsistent.

7.5   Tobit-II Model

The Tobit model considered in the previous section was a type-I model. The Tobit type-I model is characterized by the fact that the latent variable y^* decides both the censoring of an observed variable and its censored value. In the type-I model, y was censored if y^* was above or below a threshold, and it also provided a value for y in the non-censored area. The Tobit type-II model separates these two functions. One latent variable y_1^* determines when the observations will be censored, and another latent variable y_2^* provides the uncensored value. This is formulated in Equation 7-16.

$$y = \begin{cases} a_1 \text{ if } y_1^* < L_1 \\ y_2^* \text{ if } L_1 \leq y_1^* < L_2 \\ a_2 \text{ if } y_1^* \geq L_2 \end{cases}$$

$$y_1^* = \mathbf{X}\beta + \epsilon_1 \qquad\qquad (7\text{-}16)$$

$$y_2^* = \mathbf{X}\gamma + \epsilon_2$$

$$\epsilon_1 \sim N(0, \sigma_1^2)$$

$$\epsilon_2 \sim N(0, \sigma_2^2)$$

In econometrics, the Tobit type-II model is known as the **Heckman two-step model**. This is because the Heckman model involves a two-step regression. The first step is probit regression (GLM with Poisson distribution) that decides if the observation will be censored. Let us say this step has model parameters β, as shown in Equation 7-16. The next step regresses the observation in an uncensored region using another set of coefficients γ.

7.5.1   Fitting Heckman Two-Step Model

Let us fit the Tobit-II model using the Heckman two-step approach to predict the aptitude score of students in a test. The data is available from the UCLA website [48]. The data is right-censored at 800, which is the highest score that can be obtained. There is also a lower threshold of 200, but in the dataset no student reaches the lower threshold. Therefore, we cannot train a probit model for a lower threshold, and we specify only an upper threshold for right-censoring.

The explanatory variables are student scores in reading and writing, along with a categorical column indicating if a student is enrolled in "vocational," "general," or "academic" disciplines.

The code for fitting the Heckman two-step model is shown in Listing 7-6.

Listing 7-6. Implementing and Testing Heckman Two-Step Model

```
import numpy as np
import logging
import statsmodels.api as sm
import pandas as pd
import os
import seaborn as sns
import matplotlib.pyplot as plt

logging.basicConfig(level=logging.DEBUG)


class Heckman2StepModel(object):
    """ Tobit-II (censored) regression model using Heckman 2-step approach """

    def __init__(self, endog: np.ndarray, exog: np.ndarray, low=None, high=None,
                 include_constant=True, train_test_ratio=0.9, low_threshold=0.15,
                 high_threshold=0.15):
        """
        Initialize the regression model
        :param endog: y
        :param exog: X
        :param low: Low threshold for censoring
        :param high: High threshold for censoring
        :param include_constant:
        :param train_test_ratio:
        """
```

```
27    assert  (low  is  not  None) or  (high  is  not  None),  "both  low  and  high  cannot  be
          None"
28    if  (low  is  not  None) and  (high  is  not  None):
29        assert  low < high,  "low  must be  strictly  less  than  high"
30    self .low = low
31    self .high = high
32    self .lowProbitModel = None
33    self .highProbitModel = None
34    self .olsModel = None
35    self .endog = endog
36    self .exog = exog
37    self . includeConstant  =  include_constant
38    self . trainTestRatio  =  train_test_ratio
39    self .lowThreshold = low_threshold
40    self .highThreshold = high_threshold
41    self . ntraining  =  int ( self . trainTestRatio  *  self .endog.shape [0])
42    self . logger  =  logging . getLogger( self . __class__ . __name__)

44  def  fitProbit ( self ,  endog,  exog,  threshold ):
45    rows = np. where(endog >= threshold ,  1,  0)
46    model = sm. Probit (rows,  exog)
47    return  model. fit ()

49  def  fitOLS( self ,  endog,  exog):
50    model = sm.OLS(endog,  exog)
51    return  model. fit ()

53  def  fit ( self ):
54    """
55    Fit  the  Heckman  regression  model to  the  data
56    """
57    ntraining  =  self . ntraining
58    exog =  self .exog
59    if  self . includeConstant :
60        exog = sm. add_constant( self .exog,  has_constant="add")
61    olsFlag = np. ones( ntraining ,  dtype=bool)
62    if  self .low is  not  None:
63        olsFlag [ self .endog[0: ntraining ] <= self .low] = False
64        self .lowProbitModel = self . fitProbit ( self .endog[0: ntraining ],  exog[0:
              ntraining ,  :],  self .low)
65    if  self .high is  not  None:
66        olsFlag [ self .endog[0: ntraining ] >= self .high] = False
67        self .highProbitModel = self . fitProbit ( self .endog[0: ntraining ],  exog[0:
              ntraining ,  :],  self .high)
68    self .olsModel = self .fitOLS( self .endog[0: ntraining ][ olsFlag ],  exog[0: ntraining ,
          :][ olsFlag ,  :])

70  def  predict ( self ,  exog: np.ndarray  =  None) −> np.ndarray:
71    """
72    Predict  the  output  of  the  model using  exogeneous  variables  as  input .
73    :param exog: exogeneous  variables  (X)
74    : return :  output  value  from  the  model (y)
75    """
```

```
76          if exog is None:
77              exog = self .exog[ self . ntraining :,  :]
78          if self . includeConstant :
79              exog = sm. add_constant(exog, has_constant="add")
80          result  = np. zeros (exog.shape [0], dtype=np. float64 )
81          olsFlag  = np.ones(exog.shape [0], dtype=bool)
82          if self .low is not None:
83              lowProb = self .lowProbitModel. predict (exog)
84              lowVals = lowProb < (1  − self .lowThreshold)
85              olsFlag [lowVals] = False
86              result [lowVals] = self .low
87          if self .high is not None:
88              highProb = self .highProbitModel. predict (exog)
89              highVals = (highProb > self .highThreshold)
90              olsFlag [highVals] = False
91              result [highVals] = self .high
92          result [ olsFlag ] = self .olsModel. predict (exog[olsFlag ,  :])
93          return result
94
95      @staticmethod
96      def rmse(y1, y2):
97          return np. sqrt (np.mean((y1 − y2)**2))
98
99
100 if __name__ == "__main__":
101     dirname = r"C:\prog\cygwin\home\samit_000\latex\ book_stats \code\data"
102     df = pd.read_csv(os.path. join (dirname, " student_scores .csv"))
103     df.loc [:, "prog_codes"] = df.loc [:, "prog"]. astype ("category"). cat .codes
104     high = 800
105     y = df.loc [:, "apt"]. values
106     X = df.loc [:, ["read","math","prog_codes"]]. values
107     heckman = Heckman2StepModel(y, X, high=high)
108     heckman. fit ()
109     predicted  = heckman.predict ()
110     actual  = y[heckman. ntraining :]
111     rmse1 = Heckman2StepModel.rmse(actual, predicted )
112
113     # fit an OLS model
114     Xconst = sm. add_constant(X, has_constant="add")
115     ols = sm.OLS(y[0:heckman.ntraining], Xconst[0:heckman. ntraining ,  :])
116     ols = ols . fit ()
117     olsPred = ols . predict (Xconst[heckman. ntraining :,  :])
118     rmse2 = Heckman2StepModel.rmse(actual, olsPred)
119
120     logging . info ("RMSE from Heckman model: %f, OLS model: %f", rmse1, rmse2)
121
122     # plot
123     predictors  = ["Heckman"] ∗ predicted .shape[0] + ["OLS"] ∗ olsPred.shape[0] + ["
                Actual"] ∗ actual .shape[0]
124     values = np. concatenate (( predicted , olsPred, actual ), axis=0)
125     ids = df.loc [heckman.ntraining :, "id"]. values
126     ids = np. concatenate (( ids , ids , ids ))
127     df = pd.DataFrame({"Id": ids, "Aptitude Score": values, "Predictor": predictors })
```

```
128    sns. lineplot (data=df, x="Id", y="Aptitude Score", hue="Predictor")
129    plt.legend(loc="upper left")
130    plt.grid()
131    plt.savefig (os.path.join(dirname, f"heckman_v_ols.jpeg"),
132                    dpi=500)
133    plt.show()
```

In order to assess its performance, we compare the root-mean-square difference between predictions of the Heckman two-step model and actual aptitude score against the difference between an OLS fitted linear regression model and actual aptitude score using the test dataset that has not been used for training either models. The RMSE value using the Heckman two-step approach is 60.35, which is lower than that of the OLS model's value 62.85. This shows that the Heckman two-step model performs better on the test dataset. The plot of predictions from the two models and actual aptitude scores is shown in Figure 7-2.

Code Explanation
Let us step through the code in Listing 7-6 to understand the implementation of the model and the test for validation.

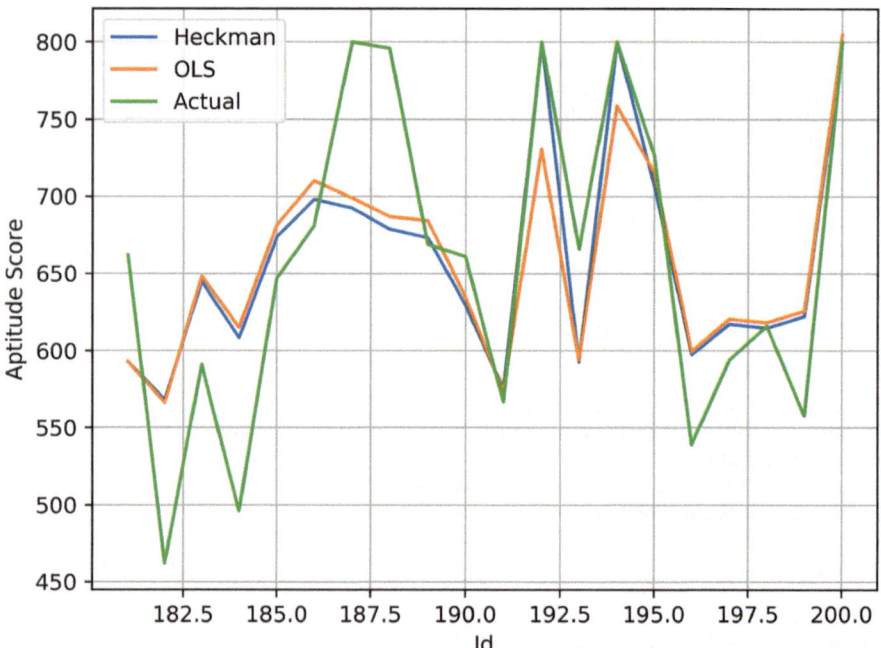

Figure 7-2. Comparison of Heckman Two-Step and OLS Models for Predicting Right-Censored Aptitude Scores

1. Class **Heckman2StepModel** implements the Tobit-II model. Its constructor in method **__init__** takes the following arguments:

   - **endog** is a one-dimensional numpy array containing the dependent variable (y).
   - **exog** is a two-dimensional numpy array containing the independent variables. This array does not have a column of ones for an intercept.
   - **low** threshold for left-censoring (optional).
   - **high** threshold for right-censoring (optional).
   - **include_constant**: A boolean flag indicating if an intercept should be added to the regression.
   - **train_test_ratio**: Proportion of the dataset to use for training the model. Remaining data is used for testing.
   - **low_threshold**: Threshold to use in **testing** to decide if the output is left-censored. This is a model hyper-parameter. A probit model predicts the probability that an endogenous variable is above the left-censoring value, specified as a **low** argument. During prediction on the test dataset, if the predicted probability is greater than **low_threshold**, the value is assumed to be not censored because it is above the limit where left-censoring is applied. If the probability is at or below this threshold, left-censoring is applied.
   - **high_threshold**: Threshold to use in **testing** to decide if the output is right-censored. This is a model hyper-parameter. A probit model predicts the probability that an endogenous variable is above the right-censoring value, specified as a **high** argument. During prediction on the test dataset, if the predicted probability is greater than **high_threshold**, the value is assumed to be censored because it is above the limit where right-censoring is applied. If the probability is at or below this threshold, right-censoring is not applied.

2. The model contains three submodels: a probit model for left-censoring, a probit model for right-censoring, and an OLS model for uncensored data.
3. Method **fitProbit** fits the probit model.
4. The **fit** method identifies the training dataset, adds a column of ones to the array of exogenous variables if the **add_constant** flag was set to **True**, and fits the probit model for left-censoring, probit model for right-censoring, and an OLS model for the uncensored part of the data.
5. The **predict** method predicts the value of the endogenous variable. If no exogenous variables are provided to this method, it uses the test dataset extracted from the original dataset passed to the constructor. This method first computes the probability that the data is left-censored using the fitted probit model for left-censoring. This is followed by a similar calculation to predict if it is right-censored using the right-censoring probit model. For censored data points, it sets the predicted value at low and high levels, respectively. These arguments were passed to the constructor. Following this, it uses the OLS model to predict the endogenous variable in the uncensored region.

6. To test the model, the data file containing student aptitude scores and exogenous variables is read.
7. Column **prog** is categorical. It is converted to a categorical data type in the pandas dataframe. This type is then cast to an integer that can be used for modeling.
8. After fitting the Heckman model, the code fits an OLS model to the dataset using training data only.
9. The models are used to predict student aptitude scores for the test dataset. For the two models, root-mean-square errors (RMSE) against actual aptitude scores are computed.
10. Predictions are plotted against actual scores using the **seaborn** plotting package.

Random Forest

8

In this chapter, we will look at random forests – a versatile, non-linear model that can be used for both regression and classification. Random forests have an impressive mathematical and statistical pedigree to be labeled as advanced statistical models, and their use of decision trees as weak learners and their reliance on bagging of data for training an ensemble of decision trees makes them among the more widely adopted machine learning tools.

Random forest is an ensemble learning model that is based on a collection of decision trees. Unlike decision trees that are prone to overfitting, random forests are robust to overfitting. They achieve this performance by growing a forest of trees using a bag of training data selected randomly from the training dataset with replacement and by randomly selecting a feature for splitting a tree node at each level. The trees are grown to full depth without pruning. With these modifications, random forests can be used for a wide variety of tasks such as classification, regression, dimensionality reduction, outlier detection, and quantifying variable importance. Because they are not susceptible to overfitting and do not require extensive problem-specific hyper-parameter tuning, they have been applied to solve varied problems with good results.

The foundational idea of random forests was proposed by [25] who investigated the impact of randomly selecting a node-splitting feature in a decision tree. [26] compared the performance of decision tree ensembles using bagging, boosting, and randomization. Random forests in their present form were introduced by [27].

Decision trees have low bias but high variance because they can fit training data well by growing in depth. However, this comes at a cost of poor generalization because decision trees learn the noise in the training dataset. Random forests address this problem by growing an ensemble of decision trees using a randomly drawn subset of training dataset (with replacement). Because individual decision trees are constructed using a differing training dataset, they have low correlation. This process of constructing trees is called bootstrap aggregation or bagging. However, if an input feature is important, i.e., has a strong influence in determining the output,

© Samit Ahlawat 2025
S. Ahlawat, *Statistical Quantitative Methods in Finance*,
https://doi.org/10.1007/979-8-8688-0962-0_8

it will get selected as a node-splitting feature in most decision trees and increase the tree correlation. This will lead to poor generalization (high variance) in the test dataset. To further reduce tree correlation, node-splitting features are selected at for each decision tree.

8.1   Strength and Correlation of Trees

Strength and correlation are two key concepts for understanding random forests' performance. An ideal model has low bias and low variance. Low bias ensures that the predicted result converges to the true result with increasing batch size. Low variance ensures that predictions are not too far from the actual but unknown result, leading to good performance in the test dataset. For random forests, strength and correlation of constituent decision trees determine bias and variance of the model.

Let us denote the vector of input features by $\mathbf{X}$ and the output of the ith decision tree as $h_i(X)$. T denotes the number of decision trees in random forest. For a classification problem, the true label for input vector $\mathbf{X}$ is Y, and there are M different classes.

Let us define margin function $m(X, Y)$ as the difference between the number of decision trees predicting the correct label Y and the number of decision trees predicting the most frequent incorrect label normalized by the number of decision trees, as shown in Equation 8-1. I is the indicator function that has a value of 1 if its argument is true and 0 otherwise.

$$m(X, Y) = \frac{1}{T} \sum_i \left[I\left(h_i(X) = Y\right) - \max_{j \neq Y} I\left(h_i(X = j)\right) \right] \tag{8-1}$$

As the number of trees, T, grows to infinity, we can write the asymptotic margin function, $\tilde{m}(X, Y)$, using probability over random forest parameters θ as shown in Equation 8-2.

$$\tilde{m}(X, Y) = P_\theta\left(h_i(X) = Y\right) - \max_{j \neq Y} P_\theta\left(h_i(X = j)\right) \tag{8-2}$$

A generalization error or prediction error, PE, is defined as the probability that an asymptotic margin is negative as shown in Equation 8-3.

$$PE = P_{X,Y}\left(\tilde{m}(X, Y) < 0\right) \tag{8-3}$$

The strength, s, of a random forest is defined as the expected value of the asymptotic margin as shown in Equation 8-4. Intuitively, the higher the margin, the lower the error and the stronger the classifier. In the extreme case, all decision trees of random forest predict the correct label, giving the highest margin.

$$s = E_{X,Y}\left[\tilde{m}(X, Y)\right] \tag{8-4}$$

Equation 8-5 is the Chebyshev's inequality. In the asymptotic case (as the number of decision trees goes to infinity), X approaches the true mean μ. This gives the expression for the generalization error of random forest as shown in Equation 8-6. s is the asymptotic strength of random forest.

$$P\left[|Z - \mu(Z)| > k\sigma(Z)\right] \leq \frac{1}{k^2} \tag{8-5}$$

$$\lim_{T \to \infty} P\left(|\tilde{m}(X, Y) - \mu| > k\sigma\right) = P\left(0 > k\sigma\right)$$

$$= P\left(s < 0\right) \text{ setting } k = \frac{s}{\sigma}$$

$$\equiv PE$$

$$\leq \frac{1}{\frac{s^2}{\sigma^2}} \text{ using Chebyshev's inequality} \tag{8-6}$$

$$\implies PE \leq \frac{\sigma^2}{s^2}$$

where $\sigma^2 = \mathrm{var}(\tilde{m}(X, Y))$

Next, using $\mathrm{var}(Z) = E[Z^2] - (E[Z])^2$, we can write $\mathrm{var}(Z) \leq E[Z^2]$. Using this, the variance of margin can be expressed as a function of the correlation of decision tree strength, ρ, as shown in Equation 8-7.

$$\mathrm{var}(\tilde{m}(X, Y)) \leq E\left[(\tilde{m}_i(X, Y))(\tilde{m}_j(X, Y))\right]$$

$$= E\left[\left(I\left(h_i(X) = Y\right) - \max_{j \neq Y} I\left(h_i(X = j)\right)\right) \right.$$

$$\left. \left(I\left(h_l(X) = Y\right) - \max_{j \neq Y} I\left(h_l(X = j)\right)\right)\right] \tag{8-7}$$

$$= \rho k^2$$

Combining Equations 8-6 and 8-7, one obtains Equation 8-8 relating the generalization error of random forest to the tree correlation ρ and strength s. As can be seen from Equation 8-8, generalization error PE increases with increasing tree correlation ρ and decreases with increasing tree strength s.

$$PE \leq \frac{\rho k^2}{s^2} \tag{8-8}$$

8.2   Building a Random Forest

Random forests can be used for classification and regression. The two problems are related, but there are minor differences in the forest construction algorithm for the two problems.

Let us denote the number of decision trees in random forest by T, the number of features by M, and the number of training data samples by N. Each decision tree is constructed as follows:

1. Sample N training data points from the dataset with replacement. This results in around $\frac{N}{3}$ data points left out of the training set for constructing a decision tree. This set is called the out-of-bag (OOB) training dataset. To see why sampling with replacement leaves around $\frac{N}{3}$ data points out-of-bag, let us look at the code in Listing 8-1. It selects N items at random from N items with replacement, counting the number of unique elements obtained. Dividing the number of unique elements by N gives the proportion of in-bag samples. Output of the code in Listing 8-1 can be seen in Listing 8-2, which confirms that the number of unique items converges to 0.64. This implies around 0.33 or one-third of items are left out-of-bag for each decision tree.
2. Select m features at random from the set of M features. Select the feature that gives the best node split using a criterion such as Gini impurity reduction or entropy decrease. The Gini impurity index measures the probability of misclas-

Listing 8-1. Unique Items Drawn by Sampling with Replacement

```
1   import numpy as np
2
3
4   class Choice(object):
5       def sim(self, N):
6           arr = np.arange(N)
7           res = np.random.choice(arr, N, replace=True)
8           return np.unique(res).shape[0] / float(N)
9
10
11  if __name__ == "__main__":
12      np.random.seed(32)
13      ch = Choice()
14      res = [ch.sim(k) for k in range(2000, 2005)]
15      print(res)
```

Listing 8-2. Number of Unique Items When Sampling with Replacement

```
1   [0.6305, 0.6481759120439781, 0.6363636363636364, 0.6370444333499751,
        0.6372255489021956]
```

sifying an element. The probability of misclassifying an element belonging to class k is $p_k(1 - p_k)$ where p_k is the probability of an element belonging to class k and $1 - p_k$ is the misclassification probability. Summing over all classes, we get the value of the Gini index, as shown in Equation 8-9.

$$\text{Gini} = \sum_k p_k(1 - p_k) = 1 - \sum_k p_k^2 \qquad (8\text{-}9)$$

The Gini index ranges from 0 to 0.5. Each time a node is split, the Gini index reduces because misclassification probability falls. Gini index reduction after a split is shown in Equation 8-10. *Gini* is the Gini index at the parent node, and *Gini$_1$* and *Gini$_2$* are the Gini indices at the two child nodes. N, N_1, and N_2 denote the number of data points in the parent node and the two child nodes.

$$\Delta\text{Gini} = \text{Gini}_{parent} - \frac{N}{N_1}\text{Gini}_1 - \frac{N}{N_2}\text{Gini}_2 \qquad (8\text{-}10)$$

Entropy, S, measures the information content or the degree of randomness. A node that has all data points belonging to one class has zero entropy, while a node with points equally distributed among the classes has maximum entropy. The definition of entropy is shown in Equation 8-11. A node partition that reduces entropy or randomness to a greater extent is better. Entropy reduction due to a node split is shown in Equation 8-12.

$$S = -\sum_k p_k \log(p_k) \qquad (8\text{-}11)$$

$$\Delta S = S_{parent} - \frac{N}{N_1}S_1 - \frac{N}{N_2}S_2 \qquad (8\text{-}12)$$

Increasing the number of splitting features, m, increases both the tree strength and correlation. As observed earlier, a higher tree strength and low correlation are needed to reduce the generalization error. Therefore, the selection of m entails a compromise. For classification problems, $m = \sqrt{M}$ has been found to be a good choice, while for regression, $m = \frac{M}{3}$ works better.

3. Grow each decision tree to its maximum depth without pruning.

During prediction, each decision tree predicts the class of input data, and the class with the maximum votes is selected.

The algorithm used to build a random forest is shown in pseudo-code 15.

Algorithm 15 Building a Random Forest for Classification

Require: Number of decision trees T, training samples N with M input features each, number of features used for node splitting m.

1: **for** each decision tree t = 0, 1,2,...,T-1 **do**
2:     Sample N samples from the training dataset with replacement. Keep track of out-of-bag (OOB) training data samples for each tree.
3:     Randomly select m distinct features from M input features.
4:     Select the feature giving highest reduction of Gini impurity or entropy. Split the node on this feature.
5:     Continue growing the tree using remaining unused features, splitting on the one giving best Gini or entropy reduction.
6: **end for**

A sample implementation for constructing a random forest is shown in List-ing 8-3. In practice, a standard library implementation is used. This implementation serves as a quick synopsis of the algorithm's salient features.

Listing 8-3. Random Forest Implementation

```
import numpy as np


class Node(object):
    def __init__(self, feature, threshold, left=None, right=None):
        self.feature = feature
        self.threshold = threshold
        self.left = left
        self.right = right


class RandomForest(object):
    ''' Construct a random forest for binary classification problem '''

    def __init__(self, ntrees, nsplits=None):
        ''' Initialize.
        :ntrees number of decision trees
        :nsplits number of features used to split tree nodes. Tree will
            have atmost nsplits height
        '''
        self.trees = [None] * ntrees
        self.nSplits = nsplits
        self.oobSamples = [None] * ntrees

    def _giniNode(self, outputs):
        pos = outputs.sum()
        prob_pos = pos / float(outputs.shape[0])
        prob_neg = 1 - prob_pos
        return prob_pos * (1 - prob_pos) + prob_neg * (1 - prob_neg)

    def _getGiniImpRed(self, inputfeature, outputs, threshold):
```

```
31        ''' Gini impurity reduction by splitting on feature at threshold
          '''
32        gini = self._giniNode(outputs)
33        left = (inputfeature < threshold)
34        right = (inputfeature >= threshold)
35        gini_left = self._giniNode(outputs[left])
36        gini_right = self._giniNode(outputs[right])
37        nobs = outputs.shape[0]
38        return gini - left.sum() / float(nobs) * gini_left - right.sum() /
              float(nobs) * gini_right
39
40    def _constructTree(self, inputs, outputs, features, splits):
41        ''' Construct a decision tree
42        :inputs 2 dimensional numpy ndarray of shape (num observations, num
              features)
43        :outputs 1 dimensional ndarray with output. Shape (num_observations
              )
44        :features list of features to split the node
45        :splits threshold on number of splits
46        '''
47        if splits <= 0:
48            return
49
50        sel_feat = None
51        reduction = None
52        sel_threshold = None
53        for feat in features:
54            threshold = np.random.choice(inputs[:, feat], size=1)
55            gini_red = self._getGiniImpRed(inputs[:, feat], outputs,
                  threshold)
56            if (reduction is None) or (reduction < gini_red):
57                reduction = gini_red
58                sel_feat = feat
59                sel_threshold = threshold
60
61        node = Node(sel_feat, sel_threshold)
62        left_data = (inputs[:, sel_feat] <= sel_threshold)
63        features_rem = [f for f in features if f != sel_feat]
64        node.left = self._constructTree(inputs[left_data, :], outputs[
              left_data], features_rem, splits - 1)
65        right_data = np.logical_not(left_data)
66        node.right = self._constructTree(inputs[right_data, :], outputs[
              right_data], features_rem, splits - 1)
67        return node
68
69    def construct(self, inputs: np.ndarray, outputs: np.ndarray) -> None:
70        ''' Construct a random forest for binary classification problem
71        :param inputs: 2 dimensional numpy ndarray of shape (num
              observations, num features)
72        :param outputs: 1 dimensional ndarray of shape (num_observations).
73        Contains booleans: True or False
74        '''
75
```

```
76    nfeat = inputs.shape[0]
77    if self.nSplits is None:
78        self.nSplits = int(np.sqrt(nfeat))
79    y_labels = sorted(list(set(outputs)))
80    assert len(y_labels) == 2
81    y_out = np.where(outputs == y_labels[0], True, False)
82    features = np.arange(inputs.shape[1])
83    for i in range(len(self.trees)):
84        sample_inputs = np.random.choice(inputs.shape[0], inputs.shape
              [0], replace=True)
85        self.trees[i] = self._constructTree(inputs[sample_inputs, :],
              y_out[sample_inputs], features)
86        self.oobSamples[i] = sample_inputs
```

For regression, random forests take an average of individual decision tree predictions instead of a majority vote. The impurity reduction criterion used for splitting a node is the mean square deviation between a known output and the mean of outputs of data points falling in a node. This is shown in Equation 8-13. N denotes the number of data points in a node, and y_i is the data point output.

$$\Delta \text{Impurity}_{\text{regression}} = \text{MSE}_{\text{parent}} - \frac{N_1}{N} \text{MSE}_{\text{child1}} - \frac{N_2}{N} \text{MSE}_{\text{child2}}$$

$$\text{MSE}_{\text{node}} = \frac{\sum_{i=1}^{N} (y_i - \bar{y})^2}{N} \tag{8-13}$$

$$\bar{y} = \frac{\sum_{i=1}^{N}}{y}$$

8.3   Performance Metrics and Error Estimates

Random forests do not need a separate cross-validation dataset to produce performance metrics. They use out-of-bag (OOB) training data for this purpose. Each decision tree has about one-third data points as out-of-bag data and is used to predict an outcome for OOB data. For classification, a majority vote from decision trees becomes the predicted outcome. There would be about one-third trees predicting an outcome for each data. For classification, a loss function such as cross-entropy loss shown in Equation 8-14 can be used to define an **OOB error estimate**. This estimate serves as an unbiased error estimate of the random forest. N is the number of data points, K is the number of classes, p_{ik} is the actual probability of data point i belonging to class k, and $\hat{p}_{ik}$ is the corresponding predicted probability.

$$\text{OOB Error} = - \sum_{i=1}^{N} \sum_{k=1}^{K} p_{ik} \log(\hat{p}_{ik}) \tag{8-14}$$

For regression, the mean square loss can be used to get an unbiased error estimate, as shown in Equation 8-15.

$$\text{OOB Error}_{regression} = \frac{\sum_{i=1}^{N}(y_i - \hat{y}_i)^2}{N} \tag{8-15}$$

Random forests can also be used to give an estimate of input feature importance. The following metrics can be used for the purpose:

1. **Permutation variable importance**: For each OOB data point, count the number of correct predictions by decision trees. Let us say this value is C_1. Now, get the range of values taken by a feature and randomly permute its value. Get the new predictions for this modified data point with one feature value randomly permuted and find the number of correct predictions C_2. Permutation variable importance can then be calculated using Equation 8-16, where T is the number of decision trees. Intuitively, if a variable is important in predicting the outcome, permuting the value of the feature will change the predicted class and give a low value of C_2. This will give a high value of permutation variable importance. Conversely, a variable that does not determine the classification of a data point will produce C_2 equal to C_1, giving 0 permutation variable importance.

$$\text{Permutation Var. Imp.} = \frac{|C_1 - C_2|}{T} \tag{8-16}$$

2. **Gini importance**: Adding up Gini impurity reduction each time a variable (feature) is used for splitting a node and normalizing the sum by the number of trees T gives Gini importance score for the variable. An important variable will have higher Gini impurity reduction on average and will get higher Gini importance.

3. **Dimensionality reduction using proximity**: Proximity measures the similarity between two data points and can be used for dimensionality reduction. To calculate this, the following steps are performed:

   - Initialize a square unit diagonal matrix A of size (N, N) where N is the number of training data points. The matrix has ones on the diagonal.
   - Get all decision trees in the forest to predict outcome for all N data points (OOB and in-the-bag training samples). Each time two data points m and n end up in the same node of a decision tree, increment $A[m, n]$ and $A[n, m]$ by one.
   - Let $R[i]$ and $C[j]$ denote the row and column averages of A matrix. μ denotes the overall average of quantities in matrix A. These can be calculated using Equation 8-17.

$$R[i] = \frac{\sum_{k=1}^{N} A[i, k]}{N}$$

$$C[j] = \frac{\sum_{k=1}^{N} A[k, j]}{N} \qquad (8\text{-}17)$$

$$\mu = \frac{\sum_{k=1}^{N} \sum_{l=1}^{N} A[k, l]}{N^2}$$

- Calculate a rescaled matrix C using row averages R, column averages C, and overall average μ using Equation 8-18.

$$C[i, j] = A[i, j] - R[i] - C[j] + \mu \qquad (8\text{-}18)$$

- Perform an eigenvalue decomposition of matrix C using SVD. Retain p most significant eigenvectors. Corresponding p eigenvectors define the new features of the reduced dimension space.

8.4   Other Applications of Random Forests

Random forests have been used in many applications besides classification and regression. Some of them are discussed below:

1. **Filling missing data in the training dataset**: In several machine learning applications, data is often unavailable. For example, in a survey, certain features like income may be missing. To fill missing values of non-categorical features, one calculates the median value of this feature for all data points belonging to a particular class. Missing feature is then filled with this median value for all data points belonging to that class. For categorical variables, the most frequent feature value is used in place of median.
2. **Filling missing data in the test dataset**: A testing data point with missing feature values cannot be supplied as an input to random forest. To overcome this problem, the testing data point is assumed to belong to class i. Using training dataset values of this feature for class i, the missing value is filled (using median for non-categorical features and most frequent value for categorical features). The testing data point with values for all features is then run as an input to the random forest and the number of votes for class i counted. This process is repeated for all classes. The class with maximum votes is selected, and the missing feature value used with that class is accepted.
3. **Clustering**: In clustering, we are interested in grouping points into clusters. Points within one cluster are closer to each other than to points from another cluster. Clustering is an example of unsupervised learning where data comes as a collection of features with no output value. Random forests tackle this problem by assigning all data points in input to class 0. New data is now generated by

randomly permuting each feature value from the set for N observed values of this feature. If the original dataset had N points, N new points are generated by this process. The new points are assigned to class 1. Random forest is trained using this two class data. If out-of-bag error estimate is less than a threshold (say 20%), it means that input features have relationships that can be used for clustering. On the other hand, if OOB error estimate is greater than threshold, it indicates that input features do not have pronounced interrelationships and data cannot be clustered. If OOB error estimate is less than a threshold, proximity is computed for original data points, i.e., those of class 0. Dimensionality reduction is performed using the method described earlier. After projecting the points to a lower dimensional feature space, the K-means algorithm can be used for clustering the data.

4. **Outlier detection**: Outliers are data points with feature values that are inconsistent with distribution observed in remaining data. To detect outliers, the proximity matrix is calculated as described earlier. Points with low average proximity to other data points can be regarded as outliers.

5. **Fixing incorrect labels**: Datasets often come with incorrect class labels. For example, a banking call center operator may have made a mistake in reporting a value. Random forests can be used to detect these data points as outliers. A new dataset is created with an output class as a new feature and outlier detection algorithm outlined earlier is applied. This will identify outliers. Outliers are then removed, and a random forest is trained on the clean dataset. Correct class labels for outliers are predicted using the trained random forest.

8.5   Financial Applications

In this section, let us look at a few practical examples to illustrate the random forest concepts learned so far.

8.5.1   Marketing a Banking Product

Banks often call potential customers to market new products such as credit cards, checking accounts, or investment products. This example is based on "Bank Marketing Data Set" which is available from the University of California, Irvine [28], and is based on research work by [29]. The dataset contains details of a Portuguese bank's term deposit marketing campaign. Term deposits are similar to certificate of deposits or CDs. The dataset contains 41,118 rows. Each row has 20 features such as client age, client job, 3-month EURIBOR, etc., and a binary output variable indicating if the client subscribed to the term deposit or not.

Steps for building a random forest prediction model are discussed below. The code uses random forest implementation from the **sklearn** library. This is a classification task, so the **sklearn.ensemble.RandomForestClassifier** class is used.

1. Identify input features that need to be excluded to prevent information leakage. **Duration** is an input feature of the data. However, duration is not known until the end of the call, at which point the outcome of the call is generally known, i.e., one knows whether a client will subscribe to the product or not. Ideally, this model will be used to screen potential customers for marketing campaign, with those that are likely to subscribe being contacted. Since **duration** of the call will not be known for the testing data, this feature is excluded. This leaves us with 19 features.
2. From the data, it can be observed that this is an imbalanced class problem - the number of respondents who do not subscribe outnumbers those who do by a ratio of 12.5 : 1.
3. Identify the testing and training datasets. In this example, let us use 90% of the data as training data and the remaining 10% as testing data.
4. Either remove data with missing feature values or assign the missing values with a special placeholder value. In this example, feature **job** has a value **unknown** for some rows. Let us retain the rows and assign it as a special value **unknown**. This is a categorical column.
5. Convert categorical columns to integers. Integers require less memory than strings in general, and most artificial intelligence models require numeric inputs.
6. Normalize the numeric columns. For a standard Gaussian distribution, 95% of probability density lies between $[-1.95, 1.95]$. Normalizing input features using $\frac{x-\mu}{2\sigma}$ where x is the feature value and (μ, σ) are the mean and standard deviation of the feature value over the training dataset gives a simple method of normalizing inputs.
7. For reproducibility of results, specify the **random_state** argument to the **RandomForestClassifier** constructor. This argument is used as a seed for random number generator.
8. Figure 8-1 shows Gini importance values for the features in the training dataset. **Age** is the most important predictor, followed by **euribor3m**, campaign, **education**, and **job**. This is confirmed by a histogram of outcome ("yes" or "no") against age and euribor3m. Certain age groups like 30–35 are more inclined to refuse, apparently due to a lack of interest in savings, as seen in Figure 8-2. When rates are high ($> 5\%$), there is less interest in saving products because of availability of other higher yielding investments (Figure 8-3).
9. Numerical features in testing data should be normalized using the same parameters applied to testing data. Categorical features in testing data need to be handled more carefully. If a category within a categorical feature has not been seen in training data, the feature value is essentially unknown because the model has not been trained on data with that categorical value of the feature. If the feature is important, the model is likely to make an incorrect prediction. All such cases should be recorded and the model retrained using new training data with those categorical feature values present. If the feature is not important in predicting the outcome, its value can be set to a default value.

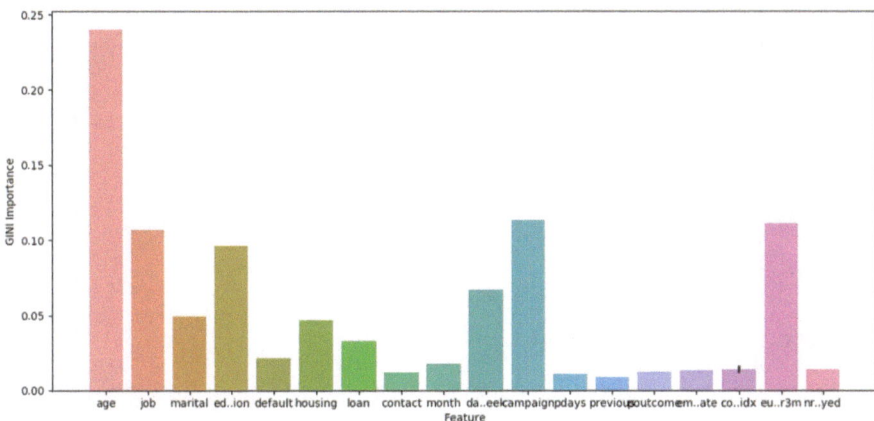

Figure 8-1. Gini Importance of Input Features

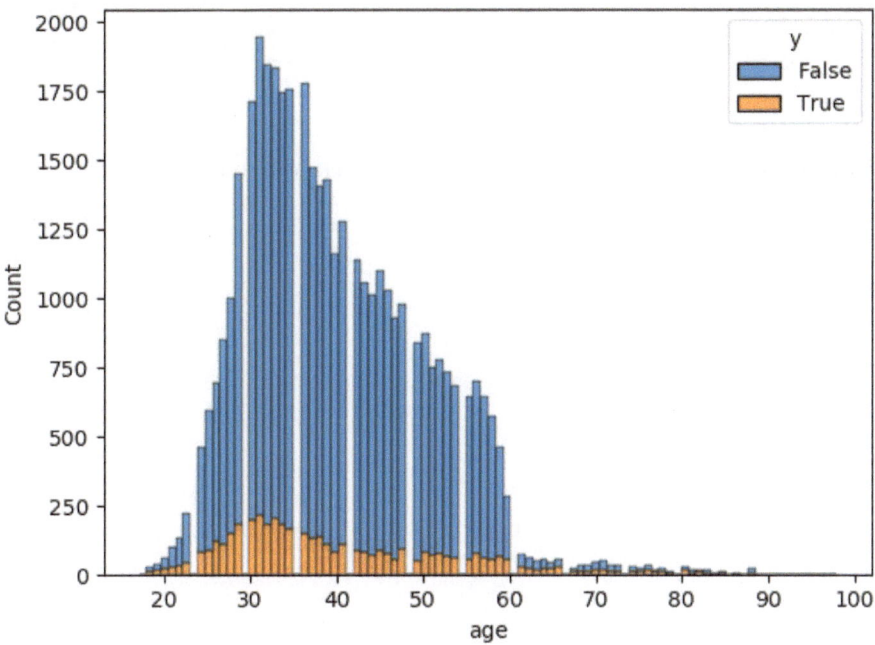

Figure 8-2. Histogram of Outcome on Age for Training Data

10. Check if the OOB score has converged to a stable value to determine the number of decision trees in the random forest. Some problems with large data variance may need larger number of trees. In this example, the plot of OOB score in Figure 8-4 against a number of trees shows that the error is stable at around 110 trees.

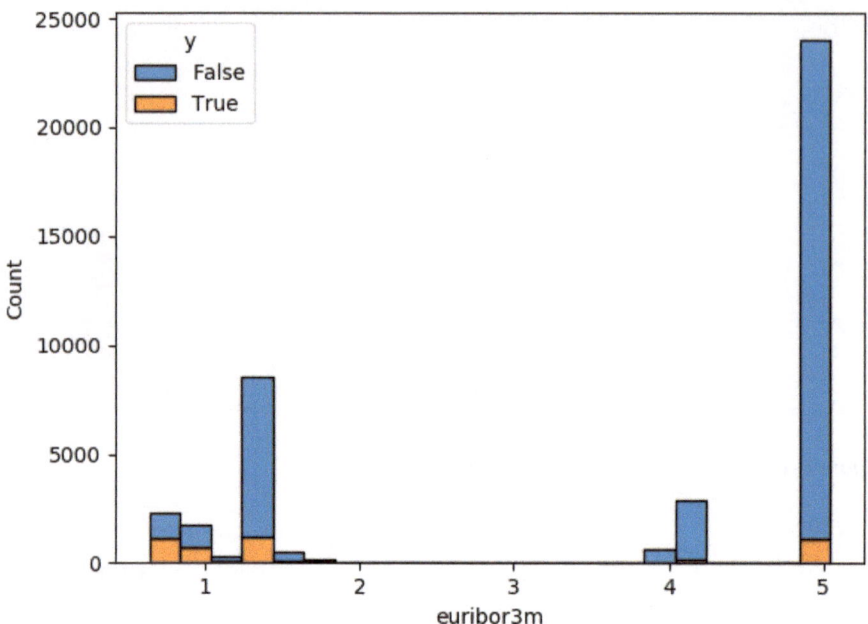

Figure 8-3. Histogram of Outcome on EURIBOR 3-Month Rate for Training Data

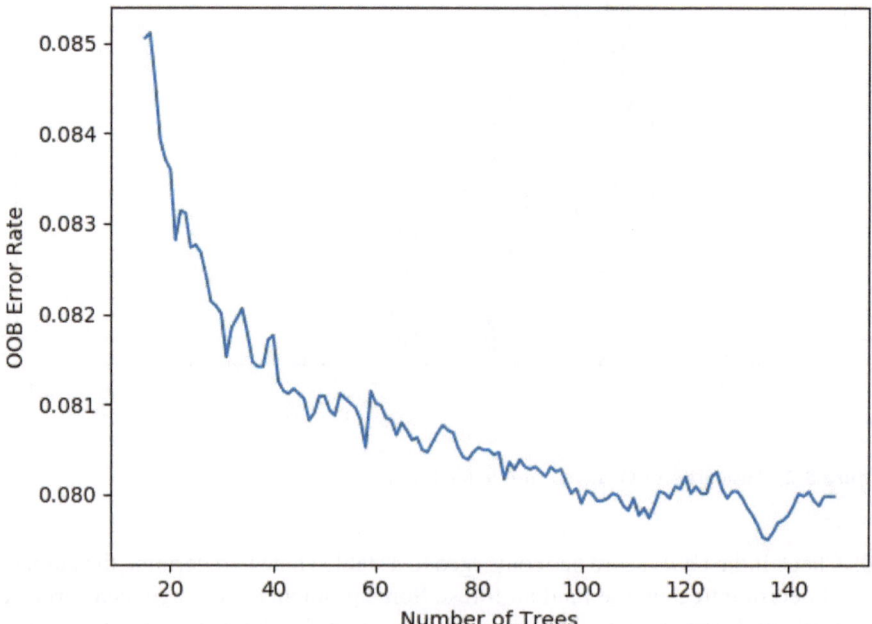

Figure 8-4. Convergence of Out-of-Bag Score with Increasing Number of Trees

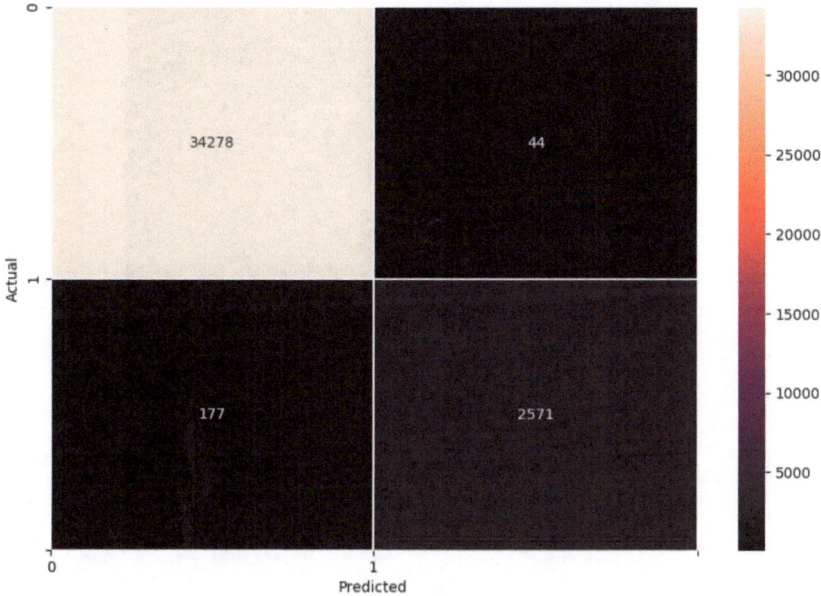

Figure 8-5. Confusion Matrix for Training Data

11. Plot the confusion matrix for both training (Figure 8-5) and testing (Figure 8-6) datasets. From the training data confusion matrix, we can see that the precision is $\frac{2571}{2571+44} \approx 98.3\%$, recall is $\frac{2571}{2571+177} \approx 93.6\%$, and accuracy is $\frac{2571+34278}{2571+34278+44+177} \approx 99.4\%$. From the testing data confusion matrix in Figure 8-6, the precision is 61.2%, recall is 16.4%, and accuracy is 56.8%. Performance metrics differ widely between training and test datasets. To understand why, let us look at the distribution of the outcome (result) across two features that have high Gini importance: **age** and **euribor3m**. From Figure 8-7, we can see that client response has changed markedly based on age between training and testing data. While in training data, respondents in age groups 30–35 were predominantly not interested in term deposits, in testing data, similar age respondents were interested in the product. Similarly, comparing the distribution of rates (**euribor3m**) and response in testing data, one can observe that rates were significantly lower in testing data, and there is a greater overall interest as seen in Figure 8-8. Perhaps the low interest rate environment observed in testing data changed preferences toward term deposit products. This illustrates the pitfall of assuming training and testing data is from the same underlying distribution, most likely, underlying distributions have changed. To mitigate this problem, additional features that capture changing customer preferences need to be included (such as growth of savings, investments, and wages). Additional training data may also be required.

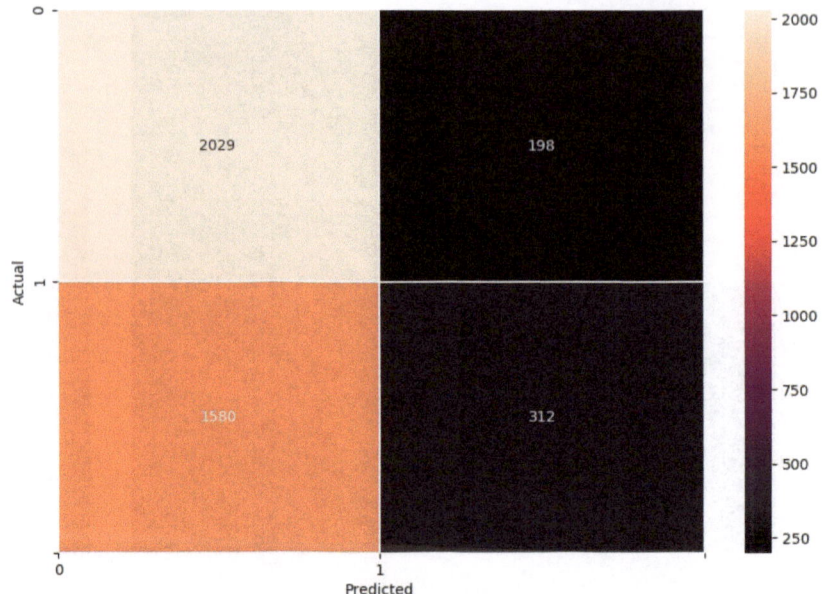

Figure 8-6. Confusion Matrix for Testing Data

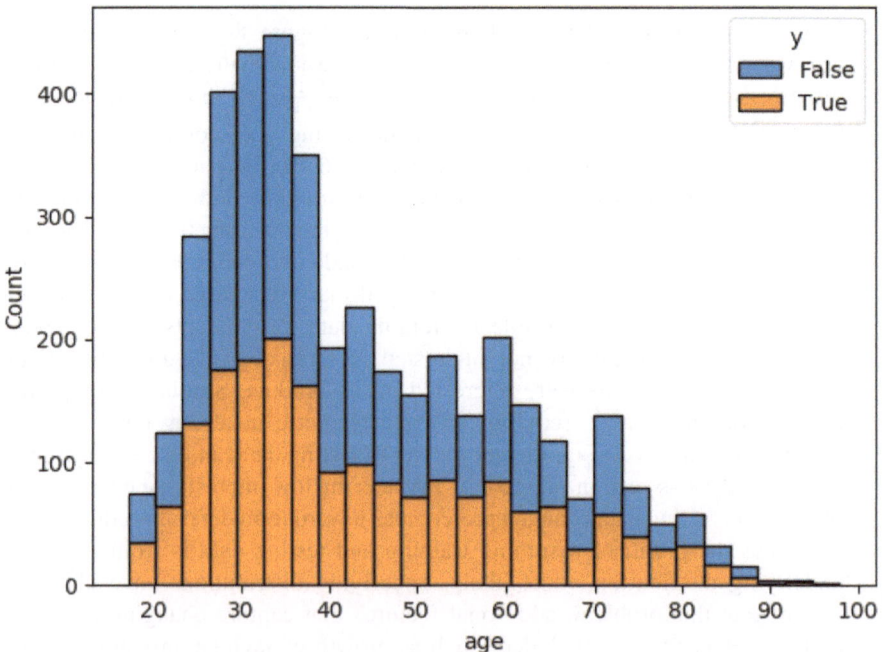

Figure 8-7. Histogram of Outcome on Age for Testing Data

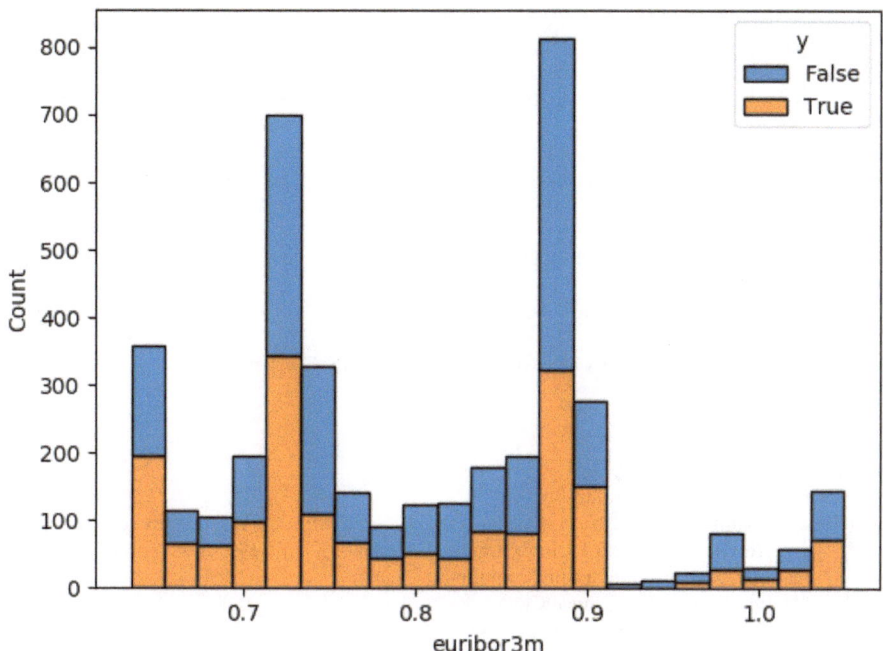

Figure 8-8. Histogram of Outcome on EURIBOR 3-Month Rate for Testing Data

The code for this example is shown in Listing 8-4.

Listing 8-4. Random Forest Classification for Banking Product

```
1   import numpy as np
2   import pandas as pd
3   import os
4   from sklearn.ensemble import RandomForestClassifier
5   import matplotlib.pyplot as plt
6   import seaborn as sns
7   from sklearn.metrics import confusion_matrix
8   import logging
9
10  logging.basicConfig(level=logging.DEBUG)
11
12
13  class BankingRF(object):
14      """ Predict if clients subscribes to a banking product (term deposit) using random
             forest """
15      LOGGER = logging.getLogger("BankingRF")
16
17      def __init__(self, datadir: str, filename: str = "bank-additional-full.csv",
             testing: float = 0.1,
18                   ntree: int = 100, fill_na_features =True, oob_score_convergence=False)
                     -> None:
```

```python
    """
    :param datadir : Data directory containing input file
    :param filename : File name containing the data
    :oaram testing : percentage of data to set aside as testing data
    :param ntree : Number of decision trees in random forest
    :param fill_na_features : Fill NA categorical features with default value
    :param oob_score_convergence: Plot OOB score convergence
    :rtype : None
    """
    df = pd.read_csv(os.path.join(datadir, filename), sep=";")
    excludeCols = ["duration"]
    df.drop(columns=excludeCols, inplace=True)
    self.resultCol = "y"
    df.loc[:, self.resultCol] = df.loc[:, self.resultCol].map({'yes': True, 'no':
        False})
    self.df = df

    nrows = self.df.shape[0]
    training = 1.0 - testing
    self.trainDf = self.df.loc[0: int(training * nrows), :].reset_index(drop=True)
    self.testDf = self.df.loc[int(training * nrows):, :].reset_index(drop=True)
    features = list(self.trainDf.columns)
    features.remove(self.resultCol)
    self.featureNames = features
    self.numericCols = []
    self.normalizeCols = {}
    self.categoricalCols = []
    self.categoricalMap = {}
    self._processColumns()
    self.fillNa = fill_na_features
    self.oobScore = oob_score_convergence
    self.model = RandomForestClassifier(n_estimators=ntree, random_state=0)
    self.trainModel()

def _processCategoricalCols(self, df: pd.DataFrame) -> None:
    """
    Process categorical columns by creating a mapping
    :param df: training dataframe
    :rtype : None
    """
    for col in self.categoricalCols:
        unique = np.sort(df.loc[:, col].unique())
        self.categoricalMap[col] = {u: i for i, u in enumerate(unique)}

def _applyCategoricalMapping(self, df: pd.DataFrame) -> pd.DataFrame:
    """
    Apply mappping to convert categorical columns to integers
    :rtype : pd.DataFrame with mapped categorical columns
    """
    for col in self.categoricalCols:
        df.loc[:, col] = df.loc[:, col].map(self.categoricalMap[col])
        return df
```

```python
def _normalizeNumericCols(self, trainingDf: pd.DataFrame) -> None:
    """
    Calclate normalizing params for numeric columns
    :param trainingDf:
    :return: None
    """
    for col in self.numericCols:
        mean = trainingDf.loc[:, col].mean()
        sd = trainingDf.loc[:, col].std()
        self.normalizeCols[col] = (mean, 2 * sd)

def _applyNormalization(self, df: pd.DataFrame) -> pd.DataFrame:
    """
    Apply normalization as col = (x-mean)/(2*sd)
    :param df:
    :return: df
    """
    for col in self.numericCols:
        mean, sd2 = self.normalizeCols[col]
        df.loc[:, col] = (df.loc[:, col].values - mean) / sd2
    return df

def _processColumns(self) -> None:
    """
    Process input columns from dataframe. Dataframe is in self.df
    :return: None
    """
    df = self.trainDf

    # identify categorical columns
    cols = list(df.columns)
    cols.remove(self.resultCol)
    for col in cols:
        if df.dtypes[col].name == "object":
            self.categoricalCols.append(col)
        else:
            self.numericCols.append(col)

    self._processCategoricalCols(df)
    self.trainDf = self._applyCategoricalMapping(self.trainDf)
    self.testDf = self._applyCategoricalMapping(self.testDf)

    self._normalizeNumericCols(df)
    self.trainDf = self._applyNormalization(self.trainDf)
    self.testDf = self._applyNormalization(self.testDf)

def _plotOOBError(self):
    estimators = range(15, 150)
    X = self.trainDf.loc[:, self.featureNames].values
    y = self.trainDf.loc[:, self.resultCol].values
    err = []
    for nest in estimators:
```

```
123        rf = RandomForestClassifier ( n_estimators =nest,  oob_score=True, random_state
                =0)
124        rf . fit (X, y)
125        err . append(1 − rf .oob_score_)
126        errdf = pd.DataFrame({"Number of Trees": list ( estimators ),  "OOB Error Rate
                ": err })
127    sns . lineplot (data=errdf, x="Number of Trees", y="OOB Error Rate")
128    plt .show()

130    def _plotConfusionMatrix( self ,  labels : np.ndarray,  predictions : np.ndarray)  −>
              None:
131        cm = confusion_matrix ( labels ,  predictions )
132        fig ,  ax = plt . subplots ( figsize =(10,  10))
133        sns .heatmap(cm, annot=True, fmt="d",  linewidths =0.25, ax=ax)
134        plt . xticks ([0,  1,  2])
135        plt . yticks ([0,  1,  2])
136        plt . ylabel ('Actual')
137        plt . xlabel ('Predicted')
138        plt .show()

140    def _calcMeasures( self )  −> None:
141        """ Calculate  and plot  measures  after   fitting   random  forest  """
142        importances = self .model. feature_importances_
143        std = np. std ([ tree . feature_importances_  for  tree  in  self .model. estimators_ ],
                axis=0)
144        shortFeatName = []
145        for  feature  in  self .featureNames:
146            if  len( feature ) > 8:
147                feature = feature [0:2] + ".." + feature [−3:]
148            shortFeatName.append( feature )
149        impdf = pd.DataFrame({"Feature":  shortFeatName,
150                                "Gini Importance":  importances ,
151                                "sd":  std })

153        ax = sns . barplot (data=impdf, x="Feature", y="Gini Importance")
154        ax. errorbar (data=impdf, x="Feature", y="Gini Importance",  ls='', lw='3', color
                ="black")
155        plt .show()

157        self ._plotOOBError()

159    def trainModel( self )  −> None:
160        """
161        Train the random forest  classifier  on training  dataset
162        : return :
163        """
164        X = self . trainDf . loc [:,  self .featureNames]. values
165        y = self . trainDf . loc [:,  self . resultCol ]. values
166        self .model. fit (X, y)
167        if  self .oobScore:
168            self ._calcMeasures()

170        ypred = self .model. predict (X)
```

```
171          self ._plotConfusionMatrix(y, ypred)
172
173          Xtest = self . testDf . loc [:, self . featureNames]. values
174          ytest = self . testDf . loc [:, self . resultCol ]. values
175
176          rowsWithNan = np.where(np.isnan(Xtest) . sum(axis=1)) [0]
177          if rowsWithNan.shape[0]:
178              self .LOGGER.info("Some categorical variables in test data were not present
                      in training !")
179              if self . fillNa :
180                  Xtest = np.nan_to_num(Xtest) # fill missing categorical variables with
                          0
181              else :
182                  rowsWithoutNan = np.array ([ i for i in range( Xtest .shape [0]) if i not in
                          set (rowsWithNan)])
183                  Xtest = Xtest [rowsWithoutNan, :]
184                  ytest = ytest [rowsWithoutNan]
185
186          testPred = self .model. predict (Xtest)
187
188          self ._plotConfusionMatrix( ytest , testPred )
189
190      def testModel( self , testDf: pd.DataFrame) −> None:
191          """
192          Test the model using provided testing data
193          :param testDf:
194          : return : None
195          """
196          if testDf is None:
197              testDf = self . testDf
198
199          X = testDf . loc [:, self . featureNames]. values
200          pred = self .model. predict (X)
201          y = testDf . loc [:, self . resultCol ]. values
202          self ._plotConfusionMatrix(y, pred)
203
204
205  if __name__ == "__main__":
206      rf = BankingRF(r"C:\prog\cygwin\home\samit_000\RLPy\data\book\bank−additional",
              ntree =110)
```

Generalized Method of Moments

9

The generalized method of moments, often abbreviated as GMM, is a method of solving a system of equations where exogenous variables are correlated with error terms and the probability distribution of data conditional on model parameters is intractable to compute. The method can be used for linear or non-linear system of equations. In the context of linear equations, we observed in an earlier chapter that the existence of correlation between exogenous variables and error term causes parameter estimates from OLS to be biased and inconsistent. This necessitated the use of instrumental variables. These variables are dependent on exogenous variables and uncorrelated with error terms. In the context of linear system of equations, GMM can be viewed as a generalization of the method of instrumental variables for estimating model parameters.

GMM was proposed by [30] in a paper titled "Large Sample Properties of Generalized Method of Moments Estimators." GMM has been widely adopted by economists and finance practitioners, as witnessed by the long list of research papers leveraging the method, for example, [31–33], and [34].

9.1 GMM for Linear Equations

Let us formulate the generalized method of moments (GMM) for a linear equation. Let us assume we have the governing equation as shown in Equation 9-1. Here, $y(t)$ is a vector of size N, $\mathbf{X}$ is a two-dimensional array of shape $N \times M$, and β is a one-dimensional vector of size M. We have some column vectors in the matrix of exogenous variables with non-zero correlation with the error term, $\epsilon(t)$. We know that estimates of β obtained using OLS are biased and inconsistent, and we must resort to using instrumental variables.

$$y(t) = \mathbf{X(t)}\beta + \epsilon(t) \text{ where } \mathbf{X(t)} = [\mathbf{X_1(t)}, \mathbf{X_2(t)}, \cdots, \mathbf{X_M(t)}]$$

$$E\left[X_m(t)\epsilon(t)\right] \neq 0 \text{ for some } m \in [1, 2, \cdots, M]$$

(9-1)

© Samit Ahlawat 2025
S. Ahlawat, *Statistical Quantitative Methods in Finance*,
https://doi.org/10.1007/979-8-8688-0962-0_9

Let us assume we have a set of K instrumental variables, $\mathbf{Z(t)}$, that are dependent on exogenous variables $\mathbf{X(t)}$ but independent of the error term $\epsilon(t)$, as shown in Equation 9-2. $\mathbf{Z(t)}$ is an $N \times K$ matrix with K columns or instrumental variables. N represents the number of observations. Let us also assume that $\mathbf{Z(t)}$ is an **ergodic process**. This means that the mean of $\mathbf{Z(t)}$ is a constant; all its auto-covariances are constant as is their sum in the limit when $t \to \infty$. This condition is shown in Equation 9-3.

$$E\left[\mathbf{Z(t)}\epsilon(t)\right] = 0 \text{ for all K instrumental variables}$$
$$E\left[\mathbf{Z(t)}\left(y(t) - \mathbf{X(t)}\boldsymbol{\beta}\right)\right] = 0 \tag{9-2}$$

$$E\left[\mathbf{Z(t)}\right] = \lim_{t \to \infty} \frac{1}{T} \sum_{t=1}^{T} \mathbf{Z(t)} = \boldsymbol{\mu}$$

$$E\left[\mathbf{Z(t)Z(t-c)}\right] = \lim_{t \to \infty} \frac{1}{T} \sum_{t=1}^{T} \left(\mathbf{Z(t)} - \boldsymbol{\mu}\right)^T \left(\mathbf{Z(t-c)} - \boldsymbol{\mu}\right) = \boldsymbol{\Sigma(c)} \tag{9-3}$$

$$\sum_{q=1}^{\infty} \boldsymbol{\Sigma(q)} = \boldsymbol{\Sigma} \text{ a finite value}$$

Expanding Equation 9-2, we can write Equation 9-4, where $E[\mathbf{Z(t)}y(t)] \equiv \boldsymbol{\Sigma_{z,y}}$ and $E[\mathbf{Z(t)X(t)}] \equiv \boldsymbol{\Sigma_{z,x}}$. Equation 9-2 comprises a set of K equations in M unknowns, namely, the elements of parameter vector β. We have N equations for each set, corresponding to the number of observations. Because we have K independent instrumental variables $\mathbf{Z}$, the rank of $\boldsymbol{\Sigma_{z,x}}$ is K.

$$E[\mathbf{Z_i(t)}y(t)] = E[\mathbf{Z_i(t)X(t)}]\boldsymbol{\beta} \text{ for i} \in (1, 2, \cdots, K)$$

$$\boldsymbol{\Sigma_{z_i}, y} = \boldsymbol{\Sigma_{z_i}, x}\boldsymbol{\beta}$$

$$\boldsymbol{\Sigma_z, y} = \boldsymbol{\Sigma_z, x}\boldsymbol{\beta} \text{ putting all equations for i as rows}$$

$$\boldsymbol{\Sigma_z, y} = (K \times 1) \text{ vector}$$

$$\boldsymbol{\Sigma_z, x} = (K \times M) \text{ matrix}$$
$$\tag{9-4}$$
$$\boldsymbol{\beta} = (M \times 1) \text{ vector}$$

$$\text{rank} (\boldsymbol{\Sigma_z, x}) = K$$

$$\text{GMM handles } K > M$$

$$\text{Use K OLS solvers for } K = M$$

$$\text{No solution for } K < M \text{ , we need more instrumental variables}$$

If $K = M$, we can solve for all M parameters in $\boldsymbol{\beta}$ by applying OLS to each of K equations. If $K < M$, we cannot determine all parameters. If $K > M$, we cannot determine all parameters uniquely. This is the case handled by GMM. Moreover, when $K = M$, the OLS solution reduces to GMM estimate, and the two methods produce identical results. Therefore, we will consider the general case of $K \geq M$ for the GMM solution.

For the case where $K \geq M$, GMM minimizes the **Mahalanobis** distance $\left(\boldsymbol{\Sigma}_{z_i,y} - \boldsymbol{\Sigma}_{z_i,x}\boldsymbol{\beta}\right)^T \mathbf{W} \left(\boldsymbol{\Sigma}_{z_i,y} - \boldsymbol{\Sigma}_{z_i,x]}\boldsymbol{\beta}\right)$ for each i. In order to calculate the Mahalanobis distance, we need a symmetric, positive-definite matrix $\mathbf{W}$ with dimensions (K, K). This is shown in Equation 9-5.

$$\min_{\boldsymbol{\beta}} \left(\boldsymbol{\Sigma}_{z_i,y} - \boldsymbol{\Sigma}_{z_i,x}\boldsymbol{\beta}\right)^T \mathbf{W} \left(\boldsymbol{\Sigma}_{z_i,y} - \boldsymbol{\Sigma}_{z_i,x}\boldsymbol{\beta}\right)$$

For each $i \in (1, 2, \cdots, K)$ $\qquad\qquad$ (9-5)

$$\therefore \min_{\boldsymbol{\beta}} \left(\boldsymbol{\Sigma}_{z,y} - \boldsymbol{\Sigma}_{z,x}\boldsymbol{\beta}\right)^T \mathbf{W} \left(\boldsymbol{\Sigma}_{z,y} - \boldsymbol{\Sigma}_{z,x}\boldsymbol{\beta}\right)$$

Differentiating Equation 9-5 with respect to $\boldsymbol{\beta}$ and setting to 0 yields the value of parameters $\boldsymbol{\beta}$, as shown in Equation 9-6.

$$\boldsymbol{\beta} = \left(\boldsymbol{\Sigma}_{z,x}^T \mathbf{W} \boldsymbol{\Sigma}_{z,x}\right)^{-1} \boldsymbol{\Sigma}_{z,x}^T \mathbf{W} \boldsymbol{\Sigma}_{z,y} \qquad\qquad (9\text{-}6)$$

Equation 9-6 has matrix $\mathbf{W}$ which we have not specified so far. GMM selects $\mathbf{W}$ in order to minimize the variance of $\boldsymbol{\beta}$. The variance of $\boldsymbol{\beta}$ can be calculated as shown in Equation 9-7.

$$\text{Variance}(\boldsymbol{\beta}) = \left(\boldsymbol{\Sigma}_{z,x}^T \mathbf{W} \boldsymbol{\Sigma}_{z,x}\right)^{-1} \boldsymbol{\Sigma}_{z,x}^T \mathbf{W} \mathbf{Z} \boldsymbol{\epsilon} \mathbf{Z} \boldsymbol{\epsilon}^T \mathbf{W} \left(\boldsymbol{\Sigma}_{z,x}^T \mathbf{W} \boldsymbol{\Sigma}_{z,x}\right)^{-1}$$

$$\min_{\mathbf{W}} \text{Variance}(\boldsymbol{\beta})$$

$$\implies \mathbf{W} = \left(\mathbf{Z} \boldsymbol{\epsilon} \boldsymbol{\epsilon}^T \mathbf{Z}^T\right)^{-1}$$

$\qquad\qquad\qquad\qquad\qquad\qquad\qquad\qquad\qquad (9\text{-}7)$

Denote $\left(\mathbf{Z} \boldsymbol{\epsilon} \boldsymbol{\epsilon}^T \mathbf{Z}^T\right) \equiv \mathbf{S}$

$$\min \text{variance}(\boldsymbol{\beta}) = \left(\boldsymbol{\Sigma}_{z,x}^T \mathbf{S}^{-1} \boldsymbol{\Sigma}_{z,x}\right)^{-1}$$

where $\boldsymbol{\epsilon} = y - \mathbf{X}\boldsymbol{\beta}$

Equations 9-6 and 9-7 describe the GMM solution for a linear system. In order to estimate the parameters, $\boldsymbol{\beta}$, one requires an estimate of $\mathbf{W}$ according to Equation 9-6. However, the estimation of $\mathbf{W}$ requires the calculation of residuals, $\boldsymbol{\epsilon} = y - \mathbf{X}\boldsymbol{\beta}$, which requires an estimate of parameters $\boldsymbol{\beta}$ that depends on $\mathbf{W}$.

This necessitates an iterative procedure where one assumes an initial value of $\mathbf{W}$, computes $\boldsymbol{\beta}$, and updates the value of $\mathbf{W}$. This iterative procedure describes GMM, as shown in pseudo-code 16.

Algorithm 16 GMM Iterative Estimation for Linear System

Require: Data y, $\mathbf{X}$, instrument variables $\mathbf{Z}$, threshold t for deciding when to terminate iterations.
1: Assume $\mathbf{W} = \mathbf{I}$ or $\boldsymbol{\Sigma}_{\mathbf{x,x}}^{-1}$.
2: Set $\delta = 1$.
3: **for** iter $= 1, 2, \ldots$, while $\delta > t$ **do**
4:     Calculate the parameter estimate, $\boldsymbol{\beta}$, using Equation 9-6.
5:     Calculate residuals, $\boldsymbol{\epsilon} = y - \mathbf{X}\boldsymbol{\beta}$.
6:     Calculate the optimum value of weighing matrix $\mathbf{W}$ that gives minimum variance for parameter estimate. Use Equation 9-7.

$$\mathbf{W} = \left(\mathbf{Z}\boldsymbol{\epsilon}\boldsymbol{\epsilon}^T \mathbf{Z}^T \right)^{-1} \tag{9-8}$$

7:     Calculate the change $\delta = \| \mathbf{W}(\mathbf{iter}) - \mathbf{W}(\mathbf{iter} - 1) \|^2$. Terminate iterations when δ falls below threshold t.
8: **end for**

A key step in pseudo-code 16 is to compute the optimum value of weighing matrix $\mathbf{W}$. This requires the calculation of $\mathbf{S} = \left(\mathbf{Z}\boldsymbol{\epsilon}\boldsymbol{\epsilon}^T \mathbf{Z}^T \right)$. If the errors, $\boldsymbol{\epsilon}$, are homoskedastic, i.e., have constant variance, a consistent estimator of $\mathbf{S}$ is shown in Equation 9-9.

$$\mathbf{S} = \sigma^2 E \left[\mathbf{Z}^T \mathbf{Z} \right]$$

$$\text{where } E \left[\boldsymbol{\epsilon}^T \boldsymbol{\epsilon} \right] = \sigma^2 \mathbf{I} \tag{9-9}$$

$$\text{if } \epsilon(t) \text{ has constant variance}$$

However, if $\epsilon(t)$ is heteroskedastic, we need to use White's heteroskedasticity consistent estimator as described by White [1]. The estimator is shown in Equation 9-10.

If ϵ is heteroskedastic with time-varying variance but zero autocorrelation

$$\mathbf{S} = \frac{1}{N} \left(\frac{1}{N} \sum_{t=1}^{N} \mathbf{Z(t)Z(t)}^T \right)^{-1} \left(\frac{1}{N} \sum_{t=1}^{N} \mathbf{Z(t)Z(t)}^T \epsilon(t)^2 \right)$$

$$\times \frac{1}{N} \left(\frac{1}{N} \sum_{t=1}^{N} \mathbf{Z(t)Z(t)}^T \right)^{-1}$$

(9-10)

If the error terms, ϵ, have heteroskedasticity and autocorrelation, we must use the Newey-West heteroskedasticity and autocorrelation (HAC) consistent estimates, shown in Equation 9-11.

$$\mathbf{S}_{HAC} = \frac{1}{N} \sum_{t=1}^{N} w_{t,n} \left(\mathbf{\Gamma}(t) + \mathbf{\Gamma}(t)^T \right)$$

(9-11)

$\mathbf{\Gamma}(t) = $ Lag t autocorrelation estimated using

heteroskedasticity consistent estimator

9.2   GMM for Non-linear Equations

If the underlying equations describing the system are non-linear in the relationship between endogenous and exogenous variables, we must use GMM for non-linear equations. The expressions are analogous to those obtained for linear equations once we linearize the non-linear moment conditions.

Let us suppose we have a non-linear relation between exogenous and endogenous variables, as shown in Equation 9-12.

$$g\left(y(t), \mathbf{X(t)}, \boldsymbol{\beta}\right) = \epsilon(t) \qquad (9\text{-}12)$$

Proceeding like we did for the case of linear equations, moment conditions can be written as shown in Equation 9-13 using a set of instrumental variables, $\mathbf{Z(t)}$.

$$E\left[\mathbf{Z(t)}\epsilon(t)\right] = 0$$

$$E\left[\mathbf{Z(t)}g\left(y(t), \mathbf{X(t)}, \boldsymbol{\beta}\right)\right] = 0$$

(9-13)

Linearizing Equation 9-13 in the neighborhood of a solution to the non-linear Equation 9-12, we obtain Equation 9-14. In Equation 9-14, β_0 is assumed to be a solution to Equation 9-13.

$$E\left[\mathbf{Z(t)}\left(g\left(y(t), \mathbf{X(t)}, \beta_0\right) + \frac{\partial g\left(y(t), \mathbf{X(t)}, \beta\right)}{\partial\beta}|_{\beta=\beta_0}\delta\beta\right)\right] = 0$$

$$\implies E\left[\mathbf{Z(t)}\left(g\left(y(t), \mathbf{X(t)}, \beta_0\right) + \right.\right.$$
$$\left.\left. \frac{\partial g\left(y(t), \mathbf{X(t)}, \beta\right)}{\partial\beta}|_{\beta=\beta_0}\left(\beta - \beta_0\right)\right)\right] = 0$$

$$E\left[\mathbf{Z(t)}\left(g\left(y(t), \mathbf{X(t)}, \beta_0\right) - \frac{\partial g\left(y(t), \mathbf{X(t)}, \beta\right)}{\partial\beta}|_{\beta=\beta_0}\beta_0 + \right.\right.$$
$$\left.\left. \frac{\partial g\left(y(t), \mathbf{X(t)}, \beta\right)}{\partial\beta}|_{\beta=\beta_0}\beta\right)\right] = 0$$

(9-14)

β_0 being a solution to non-linear system

Comparing Equation 9-14 to its linear cousin, Equation 9-4, $E[\mathbf{Z_i(t)}\left(y(t) - \mathbf{X(t)}\beta\right)] = 0$, we notice that setting y and $\mathbf{X}$ as shown in Equation 9-15, we can transform the non-linear moment equations to linear moment equations. From this point on, we can use the corresponding expressions from the last section for the GMM estimator after making the substitutions shown in Equation 9-15.

Replace $y(t)$ from the linear case with

$$g\left(y(t), \mathbf{X(t)}, \beta_0\right) - \frac{\partial g\left(y(t), \mathbf{X(t)}, \beta\right)}{\partial\beta}|_{\beta=\beta_0}\beta_0$$

(9-15)

Replace $\mathbf{X(t)}$ from the linear case with

$$-\frac{\partial g\left(y(t), \mathbf{X(t)}, \beta\right)}{\partial\beta}|_{\beta=\beta_0}$$

In the above equations, we need to know β_0 a priori. We incorporate it in the iterative procedure shown in pseudo-code 17.

9.3   Hansen J Test

The Hansen J test—also known as the Sargan-Hansen J test after [39] who initially proposed it and [30] who adapted it to GMM—is a test to establish the validity of moment conditions used in an overdetermined GMM system of equations. This test is only valid when K, or the number of moment conditions, is strictly greater than the number of free parameters, M.

Algorithm 17 GMM Iterative Estimation for Non-linear System

Require: Data y, $\mathbf{X}$, instrument variables $\mathbf{Z}$, threshold t for deciding when to terminate iterations.
1: Assume $\mathbf{W} = \mathbf{I}$ or $\mathbf{\Sigma}_{\mathbf{x},\mathbf{x}}^{-1}$.
2: Assume a value of $\boldsymbol{\beta_0}$.
3: Set $\delta = 1$.
4: **for** iter $= 1,2,...$, while $\delta > t$ **do**
5:      Compute the parameters shown in Equation 9-15 to complete the correspondence between linear and non-linear GMM equations.
6:      Calculate the parameter estimate, $\boldsymbol{\beta}$, using Equation 9-6.
7:      Calculate residuals, $\boldsymbol{\epsilon} = y - \mathbf{X}\boldsymbol{\beta}$, i.e., the linearized form of non-linear equations.
8:      Calculate the optimum value of weighing matrix $\mathbf{W}$ that gives minimum variance for the parameter estimate. Use Equation 9-7.

$$\mathbf{W} = \left(\mathbf{Z}\boldsymbol{\epsilon}\boldsymbol{\epsilon}^T\mathbf{Z}^T\right)^{-1} \tag{9-16}$$

9:      Calculate the change $\delta = \|\mathbf{W}(\text{iter}) - \mathbf{W}(\text{iter} - 1)\|^2$. Terminate iterations when δ falls below threshold t.
10: **end for**

GMM attempts to minimize the Mahalanobis norm of moment expression. Norms being non-negative, the minimum value they can take is zero. This provides the basis for the test. The test statistic is shown in Equation 9-17 and is a consistent expression for evaluating Equation 9-5.

$$J \text{ statistic} \equiv \frac{1}{N}\sum_{i=1}^{N} g\left(y_i, \mathbf{X}_i, \boldsymbol{\beta}\right)^T \hat{\mathbf{W}} \equiv \frac{1}{N}\sum_{i=1}^{N} g\left(y_i, \mathbf{X}_i, \boldsymbol{\beta}\right) \tag{9-17}$$

Under certain regularity conditions and as $N \rightarrow \infty$, J statistic converges to a χ^2 distribution with $K - M$ degrees of freedom, as shown in Equation 9-18.

$$J \text{ statistic} \sim \chi^2\left(K - M\right) \tag{9-18}$$

If values are within the range of a $\chi^2\left(K - M\right)$ distribution using a prespecified confidence interval, the null hypothesis that moment conditions are zero cannot be rejected. This is considered an evidence supporting the plausibility of moment conditions and, consequently, of the validity of parameter estimates. On the other hand, if the converse is true, the null hypothesis is rejected, and some moment conditions may not be true according to the data.

9.4   Application

In this section, let us apply GMM to estimate the parameters of the CEV model— an acronym for constant elasticity of variance. The model is frequently used in commodity markets. It posits a relationship between the evolution of an asset price as a function of drift and random fluctuations. The model is shown in Equation 9-19.

$$dS_t = \mu S_t dt + \sigma S_t^{\gamma} dW_t$$

$$W_t = N(0, 1) \text{ standard normal distribution}$$

(9-19)

In Equation 9-19, S_t denotes the asset price at time t, μ denotes the drift or diffusion coefficient, and σ is a constant that contributes to volatility. γ is another constant that contributes to volatility. The CEV model assumes that volatility is influenced by the level of asset price. For values of $\gamma > 1$, higher volatility is observed when asset price is high. Conversely, volatility declines when asset price falls. On the other hand, if $\gamma < 1$, volatility rises as prices fall. This can be seen by writing an expression for volatility, as shown in Equation 9-20, assuming the CEV model governs asset price evolution. Due to this behavior, the CEV model is sometimes referred to as the local-volatility model.

$$d \log S_t = \frac{1}{S_t} dS_t - \frac{1}{2S_t^2} (dS_t)^2 + \cdots$$

$$= \left(\mu - \sigma^2 S_t^{2(\gamma-1)} \right) dt + \sigma S_t^{\gamma-1} dW_t$$

$$\implies \text{mean} (d \log S_t) = E \left[d \log S_t \right] = \left(\mu - \sigma^2 S_t^{2(\gamma-1)} \right) dt$$

(9-20)

$$E \left[(d \log S_t - \text{mean} (d \log S_t))^2 \right] = \sigma^2 S_t^{2(\gamma-1)} dt$$

$$\text{variance} = \sigma^2 S_t^{2(\gamma-1)} dt$$

Let us estimate the values of the CEV model's parameters while fitting it to S&P 500 prices. There are three model parameters: μ, σ, and γ. Let us first derive the moment equations, $g(\theta, X)$. Rewriting the CEV model in discrete form, we get the first two equations using instrument variables 1 and S_t as shown in Equations 9-21 and 9-22.

$$S_{t+\Delta t} - S_t = \mu S_t \Delta t + \sigma S_t^{\gamma} \Delta W_t$$

$$S_{t+\Delta t} - S_t - \mu S_t \Delta t = \sigma S_t^{\gamma} \Delta W_t$$

$$E_t [S_{t+\Delta t} - S_t - \mu S_t \Delta t] = 0$$

(9-21)

$$\text{Using 1 as instrument variable}$$

$$E_t [S_t (S_{t+\Delta t} - S_t - \mu S_t \Delta t)] = E_t \left[\sigma S_t^{\gamma+1} \Delta W_t \right]$$

$$= \sigma S_t^{\gamma+1} E_t [\Delta W_t] = 0$$

(9-22)

$$\text{Using } S_t \text{ as instrument variable}$$

To derive a third moment condition, we can use the asset price at time $t-1$ as an instrument variable. The moment condition is shown in Equation 9-23.

$$E_t \left[S_{t-1} \left(S_{t+\Delta t} - S_t - \mu S_t \Delta t \right) \right] = E_t \left[\sigma S_{t-1} S_t^{\gamma} \Delta W_t \right]$$

$$= \sigma S_{t-1} S_t^{\gamma+1} E_t [\Delta W_t] = 0 \qquad (9\text{-}23)$$

Using S_{t-1} as instrument variable

It should be noted that the expectation of moment conditions 9-21, 9-22, and 9-23 is taken at time t. There, we know all the variables at time t and prior to t. Also, let μ denote the drift in price annually. Therefore, $\Delta t = \frac{1}{251}$ because a year has roughly 251 trading days. Δt represents one day because we have end-of-day prices.

We have three moment conditions, but it is still not sufficient for GMM because none of the three conditions include σ or γ. Let us derive a fourth moment condition using the expression for volatility in Equation 9-20. The derivation of the fourth moment condition is shown in Equation 9-24.

$$E \left[(d \log S_t - \text{mean}\,(d \log S_t))^2 \right] = \sigma^2 S_t^{2(\gamma-1)} dt$$

$$E \left[(d \log S_t - \text{mean}\,(d \log S_t))^2 - \sigma^2 S_t^{2(\gamma-1)} dt \right] = 0$$

$$E \left[(\log S_{t+\Delta t} - \log S_t - \text{mean}\,(d \log S_t))^2 - \sigma^2 S_t^{2(\gamma-1)} \Delta t \right] = 0 \qquad (9\text{-}24)$$

$$E \left[\left(\log \frac{S_{t+\Delta t}}{S_t} - \text{mean}\left(\log \frac{S_{t+\Delta t}}{S_t} \right) \right)^2 - \sigma^2 S_t^{2(\gamma-1)} \Delta t \right] = 0$$

Using 1 as instrument variable

Using the abovementioned four moment conditions that include the three parameters, we are in a position to use GMM to determine the parameters. The output from the model along with estimated parameter values is shown in Listing 9-1. As seen from the output, the value of γ is 0.6821. Being less than 1, it comports with the observation that market volatility increases when prices fall. Certain confidence intervals and standard errors are reported as not a number (nan) because standard errors are too small. The p-value of the Hansen J test is 0.0523. Consequently, we cannot reject the null hypothesis at 95% confidence level that moment conditions are valid and are satisfied by the data.

Listing 9-1. GMM Results for Fitting S&P 500 Returns to CEV Model

```
1  Optimization terminated successfully.
2  Current function value: 0.000621
3  Iterations: 1
4  Function evaluations: 3
5  Gradient evaluations: 3
6  INFO:CEVModel:                          CEVModel Results
```

```
7   ==================================================================
8   Dep. Variable: y              Hansen J:      3.768
9   Model:         CEVModel  Prob (Hansen J): 0.0523
10  Method:        GMM
11  Date:          Tue, 20 Aug 2024
12  Time:          16:01:04
13  No. Observations:6065
14  ==================================================================
15          coef   std err   z     P>|z|    [0.025    0.975]
16  ------------------------------------------------------------------
17  mu      0.0704 0.036   1.944  0.052   -0.001    0.141
18  sigma   0.9704 nan     nan    nan      nan      nan
19  gamma   0.6821 nan     nan    nan      nan      nan
20  ==================================================================
```

The code for fitting the CEV model using GMM is shown in Listing 9-2.

Listing 9-2. Fitting Constant Elasticity of Variance Model to S&P 500 Prices Using GMM

```
1   import pandas as pd
2   import numpy as np
3   from statsmodels.sandbox.regression.gmm import GMM
4   import logging
5   import os
6
7   logging.basicConfig(level=logging.DEBUG)
8
9
10  class CEVModel(GMM):
11      PRICE = "Close"
12      DELTAT = 1.0/251.0
13      logger = logging.getLogger("CEVModel")
14      N_MOMS = 4
15      N_PARAMS = 3
16
17      @staticmethod
18      def calculateInstrumentalVars(dirname, security):
19          CEVModel.logger = logging.getLogger("CEVModel")
20          df = pd.read_csv(os.path.join(dirname, f"{security}.csv"), parse_dates=["Date"
                ])
21          prices = df.loc[:, CEVModel.PRICE].values
22          exog = np.column_stack((prices[0:-2], prices[1:-1], prices[2:])) # (S(t-1), S(t
                ), S(t+1))
23          endog = np.zeros(exog.shape[0], dtype=np.float32) # dummy
24          const = np.ones(exog.shape[0], dtype=np.int8)
25          instruments = np.column_stack((const, prices[0:-2], prices[1:-1])) # (1, S(t
                -1), S(t))
26          return endog, exog, instruments
27
28      def momcond(self, params):
29          mu, sigma, gamma = params
30          x = self.exog
31          z = self.instrument
```

```
32      gtheta  = x [:, 2] − x [:, 1] − mu ∗ x [:, 1] ∗ CEVModel.DELTAT
33      moment = np.multiply(z, gtheta [:, np.newaxis])
34      logReturn  = np.log(x [:, 2] / x [:, 1])
35      meanLogRet = np.mean(logReturn)
36      volat  = (sigma ** 2) ∗ (x [:, 1] ∗∗ (2∗(gamma − 1))) ∗ CEVModel.DELTAT
37      fourthMoment = ((logReturn − meanLogRet) ∗∗ 2) − volat
38      moment = np.column_stack((moment, fourthMoment))
39      return  moment
40
41  def fitCEV( self ):
42      params = np.array ([0,  0.0001,  0.7])
43      result  = super (). fit (params, maxiter=100, optim_method='bfgs', weights_method=
                'hac', wargs=dict( centered=False, maxlag=1))
44      CEVModel.logger.info( result .summary(xname=['mu', 'sigma', 'gamma']))
45
46
47
48  if __name__ == "__main__":
49      dirname = r"C:\prog\cygwin\home\samit_000\latex \ book_stats \code\data"
50      security  = "SPY"
51      endog, exog,  instruments  = CEVModel.calculateInstrumentalVars (dirname,  security )
52      model1 = CEVModel(endog, exog, instruments,  k_moms=CEVModel.N_MOMS, k_params
                =CEVModel.N_PARAMS)
53      model1.fitCEV()
```

9.4.1 Code Explanation

Let us do a code walk-through for the code presented in Listing 9-2.

1. Before stepping through the execution order, let us first examine the code design:

   - Class **CEVModel** derives from the base class **GMM**. The base class is defined inside the **statsmodels** library, in the package **statsmodels.sandbox.regression.gmm**.
   - Class **GMM** requires deriving classes to implement a method **momcond** that computes the moment conditions used by the GMM estimator. It receives model parameters as an argument. In our case, we have three model parameters. So the argument will be a three-element tuple.
   - Class **CEVModel** specifies class-level variables that define the number of moment conditions **N_MOMS** and the number of parameters **N_PARAMS**.
   - The constructor of base class **GMM** accepts endogenous, exogenous, and instrument variables as arguments, but these variables are not used anywhere. They can be accessed as instance attributes. This makes them a convenient location to keep variables that will be needed in the computation of moment conditions.

2. The code begins with calling a static method **calculateInstrumentalVars** defined inside the class **CEVModel**. This method calculates and returns endogenous, exogenous, and instrument variables. These variables are required by the constructor of the base class.
3. An exogenous variable matrix is a two-dimensional matrix with $S_{t-\Delta t}$, S_t, and $S_{t+\Delta t}$ as columns.
4. An endogenous variable is a dummy array of zeros. It is not used anywhere.
5. An instrument variable is a two-dimensional matrix with 1, $S_{t-\Delta t}$, and S_t as columns.
6. Method **fitCEV** is called. This method does the following:

   - Create a three-element parameter vector and initialize it.
   - Call the **fit** method of the base class. This method is provided with the following arguments:
     - Initial values of parameters.
     - Maximum number of GMM iterations.
     - Optimization method. Here, we use the BFGS method.
     - Method for computing weight matrix. Because market returns are known to be heteroskedastic and may have autocorrelation, we use the heteroskedasticity and autocorrelation consistent weight matrix. This is specified by providing an argument "HAC."
   - The method for fitting GMM gets the moment conditions defined inside the method **momcond**. This method computes the four moment conditions defined in Equations 9-21, 9-22, 9-23, and 9-24.
   - After fitting the model, it prints summary statistics with the computed parameter values.

Benchmarking Machine Learning Models 10

Machine learning models are characterized by a rich variety of parameters that impart them their functional versatility of modeling vast domains of data with minimal handcrafted features. These models obviate the need for researchers to craft a different set of features, adjust and recalibrate the model, and reexamine the functional adequacy of the underlying model in the face of evolving data. However, this versatility often comes at a price – the pitfall of overfitting on training datasets but poor performance on test datasets. Notwithstanding the fact that some machine learning models, such as random forests, are not prone to overfitting, the vast majority of machine learning models featuring deep neural networks must be scrupulously tested to make sure they are not overfitting. This is particularly true of some of the large language models that have billions of free parameters.

Statistical models, on the other hand, are characterized by clear equations and parameters. Their properties can be studied by analytical and numerical means alike, and they often furnish confidence intervals and mathematical estimates on goodness-of-fit that can provide a higher degree of confidence in their out-of-sample performance. This modeling simplicity, however, comes at a cost. The models require extensive tuning, data processing, and often handcrafting of features (exogenous variables). They also require strong data distribution assumptions, such as the underlying data generation process conforming with a family of analytically tractable probability distributions – Gaussian, Poisson, gamma, and inverse gamma, to name a few. These assumptions may not always be true, even if they hold in training data. This could lead to unexpectedly low accuracy in certain test datasets, necessitating model redesign.

These stylistic observations regarding strengths and weaknesses of machine learning and statistical models pose the question if the two can be used collaboratively. One paradigm where such a synergistic collaboration is possible is during the phase of benchmarking machine learning models in order to provide a reference against which their performance can be judged. If the machine learning model fails to outperform the statistical model in a meaningful manner, it may highlight an

© Samit Ahlawat 2025
S. Ahlawat, *Statistical Quantitative Methods in Finance*,
https://doi.org/10.1007/979-8-8688-0962-0_10

opportunity to simplify the ML model or switch to a statistical model. On the other hand, if the machine learning model yields appreciably better fit, one may envisage using a statistical model as a tool for ensuring the prediction quality is maintained. Furthermore, the benchmark can be a useful tool during both model design and testing phases, affording a machine learning practitioner with concrete test cases and a parallel modeling framework that can be compared to establish which parts of a machine learning algorithm may benefit from improvements – feature engineering, selection of activation functions, regularization, or neural network design.

Let us look at a few specific examples where statistical models can serve as benchmarks for designing and testing machine learning models. In the ensuing examples, the **tensorflow** library has been used to implement machine learning models.

10.1   Predicting Asset Returns

Let us create a deep neural network to predict daily returns of the S&P 500 index. While working on this example, we will focus on the following two objectives:

1. Deciding if a simple statistical model suffices for a modeling task or if it is imperative to use a deep neural network for higher prediction accuracy
2. How to select the number of epochs for training a deep neural network

The endogenous variable is the daily return of the S&P 500 index, defined as shown in Equation 10-1. In this equation, $P(t)$ denotes the closing price on day t.

$$r(t) = \frac{P(t+1)}{P(t)} - 1 \tag{10-1}$$

Let us use the following exogenous variables:

1. One-day lagged return, defined as shown in Equation 10-2.

$$r_{1DayLag}(t) = \frac{P(t-1)}{P(t-2)} - 1 \tag{10-2}$$

2. Two-day lagged return, defined as shown in Equation 10-3.

$$r_{2DayLag}(t) = \frac{P(t-2)}{P(t-3)} - 1 \tag{10-3}$$

3. Three-day lagged return, defined as shown in Equation 10-4.

$$r_{3DayLag}(t) = \frac{P(t-3)}{P(t-4)} - 1 \tag{10-4}$$

4. Three-day moving average of price subtracted from five-day moving average: This feature measures price momentum. If three-day moving average is above five-day moving average, it indicates recent price changes have been on the upside relative to price changes over the past five days.
5. Relative volatility: This is defined as the ratio of 5-day (or 1-week) volatility of returns to 21-day (or 1-month) volatility of returns.
6. Relative volume: This is defined as the ratio of three-day moving average of volume to five-day moving average of volume. Volume represents the number of shares traded on a day.

A deep neural network is constructed and is trained using stochastic gradient descent with a batch size of 32. Figure 10-1 shows the training and testing RMS errors obtained using the deep neural network as a function of training epochs.

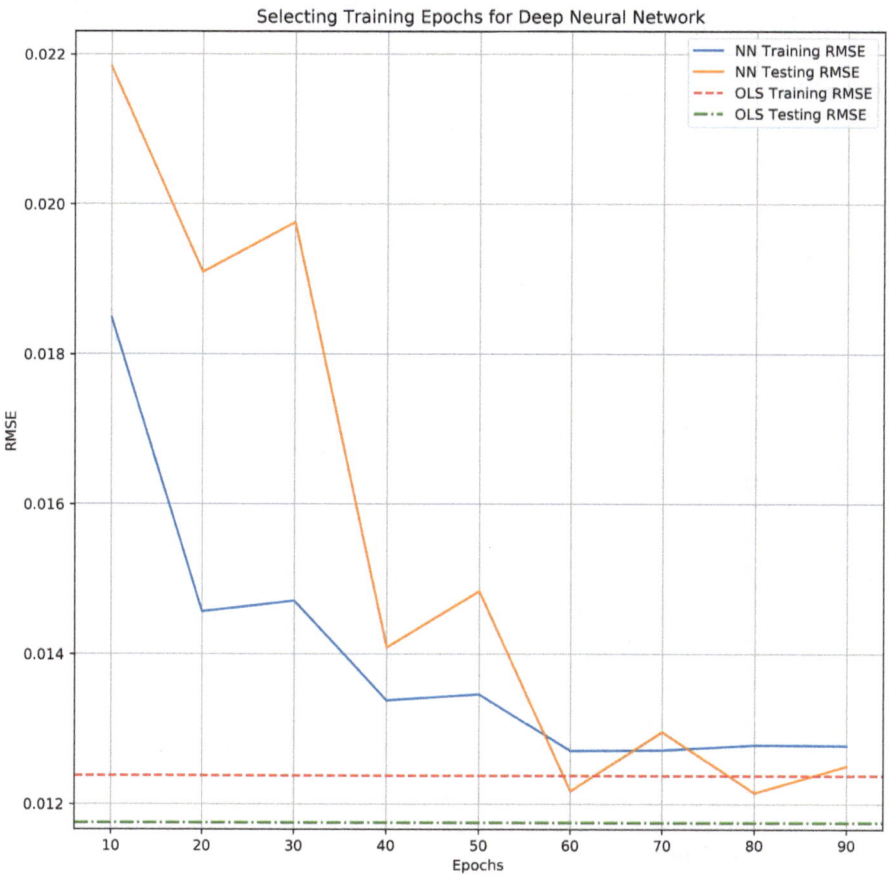

Figure 10-1. RMS Errors on Training and Test Datasets of Deep Neural Network and OLS Model

Two horizontal lines indicate the RMS errors obtained using the OLS model that is trained on the same training dataset and tested on the test dataset – identical to the datasets used for the neural network.

From Figure 10-1, we can observe that a simpler OLS model performs better than the deep neural network. This illustrates a useful principle: when a simpler statistical model can do a task just as well, there is little advantage to using a more complex neural network model. OLS RMSE of 0.012378 on the training dataset and 0.011753 on the test dataset is lower than that obtained using the neural network for all training epochs.

Secondly, training epochs of around 60 seem sufficient to train the network. Beyond this, increasing the number of training epochs does not improve training RMSE, while testing RMSE begins to increase. This is the pitfall of overfitting, where slight gains in training accuracy come at the cost of falling accuracy on the test dataset.

The code for fitting and benchmarking the deep neural network model using an OLS model is shown in Listing 10-1.

Listing 10-1. Using OLS Model to Benchmark the Performance of a Neural Network

```
import numpy as np
import pandas as pd
import logging
import matplotlib.pyplot as plt
import tensorflow as tf
import os
import statsmodels.api as sm

logging.basicConfig(level=logging.DEBUG)


class AssetReturnPredictor:
    PERIOD = 1
    PRICE_COL = "Close"
    VOLUME_COL = "Volume"

    def __init__(self, dirname, security, trainTestRatio=0.9, maxEpochs
        =100, batchSize=32):
        self.logger = logging.getLogger(self.__class__.__name__)
        self.dirname = dirname
        self.security = security
        self.maxEpochs = maxEpochs
        self.batchSize = batchSize
        self.df = pd.read_csv(os.path.join(dirname, f"{security}.csv"),
            parse_dates=["Date"])
        self.endog, self.exog = None, None
        self.beginIndex = None
        self.endIndex = None
        self.calculateEndogExogVars()
        self.ntraining = int(trainTestRatio * self.df.shape[0])
        self.nn = None
```

```
30          self.ols = self.createOLSModel()
31          self.olsFitted = False
32
33      def movingAverage(self, arr, period):
34          result = np.zeros(len(arr), dtype=np.float32)
35          sum1 = np.sum(arr[0:period])
36          for i in range(period, len(arr), 1):
37              result[i] = sum1 / period
38              sum1 += arr[i] - arr[i-period]
39          return result
40
41      def volatility(self, arr, lookback):
42          result = np.zeros(len(arr), dtype=np.float32)
43          sumsq = np.sum(arr[0:lookback] ** 2)
44          for i in range(lookback, len(arr), 1):
45              result[i] = sumsq / lookback
46              sumsq += arr[i]*arr[i] - arr[i-lookback]*arr[i-lookback]
47          return result
48
49      def calculateEndogExogVars(self):
50          prices = self.df.loc[:, self.PRICE_COL].values
51          returns = prices[self.PERIOD:] / prices[0:-self.PERIOD] - 1
52          self.df.loc[:, "returns"] = 0
53          self.df.loc[0:self.df.shape[0] - 1 - self.PERIOD, "returns"] =
                  returns
54          self.endog = "returns"
55
56          self.df.loc[:, "lag1Return"] = 0
57          self.df.loc[self.PERIOD+1:, "lag1Return"] = returns[0:self.df.shape
                  [0]-self.PERIOD-1]
58
59          self.df.loc[:, "lag2Return"] = 0
60          self.df.loc[self.PERIOD+2:, "lag2Return"] = returns[0:self.df.shape
                  [0]-self.PERIOD-2]
61
62          self.df.loc[:, "lag3Return"] = 0
63          self.df.loc[self.PERIOD+3:, "lag3Return"] = returns[0:self.df.shape
                  [0]-self.PERIOD-3]
64
65          self.df.loc[:, "ma3m5"] = 0
66          ma3 = self.movingAverage(prices, 3)
67          ma5 = self.movingAverage(prices, 5)
68          self.df.loc[5:, "ma3m5"] = ma3[5:] - ma5[5:]
69
70          volatility = self.volatility(returns, lookback=5)
71          moVolatility = self.volatility(returns, lookback=21)
72          relVolat = volatility[21:] / moVolatility[21:]
73          self.df.loc[:, "relVolatility"] = 0
74          self.df.loc[21:self.df.shape[0] - 1 - self.PERIOD, "relVolatility"]
                  = relVolat
75
76          volume = self.df.loc[:, self.VOLUME_COL].values
77          vol3 = self.movingAverage(volume, 3)
```

```python
78      vol5 = self.movingAverage(volume, 5)
79      relVolume = vol3[5:] / vol5[5:]
80      self.df.loc[:, "relVolume"] = 0
81      self.df.loc[5:, "relVolume"] = relVolume
82
83      self.exog = ["lag1Return", "lag2Return", "lag3Return", "ma3m5", "
            relVolatility", "relVolume"]
84      self.beginIndex = 21
85      self.endIndex = self.df.shape[0] - self.PERIOD
86
87  def createNN(self):
88      nn = tf.keras.models.Sequential()
89      nn.add(tf.keras.layers.BatchNormalization())
90      nn.add(tf.keras.layers.Dense(30, activation=tf.keras.activations.
            tanh))
91      nn.add(tf.keras.layers.Dense(10, activation=tf.nn.leaky_relu))
92      nn.add(tf.keras.layers.Dense(30, activation=tf.nn.leaky_relu))
93      nn.add(tf.keras.layers.Dense(20))
94      nn.add(tf.keras.layers.Dense(5))
95      nn.add(tf.keras.layers.Dense(1, activation=tf.keras.activations.
            tanh))
96      nn.compile(optimizer=tf.keras.optimizers.Adam(learning_rate=0.0001)
            , loss=tf.keras.losses.MeanSquaredError())
97      return nn
98
99  def fitNN(self, nepochs):
100     y = self.df.loc[self.beginIndex:self.ntraining, self.endog].values
101     X = self.df.loc[self.beginIndex:self.ntraining, self.exog].values
102     Xy = np.concatenate((X, y[:, np.newaxis]), axis=1)
103     np.random.shuffle(Xy)
104     X = Xy[:, 0:-1]
105     y = Xy[:, -1]
106     self.nn = self.createNN()
107     return self.nn.fit(X, y, batch_size=self.batchSize, epochs=nepochs)
108
109 def createOLSModel(self):
110     y = self.df.loc[self.beginIndex:self.ntraining, self.endog].values
111     X = self.df.loc[self.beginIndex:self.ntraining, self.exog].values
112     X = sm.add_constant(X, has_constant="add")
113     return sm.OLS(endog=y, exog=X)
114
115 def fitOLS(self):
116     if self.olsFitted:
117         return self.ols
118     self.ols = self.ols.fit()
119     self.olsFitted = True
120     return self.ols
121
122 def fit(self, nepochs):
123     self.fitOLS()
124     nnFitHistory = self.fitNN(nepochs)
125     return np.sqrt(nnFitHistory.history["loss"][-1])
126
```

```python
127      def testNN(self, y=None, X=None):
128          if y is None:
129              y = self.df.loc[self.ntraining:self.endIndex-1, self.endog].
                     values
130              X = self.df.loc[self.ntraining:self.endIndex-1, self.exog].
                     values
131          yhatNN = self.nn.predict(X)
132          rmseNN = np.sqrt(np.mean((y - yhatNN) ** 2))
133          return rmseNN
134
135      def testOLS(self, y=None, X=None):
136          if y is None:
137              y = self.df.loc[self.ntraining:self.endIndex-1, self.endog].
                     values
138              X = self.df.loc[self.ntraining:self.endIndex-1, self.exog].
                     values
139          Xols = sm.add_constant(X, has_constant="add")
140          yhatOls = self.ols.predict(exog=Xols)
141          rmseOLS = np.sqrt(np.mean((y - yhatOls) ** 2))
142          return rmseOLS
143
144      def trainingDatasetTestNN(self):
145          y = self.df.loc[self.beginIndex:self.ntraining, self.endog].values
146          X = self.df.loc[self.beginIndex:self.ntraining, self.exog].values
147          return self.testNN(y=y, X=X)
148
149      def trainingDatasetTestOLS(self):
150          y = self.df.loc[self.beginIndex:self.ntraining, self.endog].values
151          X = self.df.loc[self.beginIndex:self.ntraining, self.exog].values
152          return self.testOLS(y=y, X=X)
153
154      def plot(self, epochs, trainError, testError, olsErrorTrain,
             olsErrorTest):
155          fig, axs = plt.subplots(1, 1, figsize=(10, 10))
156          axs.plot(epochs, trainError, label="NN Training RMSE")
157          axs.plot(epochs, testError, label="NN Testing RMSE")
158          axs.axhline(y=olsErrorTrain, color='r', linestyle='dashed', label="
                 OLS Training RMSE")
159          axs.axhline(y=olsErrorTest, color='g', linestyle='dashdot', label="
                 OLS Testing RMSE")
160          axs.set(title="Selecting Training Epochs for Deep Neural Network")
161          axs.legend()
162          axs.grid()
163          axs.set_xlabel("Epochs")
164          axs.set_ylabel("RMSE")
165          plt.savefig(os.path.join(self.dirname, f"AssetReturn_{self.security
                 }.jpeg"),
166                      dpi=500)
167          plt.show()
168
169      def findOptimalTrainingEpochs(self):
170          epochs = list(range(10, self.maxEpochs, 10))
171          testError = []
```

```
172    trainError = []
173    self.fitOLS()
174    olsErrorTrain = self.trainingDatasetTestOLS()
175    olsErrorTest = self.testOLS()
176    for epoch in epochs:
177        nnerror = self.fit(nepochs=epoch)
178        self.logger.info("Epoch: %d, Fitting RMSE: %f", epoch, nnerror)
179        nnErrorTrain = self.trainingDatasetTestNN()
180        nnErrorTest = self.testNN()
181        testError.append(nnErrorTest)
182        trainError.append(nnErrorTrain)
183    self.plot(epochs, trainError, testError, olsErrorTrain,
               olsErrorTest)
184    self.logger.info("OLS RMS error on training dataset: %f, test
               dataset: %f", olsErrorTrain, olsErrorTest)


187 if __name__ == "__main__":
188    dirname = r"C:\prog\cygwin\home\samit_000\latex\book_stats\code\data"
189    pred = AssetReturnPredictor(dirname, "SPY")
190    np.random.seed(32)
191    tf.random.set_seed(32)
192    pred.findOptimalTrainingEpochs()
```

10.1.1 Code Explanation

A code walk-through is presented below:

1. An instance of class **AssetReturnPredictor** is instantiated, passing the directory name containing the data file and the name of the S&P 500 file as arguments.
2. Inside the constructor, the data file is read, and endogenous and exogenous variables are computed, as described earlier. It should be noted that the code avoids in-sample bias. That is, on day t, all exogenous variables can be calculated using data available prior to day t.
3. Seeds for numpy and tensorflow random number generators are set for reproducibility.
4. Method **findOptimalTrainingEpochs** is invoked on the object of class **AssetReturnPredictor**. This method does the following:

   - The OLS model is fitted using the training dataset. This dataset consists of the first 90% of daily data from the security file for S&P 500. Since OLS model fitting does not depend on training epochs, the OLS model is fitted just once. The OLS model includes a constant because the neural network model uses layers with bias.
   - A neural network is created. This is a **Sequential** network consisting of seven layers. Details of the neural network can be found in method **createNN**.
   - The neural network uses an **Adam** stochastic gradient descent optimizer. This optimizer was proposed by [35]. It takes the first moment (gradient) and the

second moment (gradient-square) into account while computing parameter updates. The inclusion of the second moment in the computation of parameter correction makes Adam akin to Newton's iterations that use gradient and Hessian. A learning rate of 0.0001 is used in the Adam optimizer.

- The loss function is set to **tf.keras.losses.MeanSquaredError**.
- The neural network is fitted to the training dataset using the Adam optimizer. A batch size of 32, along with a specified number of training epochs, is used. This is performed inside method **fitNN**. Before fitting, the training data is shuffled so that time-series correlation is broken. This is important when stochastic gradient descent is used for updating parameters.
- The fitted neural network model is used to predict using training and test datasets and RMSE computed using the actual value of S&P 500 returns.
- The above steps are repeated for a different number of training epochs, and a plot of RMSE observed on training and test datasets is plotted as a function of the number of epochs.
- RMSE for the OLS model using the same training and test datasets is overlaid as horizontal lines on the plot.

10.2   Word2Vec

In the field of natural language processing (NLP), Word2Vec is a cornerstone algorithm that marked the onset of an era characterized by rapid and unheralded improvements in the quality of predictions in natural language tasks. The algorithm was proposed in a seminal paper by [36] and demonstrated the principle that dense word vector representations obtained from local word contexts could extract meaningful word vector representations quantifying word similarity. The algorithm is implemented in one of two ways:

1. **Skipgram**: This involves predicting context words using the center word in a word-window.
2. **CBOW**: Acronym for continuous-bag-of-words, the model predicts the center word of a window using context words (surrounding words).

Word2Vec does not use relative positional information of words in a sentence into account. As long as a word is within a window, it has the same impact as other words.

Statistical concepts such as SVD and cosine similarity can be used to benchmark the quality of word vector representations produced by Word2Vec. In this section, let us look at how we can use SVD for this purpose, in addition to using cosine similarity for quantifying word similarity.

SVD, or singular value decomposition, was explained in the chapter on linear regression. Use of SVD for obtaining word vector representations predates Word2Vec.

Word2Vec (skipgram) learns dense vector representations of words by maximizing the objective function shown in Equation 10-5.

$$L(\boldsymbol{\theta}) = \frac{\exp\left(\boldsymbol{\theta_0}^T \boldsymbol{\theta_c}\right)}{\sum_{w \in \mathcal{V}} \exp\left(\boldsymbol{\theta_0}^T \boldsymbol{\theta_w}\right)}$$

$$\mathcal{V} \equiv \text{Vocabulary} \qquad\qquad (10\text{-}5)$$

$$\boldsymbol{\theta_0} \equiv \text{Word vector for center word}$$

$$\boldsymbol{\theta_c} \equiv \text{Word vector for context word}$$

By maximizing the likelihood expression in Equation 10-5, word vector representations of a center word, θ_0, and a context word, θ_c, become similar, while those of center word with words that do not occur in its context become dissimilar. Similarity is measured by the dot product between the word vectors. A key implementation impediment to a straightforward computation of Equation 10-5 involves computation of the denominator which includes all words in the vocabulary. Word2Vec uses **negative sampling** to make the computational cost more tractable. Negative sampling involves randomly sampling a few words from the vocabulary. For decent-sized vocabularies, random sampling will most likely pick words that do not occur in the context of the center word. Word2Vec then appends one context word to this list and computes the denominator using $1 + N$ words instead of using all words in the vocabulary. N represents the number of negative samples selected at random.

Table 10-1 shows the top five similar words to a group of ten randomly selected words from the vocabulary. The code uses a rather limited vocabulary size of 1024. Even with this small-sized vocabulary, Word2Vec word vectors reflect meaningful word associations. For example, the word "may" is found to be semantically similar to months of April, July, and June. At the same time, "may" can be used in another

Table 10-1. Words Similar to Ten Randomly Chosen Words from Vocabulary

word	Similar Word 1	Similar Word 2	Similar Word 3	Similar Word 4	Similar Word 5
may	april	will	july	june	would
seeking	seek	approval	want	who	move
meeting	meet	council	ministers	conference	statement
mths	oper	vs	qtr	3rd	shrs
value	150	volume	share	because	change
city	trust	inc	industries	lt	corp
control	seek	acquire	take	way	division
prime	lending	base	cuts	cut	point
stocks	liquidity	tonnes	tons	sugar	output
seasonally	adjusted	february	rises	index	unemployment

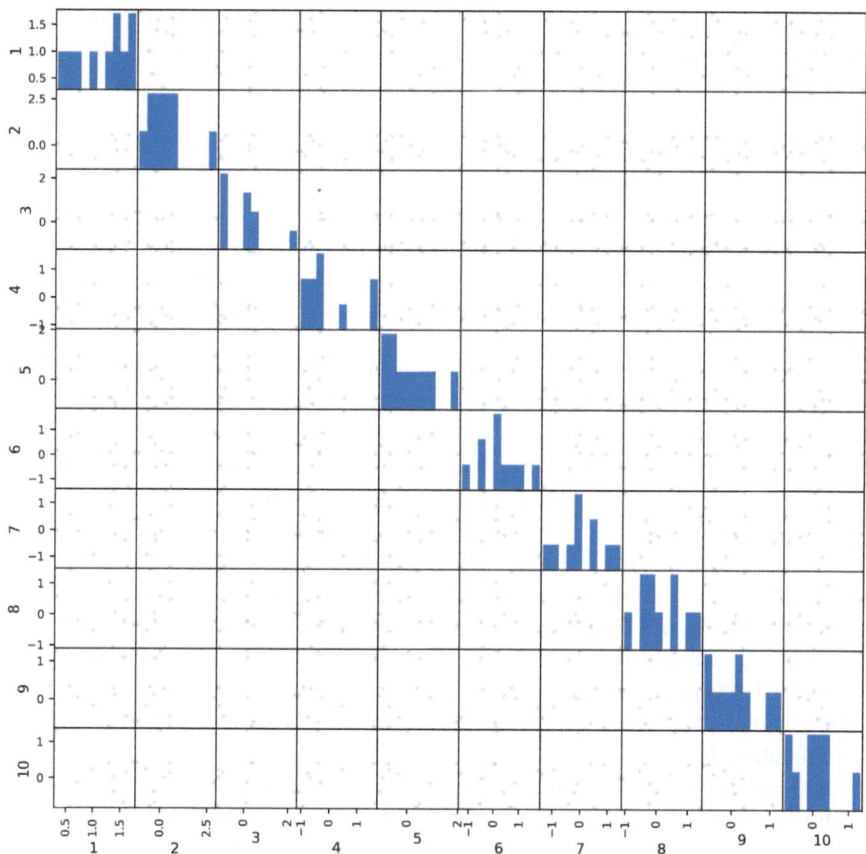

Figure 10-2. Visualizing 128-Dimensional Word Vectors from Word2Vec Using SVD

word sense – to connote the ability to do something. In this second sense, it is found similar to words "will" and "would."

Similarly, the word "meeting" is found to be similar to words "meet" and "conference." Because word vectors are learned using local context windows, words that occur together frequently will show up with similar word vector representations, even though they may have different meanings. This can be seen from the word "seasonally" which is found to be most similar to the word "adjusted." Even though the two words mean something different, they frequently occur together in news related to economics. The training corpus used in the code is derived from business news, which explains this observation.

Figure 10-2 shows reduced dimensional visualization of ten most frequently occurring words in the corpus. We have chosen a 128-dimensional word vector space in this Word2Vec implementation. Visualizing word vectors in such high-dimensional spaces is not easy. We use the statistical method of SVD to reduce 128-dimensional vectors to 10-dimensional vectors and visualize each of those vectors in 2 dimensions at a time. In Figure 10-2, subplots on the diagonal show a histogram distribution of word vectors along each of the ten reduced dimensions. We can observe that word vectors cluster into groups.

10.2.1 Code Explanation

Before performing a code walk-through of Word2Vec implementation in Listing 10-2, let us briefly recapitulate the training corpus used in the code along with parameter settings:

1. We use the reuters corpus available using the **nltk** (Natural Language Toolkit) API. This is a corpus comprising business news published by the Reuters news agency.
2. A vocabulary size of 1024 is used. This is rather restrictive; in a practical application, one would typically use a larger vocabulary size. The corpus has 30,952 unique words.
3. A window size of 3 is used. This means that three words before and three words after a word are considered to be within the context of the center word. The window shrinks if it oversteps the boundaries of a sequence.
4. A sequence length of 10 is used. This means that sentences are composed of ten-word groups.
5. Five negative samples per positive sample are used to evaluate the denominator in Equation 10-5.
6. Word vectors have 128 dimensions. This is a hyper-parameter that needs to be adjusted to ascertain a good value for the application. Low dimensionality gives greater bias (i.e., lower accuracy in both training and test datasets), while high dimensionality gives high variance (good accuracy on the training dataset but poor accuracy on the test dataset).

Listing 10-2. Fitting Word2Vec Model to Reuters Corpus Using a Vocabulary Size of 1024

```
1   import numpy as np
2   import tensorflow as tf
3   import tqdm
4   import string
5   import re
6   import logging
7   import pandas as pd
8   import matplotlib.pyplot as plt
9   from sklearn.decomposition import TruncatedSVD
10  import nltk
11  import os
12
13  logging.basicConfig(level=logging.DEBUG)
14
15
```

```
16   class TextRetriever(object):
17       """ Utility class to read corpus """
18
19       @staticmethod
20       def standardize_text(input_text):
21           return re.sub("[%s]" % re.escape(string.punctuation), "",
22               input_text.lower())
23
23       @staticmethod
24       def tf_standardize_text(input_text):
25           lowercase = tf.strings.lower(input_text)
26           return tf.strings.regex_replace(lowercase,
27                                   '[%s]' % re.escape(string.punctuation),
                                        ' ')
28
29       @staticmethod
30       def read_file(path_to_file):
31           vocab_size = 0
32           with open(path_to_file, "r") as f:
33               vocab_size = len(set(f.read().lower().split()))
34           return path_to_file, vocab_size
35
36       @staticmethod
37       def write_corpus_file(corpus_name, dirname):
38           corpus = getattr(nltk.corpus, corpus_name)
39           files = corpus.fileids()
40           filename = os.path.join(dirname, f"{corpus_name}.txt")
41           with open(filename, "w") as fp:
42               for f in files:
43                   words = corpus.words(f)
44                   for i in range(0, len(words), 10):
45                       fp.write(' '.join(words[i:min(i+10, len(words))]) + '\n')
46           return filename
47
48
49       @staticmethod
50       def read_corpus(corpus_name, dirname, sequence_len, batch_size,
                         vocab_size=None):
51           file_name = TextRetriever.write_corpus_file(corpus_name, dirname)
52           path_to_file, vsz = TextRetriever.read_file(path_to_file=file_name)
53           if vocab_size is None:
54               vocab_size = vsz
55           text_ds = tf.data.TextLineDataset(path_to_file).filter(lambda x: tf
                 .cast(tf.strings.length(x), bool))
56           vectorize_layer = tf.keras.layers.TextVectorization(standardize=
                 TextRetriever.tf_standardize_text, max_tokens=vocab_size,
                 output_mode="int", output_sequence_length=sequence_len)
57           vectorize_layer.adapt(text_ds.batch(batch_size))
58           # returns vocabulary sorted in descending order by frequency
59           text_vector_ds = text_ds.batch(batch_size).prefetch(tf.data.
                 AUTOTUNE).map(vectorize_layer).unbatch()
60           sequences = list(text_vector_ds.as_numpy_iterator())
61           inverse_vocab = vectorize_layer.get_vocabulary()
```

```python
            TextRetriever.inspect_dataset(sequences, inverse_vocab, 10)
            return sequences, inverse_vocab, vocab_size

    @staticmethod
    def inspect_dataset(sequences, inverse_vocab, num_to_inspect):
        logging.info(len(sequences))
        end = min(num_to_inspect, len(sequences))
        for seq in sequences[:end]:
            logging.info(f"{seq} => {[inverse_vocab[i] for i in seq]}")


class Plotter(object):
    @staticmethod
    def plot_weights(weights, size, labels=None, dirname=None):
        if size < weights.shape[1]:
            weights = Plotter.reduce_to_k_dim(weights, size)
        if labels is None:
            labels = ["%d" % (i + 1) for i in range(size)]

        data = pd.DataFrame(weights, columns=labels)
        pd.plotting.scatter_matrix(data, alpha=0.2, diagonal='hist',
            figsize=(10, 10))
        if dirname:
            plt.savefig(os.path.join(dirname, f"ReducedWts_Word2Vec.jpeg"),
                dpi=500)
        plt.show()


    @staticmethod
    def reduce_to_k_dim(M, k=2, n_iter=10):
        """ Reduce a matrix M (n, m) to a matrix of dimensionality (n, k)
            using the
            following SVD function from Scikit-Learn:
                - http://scikit-learn.org/stable/modules/generated/sklearn.
                    decomposition.TruncatedSVD.html

            Params:
                M (n,m): co-occurence matrix of word counts
                k (int): embedding size of each word after dimension
                    reduction
            Return:
                M_reduced (numpy matrix of shape (number of corpus words, k)
                    ): matrix of k-dimensioal word embeddings.
                        In terms of the SVD from math class, this actually
                            returns U * S
        """
        svd = TruncatedSVD(n_components=k, n_iter=n_iter)
        return svd.fit_transform(M)


class Word2Vec(tf.keras.Model):
    """ Skipgram model """
```

```python
def __init__(self, embedding_dim, num_neg_samples, window_size,
        corpus_name, dirname,
            batch_size=1024, seed=10, vocab_size=None, sequence_len=10,
            buffer_size=10000):
    super(Word2Vec, self).__init__()
    self.dirname = dirname

    self.sequences, self.inverse_vocab, self.vocab_size = TextRetriever
            .read_corpus(corpus_name, dirname, sequence_len, batch_size,
            vocab_size)
    self.embedding_dim = embedding_dim
    self.num_neg_samples = num_neg_samples
    self.window_size = window_size
    self.batch_size = batch_size
    self.buffer_size = buffer_size
    self.seed = seed
    self.word_to_index_dict = {v: i for i, v in enumerate(self.
            inverse_vocab)}
    self.target_embedding = tf.keras.layers.Embedding(vocab_size,
            embedding_dim, input_length=1, name="target_emb")
    self.context_embedding = tf.keras.layers.Embedding(vocab_size,
            embedding_dim, input_length=num_neg_samples + 1, name="
            context_softmax_emb")
    self.compile(optimizer="adam", loss=tf.keras.losses.
            CategoricalCrossentropy(from_logits=True), metrics=["accuracy"
            ])

def call(self, pair):
    target, context = pair
    if len(target.shape) == 2:
        target = tf.squeeze(target, axis=1)
    word_embed = self.target_embedding(target)
    context_embed = self.context_embedding(context)
    dotprod = tf.einsum("ik,ijk->ij", word_embed, context_embed)
    return dotprod

def generate_training_data(self):
    """
    Generates skip-gram pairs with negative sampling for a list of
        sequences
    (int-encoded sentences) based on window size, number of negative
        samples
    and vocabulary size.
    """
    targets, contexts, labels = [], [], []
    sampling_table = tf.keras.preprocessing.sequence.
            make_sampling_table(self.vocab_size)
    for sequence in tqdm.tqdm(self.sequences):
        positive_skipgrams, _ = tf.keras.preprocessing.sequence.
                skipgrams(sequence, vocabulary_size=self.vocab_size,
                sampling_table=sampling_table, window_size=self.window_size
                , negative_samples=0)
        for target_word, context_word in positive_skipgrams:
```

```
145         context_class = tf.expand_dims(tf.constant([context_word],
                dtype="int64"), 1)
146         neg_samples, _, _ = tf.random.log_uniform_candidate_sampler(
                true_classes=context_class, num_true=1, num_sampled=self
                .num_neg_samples, unique=True, range_max=self.vocab_size
                , seed=self.seed, name="neg_sampling")
147         context = tf.concat([tf.squeeze(context_class, 1),
                neg_samples], 0)
148         label = tf.constant([1] + [0] * self.num_neg_samples, dtype=
                "int64")
149         targets.append(target_word)
150         contexts.append(context)
151         labels.append(label)

153     return np.array(targets), np.array(contexts), np.array(labels)

155 def fit(self, epochs=20):
156     targets, contexts, labels = self.generate_training_data()
157     dataset = tf.data.Dataset.from_tensor_slices(((targets, contexts),
            labels))
158     dataset = dataset.shuffle(self.buffer_size).batch(self.batch_size,
            drop_remainder=True)
159     super().fit(dataset, epochs=epochs)

161 def write_weights(self, file_name):
162     weights = self.target_embedding.get_weights()[0]
163     with open(file_name, "w") as fp:
164         for index, word in enumerate(self.inverse_vocab):
165             if index == 0:
166                 continue # skip 0, it's padding.
167             vec = weights[index]
168             fp.write(word + "," + ",".join([str(x) for x in vec]) + "\n"
                    )

170 def get_weights(self, top_n=None, word_list=None):
171     if top_n:
172         word_list = self.inverse_vocab[1:top_n + 1]
173     weights = self.target_embedding.get_weights()[0]
174     indices = np.array([self.word_to_index_dict.get(w, 0) for w in
            word_list])
175     return weights[indices, :], word_list

177 def cosine_similarity(self, top_n=None, word_list=None):
178     if top_n:
179         word_list = self.inverse_vocab[1:top_n + 1]
180     nwords = len(word_list)
181     weights = self.target_embedding.get_weights()[0]
182     indices = np.array([self.word_to_index_dict.get(w, 0) for w in
            word_list])
183     wts = weights[indices, :]
184     lengths = np.sum(np.multiply(wts, wts), axis=1)
185     cosine = np.zeros((nwords, nwords), dtype=np.float64)
186     for i in range(nwords):
```

```
187            cosine[i, i] = 1.0
188            for j in range(i):
189                cosine[i, j] = np.dot(wts[i, :], wts[j, :]) / np.sqrt(
                       lengths[i] * lengths[j])
190                cosine[j, i] = cosine[i, j]
191        return cosine, word_list
192
193    def length_similarity(self, top_n=None, word_list=None):
194        if top_n:
195            word_list = self.inverse_vocab[1:top_n + 1]
196        nwords = len(word_list)
197        weights = self.target_embedding.get_weights()[0]
198        indices = np.array([self.word_to_index_dict.get(w, 0) for w in
                   word_list])
199        wts = weights[indices, :]
200
201        lengths = np.zeros((nwords, nwords), dtype=np.float64)
202        for i in range(nwords):
203            for j in range(i):
204                dist = np.subtract(wts[i, :], wts[j, :])
205                lengths[i, j] = np.sqrt(np.dot(dist, dist))
206                lengths[j, i] = lengths[i, j]
207        return lengths, word_list
208
209    @staticmethod
210    def get_similar_words(weights_file, topN=6):
211        np.random.seed(64)
212        df = pd.read_csv(weights_file, header=None)
213        words = np.random.choice(df.shape[0], 10, replace=False)
214        similarWords = [[]]
215        for i in range(topN):
216            similarWords.append([])
217
218        for iword in words:
219            word = df.loc[iword, 0]
220            vec = df.loc[iword, 1:].values
221            l1 = np.dot(vec, vec)
222            cosineArr = []
223            for j in range(df.shape[0]):
224                if j == iword:
225                    continue
226                word2 = df.loc[j, 0]
227                vec2 = df.loc[j, 1:].values
228                l2 = np.dot(vec2, vec2)
229                cosineSim = np.dot(vec, vec2) / np.sqrt(l1 * l2)
230                cosineArr.append((cosineSim, word2))
231            cosineArr.sort(key=lambda x: x[0], reverse=True)
232            similarWords[0].append(word)
233            for i in range(topN):
234                similarWords[i+1].append(cosineArr[i][1])
235
236        columns = ["word"] + ["SimWord%d" % (i+1) for i in range(topN)]
237        data = {c:arr for c,arr in zip(columns, similarWords)}
```

```
238         df2 = pd.DataFrame(data=data)
239         logging.info(df2.to_latex(index=False))
240
241
242  if __name__ == "__main__":
243      embedding_dim = 128
244      num_neg_samples = 5
245      window_size = 3
246      corpus_name = "reuters"
247      vocab_size = 1024
248      sequence_len = 10
249      dirname = r"C:\prog\cygwin\home\samit_000\latex\book_stats\code\data"
250      word2vec = Word2Vec(embedding_dim, num_neg_samples, window_size,
              corpus_name, dirname,
251                          vocab_size=vocab_size,
252                          sequence_len=sequence_len)
253      word2vec.fit()
254      weights_file = os.path.join(dirname, "weights.csv")
255      word2vec.write_weights(weights_file)
256
257      weights, words = word2vec.get_weights(top_n=10)
258      Plotter.plot_weights(weights, size=10, dirname=dirname)
259      logging.info(",".join(words))
260      logging.info(weights)
261
262      Word2Vec.get_similar_words(weights_file, topN=5)
```

With these assumptions in mind, let us look at the code following the order of
execution:

1. An object of class Word2Vec is instantiated with the following constructor
   arguments:

   - **embedding_dim** representing the word vector dimension. This is set to 128.
   - **num_neg_samples** denoting the number of negative samples for each positive
     sample for a context word and center word pair. This is set to 5.
   - **window_size** of 3 representing context window length on both sides of the
     center word.
   - **corpus_name** set to "reuters" representing the name of the training corpus.
   - **dirname** pointing to the directory name where the corpus file will be written
     after being downloaded using the **nltk** API. This directory is also used for
     writing weight vectors.
   - **vocab_size** representing the vocabulary size of 1024. The corpus used
     has 30,952 unique words. This means that the top 1024 most frequently
     occurring words are considered to be in the vocabulary with the rest being
     assigned a "UNK" (unknown) tag. Two special words "UNK" and "PAD"are
     automatically added to the vocabulary. This means that the actual vocabulary
     size is 1026. "PAD" is the word for padding. It is added to sequences that are
     shorter than the specified sequence length to make them uniform in length.

- **sequence_len** denotes the length of sequence and is set to ten. All word sequences are of this length, i.e., ten words long. Sequence may end in the middle of an actual sentence.
- Remaining arguments to the constructor use default values, such as **batch_size** and **buffer_size**.

2. Inside the constructor, the following steps are performed:

- The training corpus is read inside the static method **read_corpus** of utility class **TextRetriever**. This method downloads the corpus, chops the sentences to ten-word long sequences, and writes them to a file.
- It reads the corpus data from the file and creates a tensorflow dataset using the **tf.data.TextLineDataset** class.
- It creates an embedding layer of target word (or center word) and an embedding for context words. The **tf.keras.layers.Embedding** layer is used for the purpose. The target layer is applied to the center word, while the context layer is applied to context words. After training has converged, the two embeddings should be similar. We need separate embeddings for the two because the context embedding layer is also applied to negatively sampled words, i.e., randomly selected words that will most likely not occur in the context of a center word.
- The loss function used is **tf.keras.losses.CategoricalCrossentropy**. This calculates the expression shown in Equation 10-5 after providing the **from_logits** argument with value **True**.

3. The **fit** method is called to train the model. This method does the following:

- Training data consisting of positive and negative skipgrams is constructed inside the method **generate_training_data**.
- A dataset consisting of positive skipgrams (word-context pairs) and negative skipgrams (one positive word-context pair along with five negative word-context pairs), along with a set of labels with one indicating the pair occur together and zero indicating they do not occur together in the window, is created. The **from_tensor_slices** method is used.
- The model is fitted to training data using 20 epochs.

4. Trained word vectors are written to a file weights.csv.
5. Word vectors are visualized using SVD by reducing their dimensionality to ten.
6. To evaluate the quality of word vectors, ten random words are selected from the vocabulary, and top five semantically similar words for each are reported. Cosine similarity is used to measure the degree of similarity. This measure is defined in Equation 10-6, where θ denotes a word vector.

$$\text{cosine sim.}(\boldsymbol{\theta}_1, \boldsymbol{\theta}_2) = \frac{\boldsymbol{\theta}_1{}^T \boldsymbol{\theta}_2}{\sqrt{\left(\|\boldsymbol{\theta}_1\|^2 \, \|\boldsymbol{\theta}_2\|^2\right)}} \tag{10-6}$$

$$\|\boldsymbol{\theta}\|^2 = \boldsymbol{\theta}^T \boldsymbol{\theta}$$

10.3   Glove

Glove is an acronym for **glo**bal **ve**ctors for word representation. It was proposed by [37] as an algorithm to bridge the apparent difference in word vector quality obtained using statistical methods such as LSA (latent semantic analysis) and word-window-based methods such as Word2Vec. Word vectors produced from Glove were reported to perform better than LSA and Word2Vec in word analogy tasks. More recent work has attempted to explain the performance of Pointwise Mutual Information (PMI)-based methods such as LSA and context-based word vectors such as Word2Vec and Glove in a unified framework and has attributed some of the observed out-performance of Glove to the choice of hyper-parameters. Interested readers are referred to [38].

The Glove algorithm for word vectors has the following salient features:

1. The algorithm uses a co-occurrence matrix computed by taking the logarithm of word co-occurrence counts in context windows. It should be noted that LSA (latent semantic analysis) uses a co-occurrence matrix for the whole document, where the context spans the length of the entire document. Word2Vec, on the other hand, considers a local context that is typically just a few words long. It does not use a co-occurrence matrix.

2. The algorithm avoids computation of word vector products with all vocabulary words needed in the denominator of Equation 10-5 by using co-occurrence counts instead of negative sampling used in Word2Vec. The objective function optimized in Glove is shown in Equation 10-7.

$$L(\boldsymbol{\theta}) = \sum_{i,j=1}^{\|\mathcal{V}\|} f(X_{i,j}) \left(\boldsymbol{\theta}_i{}^T \boldsymbol{\theta}_j + b_i + b_j - \log\left(X_{i,j} + 1\right) \right)$$

b_i, b_j are biases or intercepts for weight vectors $\boldsymbol{\theta}_i, \boldsymbol{\theta}_j$ \hfill (10-7)

$$f(X_{i,j}) = \begin{cases} \left(\frac{X_{i,j}}{100}\right)^{0.75} & \text{if } X_{i,j} < 100 \\ 1 \text{ otherwise} \end{cases}$$

In Equation 10-7, $X_{i,j}$ is the co-occurrence matrix formed by counting the number of times word j occurs in the context of word i. An addition of one is required before taking the logarithm so that $\log\left(X_{i,j} + 1\right)$ stays well defined for entries where the $x_{i,j}$ count is zero.

Table 10-2. Words Similar to Ten Randomly Chosen Words from Vocabulary Found Using Glove

word	Similar Word 1	Similar Word 2	Similar Word 3	Similar Word 4	Similar Word 5
may	april	july	june	march	shipment
seeking	seek	financing	currently	usair	proposals
meeting	review	special	here	report	talks
mths	months	oper	89	39	12
value	volume	remaining	australian	number	600
city	equipment	computer	southern	europe	chemical
control	acquisitions	computer	impact	steel	usair
prime	finance	discount	cuts	base	cut
stocks	pipeline	ships	shipments	sharply	supplies
seasonally	unemployment	adjusted	69	discount	rises

Let us implement Glove and analyze semantically similar words using cosine similarity. As before, we will also use SVD to visualize the word vectors in addition to using cosine similarity.

We select ten words at random from the vocabulary and use a cosine similarity measure to determine the top five most similar words. Results are shown in Table 10-2. We can observe improvement in the quality of word vectors obtained using Glove, compared with Word2Vec. The word "may" has neighboring four months as the top four most similar words. The algorithm is also able to identify commonly used abbreviated words. For example, the word "months" is found to be most similar to the abbreviated word "mths." In other examples, we observe Glove selecting similar words as those selected by Word2Vec. For example, "seasonally" is adjudged to be similar to the word "adjusted." This shows that Glove shares a few shortcomings with Word2Vec owing to the fact that both use context windows to determine word vectors and word similarity.

Figure 10-3 shows reduced dimensional visualization of ten most frequently occurring words in the corpus. In Figure 10-3, subplots on the diagonal show a histogram distribution of word vectors along each of the ten reduced dimensions. As before, we can observe that word vectors cluster into groups.

The code for implementing and fitting the Glove model on the same corpus used for training Word2Vec is shown in Listing 10-3.

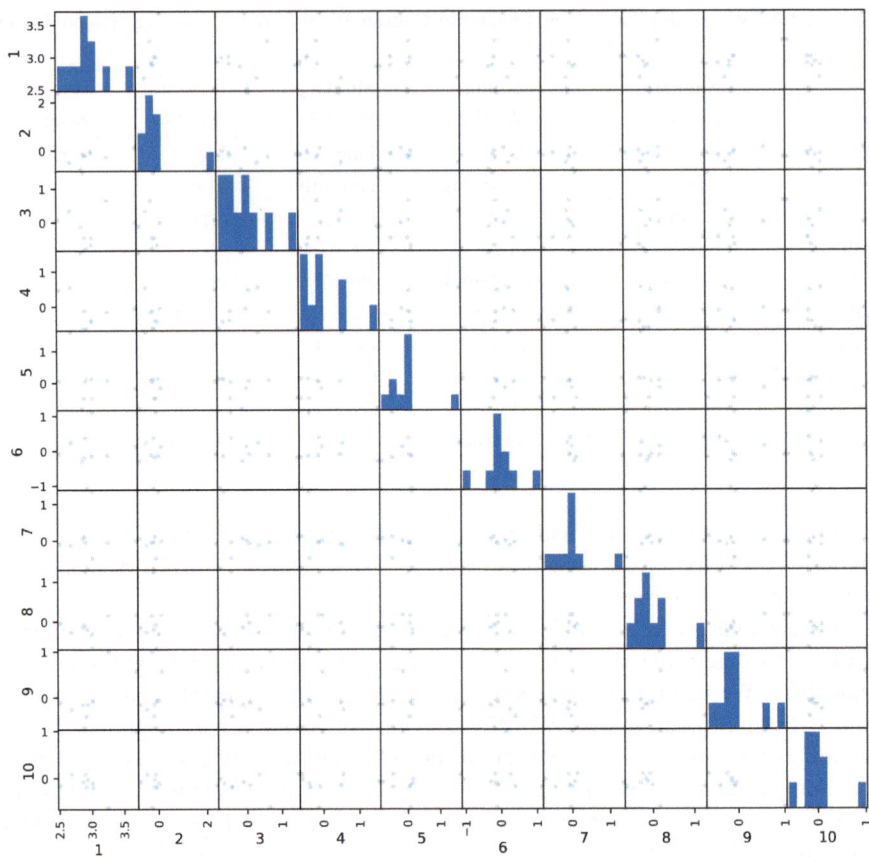

Figure 10-3. Visualizing 128-Dimensional Word Vectors from Glove Using SVD

Listing 10-3. Fitting Glove Model to Reuters Corpus Using a Vocabulary Size of 1024

```
import numpy as np
import tensorflow as tf
import tqdm
import logging
import pandas as pd
import matplotlib.pyplot as plt
from sklearn.decomposition import TruncatedSVD
from src.Word2Vec import TextRetriever
import os


logging.basicConfig(level=logging.DEBUG)



class WeightedMeanSquaredError(tf.keras.losses.Loss):
    def __init__(self, xmax=100.0, power=0.75):
        super().__init__()
```

```
17          self.xmax = xmax
18          self.power = power
19
20      def call(self, y_true, y_pred):
21          xij = tf.exp(y_true) - 1
22          sample_weight = tf.where(xij < self.xmax, tf.pow(xij / self.xmax,
                self.power), 1)
23          val = tf.math.square(y_true - y_pred)
24          return tf.math.reduce_mean(sample_weight * val)
25
26
27  class Glove(tf.keras.Model):
28      """ Word co-occurrence model """
29
30      def __init__(self, embedding_dim, window_size, corpus_name, dirname,
            xmax=100, power=0.75,
31                  skip_words=None, batch_size=1024, seed=10, vocab_size=None,
                    sequence_len=10,
32                  buffer_size=10000):
33          super(Glove, self).__init__()
34          self.dirname = dirname
35
36          self.sequences, self.inverse_vocab, self.vocab_size = TextRetriever
                .read_corpus(corpus_name, dirname, sequence_len, batch_size,
                vocab_size)
37          self.embedding_dim = embedding_dim
38          self.window_size = window_size
39          self.batch_size = batch_size
40          self.buffer_size = buffer_size
41          self.seed = seed
42          self.skip_words = {}
43          if skip_words:
44              self.skip_words = set(skip_words)
45          self.co_occurrence_matrix = np.zeros((self.vocab_size, self.
                vocab_size), dtype=np.int32)
46          self.word_to_index_dict = {v: i for i, v in enumerate(self.
                inverse_vocab)}
47          self.target_embedding = tf.keras.layers.Embedding(vocab_size,
                embedding_dim, input_length=1, name="target_emb")
48          self.context_embedding = tf.keras.layers.Embedding(vocab_size,
                embedding_dim, input_length=1, name="context_emb")
49          self.bias_a = tf.keras.layers.Embedding(vocab_size, 1, input_length
                =1, name="bias1")
50          self.bias_b = tf.keras.layers.Embedding(vocab_size, 1, input_length
                =1, name="bias2")
51          self.compile(optimizer="adam", loss=WeightedMeanSquaredError(xmax=
                xmax, power=power), metrics=["accuracy"])
52
53      def call(self, pair):
54          target, context = pair
55          if len(target.shape) == 2:
56              target = tf.squeeze(target, axis=1)
57          if len(context.shape) == 2:
```

```
58              context = tf.squeeze(context, axis=1)
59          word_embed = self.target_embedding(target)
60          context_embed = self.context_embedding(context)
61          bias_a = tf.squeeze(self.bias_a(target), axis=1)
62          bias_b = tf.squeeze(self.bias_b(context), axis=1)
63          dotprod = tf.einsum("ij,ij->i", word_embed, context_embed)
64          return dotprod + bias_a + bias_b
65
66      def generate_training_data(self):
67          """
68          Generates skip-gram pairs for a list of sequences
69          (int-encoded sentences) based on window size, number of negative
                samples
70          and vocabulary size.
71          """
72          targets, contexts = [], []
73          self.co_occurrence_matrix[:, :] = 0
74          for sequence in tqdm.tqdm(self.sequences):
75              positive_skipgrams, _ = tf.keras.preprocessing.sequence.
                    skipgrams(sequence, vocabulary_size=self.vocab_size,
                    window_size=self.window_size)
76              for target_word, context_word in positive_skipgrams:
77                  if (self.inverse_vocab[target_word] in self.skip_words) or (
78                          self.inverse_vocab[context_word] in self.skip_words):
79                      continue
80                  self.co_occurrence_matrix[target_word, context_word] += 1
81                  self.co_occurrence_matrix[context_word, target_word] += 1
82                  targets.append(target_word)
83                  contexts.append(context_word)
84
85          return np.array(targets), np.array(contexts)
86
87      def fit(self, epochs=20):
88          targets, contexts = self.generate_training_data()
89          output = np.log(self.co_occurrence_matrix + 1)
90          dataset = tf.data.Dataset.from_tensor_slices(((targets, contexts),
                output[targets, contexts]))
91          dataset = dataset.shuffle(self.buffer_size).batch(self.batch_size,
                drop_remainder=True)
92          super().fit(dataset, epochs=epochs)
93
94      def write_weights(self, file_name):
95          weights = self.target_embedding.get_weights()[0]
96          with open(file_name, "w") as fp:
97              for index, word in enumerate(self.inverse_vocab):
98                  if index == 0:
99                      continue # skip 0, it's padding.
100                 vec = weights[index]
101                 fp.write(word + "," + ",".join([str(x) for x in vec]) + "\n"
                        )
102
103     def get_weights(self, top_n=None, word_list=None):
104         if top_n:
```

```
105                word_list = self.inverse_vocab[1:top_n + 1]
106            weights = self.target_embedding.get_weights()[0]
107            indices = np.array([self.word_to_index_dict.get(w, 0) for w in
                   word_list])
108            return weights[indices, :], word_list
109
110        def cosine_similarity(self, top_n=None, word_list=None):
111            if top_n:
112                word_list = self.inverse_vocab[1:top_n + 1]
113            nwords = len(word_list)
114            weights = self.target_embedding.get_weights()[0]
115            indices = np.array([self.word_to_index_dict.get(w, 0) for w in
                   word_list])
116            wts = weights[indices, :]
117            lengths = np.sum(np.multiply(wts, wts), axis=1)
118            cosine = np.zeros((nwords, nwords), dtype=np.float64)
119            for i in range(nwords):
120                cosine[i, i] = 1.0
121                for j in range(i):
122                    cosine[i, j] = np.dot(wts[i, :], wts[j, :]) / np.sqrt(
                       lengths[i] * lengths[j])
123                    cosine[j, i] = cosine[i, j]
124            return cosine, word_list
125
126        def length_similarity(self, top_n=None, word_list=None):
127            if top_n:
128                word_list = self.inverse_vocab[1:top_n + 1]
129            nwords = len(word_list)
130            weights = self.target_embedding.get_weights()[0]
131            indices = np.array([self.word_to_index_dict.get(w, 0) for w in
                   word_list])
132            wts = weights[indices, :]
133
134            lengths = np.zeros((nwords, nwords), dtype=np.float64)
135            for i in range(nwords):
136                for j in range(i):
137                    dist = np.subtract(wts[i, :], wts[j, :])
138                    lengths[i, j] = np.sqrt(np.dot(dist, dist))
139                    lengths[j, i] = lengths[i, j]
140            return lengths, word_list
141
142    @staticmethod
143    def get_similar_words(weights_file, topN=6):
144        np.random.seed(64)
145        df = pd.read_csv(weights_file, header=None)
146        words = np.random.choice(df.shape[0], 10, replace=False)
147        similarWords = [[]]
148        for i in range(topN):
149            similarWords.append([])
150
151        for iword in words:
152            word = df.loc[iword, 0]
153            vec = df.loc[iword, 1:].values
```

```python
            l1 = np.dot(vec, vec)
            cosineArr = []
            for j in range(df.shape[0]):
                if j == iword:
                    continue
                word2 = df.loc[j, 0]
                vec2 = df.loc[j, 1:].values
                l2 = np.dot(vec2, vec2)
                cosineSim = np.dot(vec, vec2) / np.sqrt(l1 * l2)
                cosineArr.append((cosineSim, word2))
            cosineArr.sort(key=lambda x: x[0], reverse=True)
            similarWords[0].append(word)
            for i in range(topN):
                similarWords[i+1].append(cosineArr[i][1])

        columns = ["word"] + ["SimWord%d" % (i+1) for i in range(topN)]
        data = {c:arr for c,arr in zip(columns, similarWords)}
        df2 = pd.DataFrame(data=data)
        logging.info(df2.to_latex(index=False))


class Plotter(object):
    @staticmethod
    def plot_weights(weights, size, labels=None, dirname=None):
        if size < weights.shape[1]:
            weights = Plotter.reduce_to_k_dim(weights, size)
        if labels is None:
            labels = ["%d" % (i + 1) for i in range(size)]

        data = pd.DataFrame(weights, columns=labels)
        pd.plotting.scatter_matrix(data, alpha=0.2, diagonal='hist',
            figsize=(10, 10))
        if dirname:
            plt.savefig(os.path.join(dirname, f"ReducedWts_Glove.jpeg"), dpi
                =500)
        plt.show()


    @staticmethod
    def reduce_to_k_dim(M, k=2, n_iter=10):
        """ Reduce a matrix M (n, m) to a matrix of dimensionality (n, k)
            using the
            following SVD function from Scikit-Learn:
                - http://scikit-learn.org/stable/modules/generated/sklearn.
                    decomposition.TruncatedSVD.html

            Params:
                M (n,m): co-occurence matrix of word counts
                k (int): embedding size of each word after dimension
                    reduction
            Return:
                M_reduced (numpy matrix of shape (number of corpus words, k)
                    ): matrix of k-dimensioal word embeddings.
```

```
                         In terms of the SVD from math class, this actually
                             returns U * S
         """
         svd = TruncatedSVD(n_components=k, n_iter=n_iter)
         return svd.fit_transform(M)


if __name__ == "__main__":
    embedding_dim = 128
    window_size = 3
    corpus_name = "reuters"
    dirname = r"C:\prog\cygwin\home\samit_000\latex\book_stats\code\data"
    vocab_size = 1024
    sequence_len = 10
    glove = Glove(embedding_dim, window_size, corpus_name, dirname,
            vocab_size=vocab_size, sequence_len=sequence_len)
    glove.fit(epochs=2)
    weights_file = os.path.join(dirname, "weights_glove.csv")
    glove.write_weights(weights_file)

    weights, words = glove.get_weights(top_n=10)
    Plotter.plot_weights(weights, size=10, dirname=dirname)
    logging.info(",".join(words))
    logging.info(weights)

    Glove.get_similar_words(weights_file, topN=5)
```

10.3.1 Code Explanation

The structure of the code is very similar to the code for Word2Vec. Therefore, only portions of code with material difference from Word2Vec have been explained at length. The training corpus and most parameter settings are the same as in Word2Vec and have been summarized in Table 10-3.

1. The code begins with the instantiation of an object of class **Glove**. This class derives from **tf.models.Model** which is the base class of all tensorflow neural network models.

Table 10-3.
Hyper-parameters Used to
Train Glove Model

Name	Value
Training Corpus	reuters (NLTK)
Vocabulary Size	1024
Context Window Size	3
Embedding Dimension	128
Sequence Length	10
Batch Size	1024
Training Epochs	2

2. Inside the constructor, following steps are performed:

- The training corpus is read, and a vocabulary is constructed. Sequences conforming to the provided sequence length of ten are created.
- A word co-occurrence matrix is computed using local context windows. Around each word (center word), neighboring words within the window are considered as context words as the co-occurrence count is updated. This is done for all words in the corpus.
- Target embedding and context embedding are created as embedding layers of type **tf.keras.layers.Embedding**. Because Glove does not use negative sampling, the context embedding layer is of the same dimensions as the target embedding layer: accepting an input of 1026 dimensions (vocabulary size + 2) and producing a 128-dimensional word vector. The input dimension is 1026 because two special words "UNK" for out-of-vocabulary words and "PAD" for padding word are appended to the vocabulary size, which has been specified as 1024.
- Two bias layers of type **tf.keras.layers.Embedding** are created. These layers accept a vocabulary-sized vector and transform it to a scalar. There are two biases: one applied to the target word and another applied to the context word.
- An optimizer is created. The Adam optimizer is used.
- The loss function used is shown in Equation 10-7. This is implemented as a weighted mean square error inside the class **WeightedMeanSquaredError**. This class derives from **tf.keras.losses.Loss** and calculates the loss function.

3. Following this, the model is trained. SVD is used for visualizing the word vectors in a reduced ten-dimensional space.
4. As before, a cosine similarity measure is used to identify top five similar words for ten randomly chosen words.

10.4   Regression Using Random Forest

Random forests were studied extensively in an earlier chapter where we observed that they can be used for classification, regression, and clustering. In this section, let us use random forests for regression and look at how we can select the number of estimators or trees. Further, we will look at benchmarking the regression using the OLS model.

The setup of this model is very similar to that used for predicting asset returns. We predict daily S&P 500 returns using the same set of features as used in Section 10.1.

In random forests, a key hyper-parameter is the number of estimators or trees. We will select an optimum value for this parameter.

Figure 10-4 shows the variation of RMSE on testing and training datasets for random forest as a function of estimators or trees. Overlaid on top are two horizontal lines showing RMSE for the OLS model in training and test datasets.

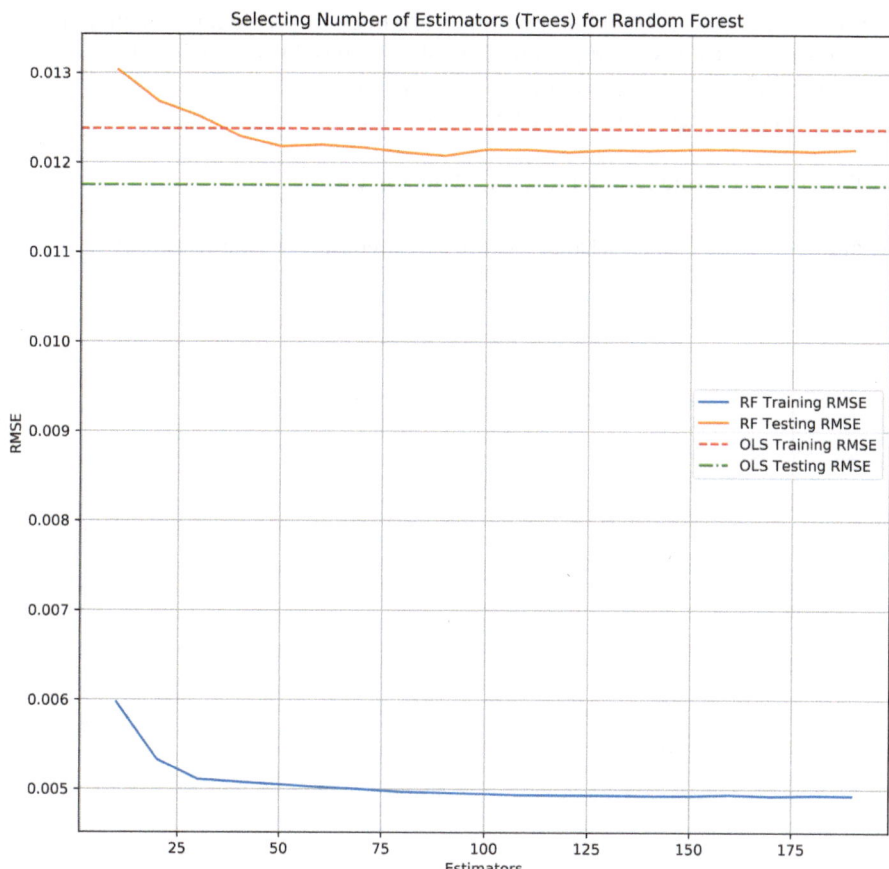

Figure 10-4. RMS Errors on Training and Test Datasets of Random Forest and OLS Model

We can make the following observation from Figure 10-4:

1. Random forest is not overfitting. This can be seen from the fact that RMSE does not go up on either testing or training datasets with increasing number of estimators or decision trees in the forest.

2. Around 75 trees seem to give optimal error reduction in training and test datasets. Beyond this, the improvement in accuracy is minimal.

3. Benchmarking random forest with the OLS model, we observe that random forest performs significantly better in the training dataset. In the test dataset, the OLS model performs marginally better. However, because we expect testing errors to be generally higher than training errors, there is a reason to believe that testing errors obtained using the OLS model will go up as new data arrives. Using existing data, RMSE on the test dataset for random forest is only marginally higher than that of OLS. In light of the previous observation, it is advisable to

use random forest for obtaining higher accuracy in predictions for this task as compared with OLS. Furthermore, it would be advisable to use the OLS model as a benchmarking tool on a continuous basis to monitor the accuracy of predictions from random forest as new data becomes available.

The code for fitting random forest is shown in Listing 10-4.

Listing 10-4. Using OLS Model to Benchmark the Performance of a Random Forest

```python
import numpy as np
import pandas as pd
import logging
import matplotlib.pyplot as plt
from sklearn.ensemble import RandomForestRegressor
import os
import statsmodels.api as sm

logging.basicConfig(level=logging.DEBUG)


class RandomForestPredictor:
    PERIOD = 1
    PRICE_COL = "Close"
    VOLUME_COL = "Volume"

    def __init__(self, dirname, security, trainTestRatio=0.9, maxTrees=200,
            batchSize=32):
        self.logger = logging.getLogger(self.__class__.__name__)
        self.dirname = dirname
        self.security = security
        self.maxTrees = maxTrees
        self.batchSize = batchSize
        self.df = pd.read_csv(os.path.join(dirname, f"{security}.csv"),
            parse_dates=["Date"])
        self.endog, self.exog = None, None
        self.beginIndex = None
        self.endIndex = None
        self.calculateEndogExogVars()
        self.ntraining = int(trainTestRatio * self.df.shape[0])
        self.nn = None
        self.ols = self.createOLSModel()
        self.rf = None

    def movingAverage(self, arr, period):
        result = np.zeros(len(arr), dtype=np.float32)
        sum1 = np.sum(arr[0:period])
        for i in range(period, len(arr), 1):
            result[i] = sum1 / period
            sum1 += arr[i] - arr[i-period]
        return result

    def volatility(self, arr, lookback):
```

```
42        result = np.zeros(len(arr), dtype=np.float32)
43        sumsq = np.sum(arr[0:lookback] ** 2)
44        for i in range(lookback, len(arr), 1):
45            result[i] = sumsq / lookback
46            sumsq += arr[i]*arr[i] - arr[i-lookback]*arr[i-lookback]
47        return result
48
49    def calculateEndogExogVars(self):
50        prices = self.df.loc[:, self.PRICE_COL].values
51        returns = prices[self.PERIOD:] / prices[0:-self.PERIOD] - 1
52        self.df.loc[:, "returns"] = 0
53        self.df.loc[0:self.df.shape[0] - 1 - self.PERIOD, "returns"] =
                 returns
54        self.endog = "returns"
55
56        self.df.loc[:, "lag1Return"] = 0
57        self.df.loc[self.PERIOD+1:, "lag1Return"] = returns[0:self.df.shape
                 [0]-self.PERIOD-1]
58
59        self.df.loc[:, "lag2Return"] = 0
60        self.df.loc[self.PERIOD+2:, "lag2Return"] = returns[0:self.df.shape
                 [0]-self.PERIOD-2]
61
62        self.df.loc[:, "lag3Return"] = 0
63        self.df.loc[self.PERIOD+3:, "lag3Return"] = returns[0:self.df.shape
                 [0]-self.PERIOD-3]
64
65        self.df.loc[:, "ma3m5"] = 0
66        ma3 = self.movingAverage(prices, 3)
67        ma5 = self.movingAverage(prices, 5)
68        self.df.loc[5:, "ma3m5"] = ma3[5:] - ma5[5:]
69
70        volatility = self.volatility(returns, lookback=5)
71        moVolatility = self.volatility(returns, lookback=21)
72        relVolat = volatility[21:] / moVolatility[21:]
73        self.df.loc[:, "relVolatility"] = 0
74        self.df.loc[21:self.df.shape[0] - 1 - self.PERIOD, "relVolatility"]
                 = relVolat
75
76        volume = self.df.loc[:, self.VOLUME_COL].values
77        vol3 = self.movingAverage(volume, 3)
78        vol5 = self.movingAverage(volume, 5)
79        relVolume = vol3[5:] / vol5[5:]
80        self.df.loc[:, "relVolume"] = 0
81        self.df.loc[5:, "relVolume"] = relVolume
82
83        self.exog = ["lag1Return", "lag2Return", "lag3Return", "ma3m5", "
                 relVolatility", "relVolume"]
84        self.beginIndex = 21
85        self.endIndex = self.df.shape[0] - self.PERIOD
86
87    def fitRF(self, ntrees):
```

```
88      self.rf = self.rf = RandomForestRegressor(n_estimators=ntrees,
            random_state=0)
89      y = self.df.loc[self.beginIndex:self.ntraining, self.endog].values
90      X = self.df.loc[self.beginIndex:self.ntraining, self.exog].values
91      self.rf = self.rf.fit(X, y)
92      yhat = self.rf.predict(X)
93      rmseRF = np.sqrt(np.mean((y - yhat) ** 2))
94      return rmseRF
95
96  def createOLSModel(self):
97      y = self.df.loc[self.beginIndex:self.ntraining, self.endog].values
98      X = self.df.loc[self.beginIndex:self.ntraining, self.exog].values
99      X = sm.add_constant(X, has_constant="add")
100     return sm.OLS(endog=y, exog=X)
101
102 def fitOLS(self):
103     self.ols = self.ols.fit()
104     return self.ols
105
106 def testRF(self, y, X):
107     yhatRF = self.rf.predict(X)
108     rmseRF = np.sqrt(np.mean((y - yhatRF) ** 2))
109     return rmseRF
110
111 def testOLS(self, y, X):
112     Xols = sm.add_constant(X, has_constant="add")
113     yhatOls = self.ols.predict(exog=Xols)
114     rmseOLS = np.sqrt(np.mean((y - yhatOls) ** 2))
115     return rmseOLS
116
117 def plot(self, trees, trainError, testError, olsErrorTrain,
            olsErrorTest):
118     fig, axs = plt.subplots(1, 1, figsize=(10, 10))
119     axs.plot(trees, trainError, label="RF Training RMSE")
120     axs.plot(trees, testError, label="RF Testing RMSE")
121     axs.axhline(y=olsErrorTrain, color='r', linestyle='dashed', label="
            OLS Training RMSE")
122     axs.axhline(y=olsErrorTest, color='g', linestyle='dashdot', label="
            OLS Testing RMSE")
123     axs.set(title="Selecting Number of Estimators (Trees) for Random
            Forest")
124     axs.legend()
125     axs.grid()
126     axs.set_xlabel("Estimators")
127     axs.set_ylabel("RMSE")
128     plt.savefig(os.path.join(self.dirname, f"AssetReturnRF_{self.
            security}.jpeg"),
                dpi=500)
130     plt.show()
131
132 def findOptimalTrainingEstimators(self):
133     ntrees = list(range(10, self.maxTrees, 10))
134     testError = []
```

```
135    trainError = []
136    self.fitOLS()
137    ytrain = self.df.loc[self.beginIndex:self.ntraining, self.endog].
           values
138    Xtrain = self.df.loc[self.beginIndex:self.ntraining, self.exog].
           values
139    ytest = self.df.loc[self.ntraining:self.endIndex - 1, self.endog].
           values
140    Xtest = self.df.loc[self.ntraining:self.endIndex - 1, self.exog].
           values
141    olsErrorTrain = self.testOLS(ytrain, Xtrain)
142    olsErrorTest = self.testOLS(ytest, Xtest)
143    for ntree in ntrees:
144        nnerror = self.fitRF(ntrees=ntree)
145        self.logger.info("Estimators: %d, Fitting RMSE: %f", ntree,
               nnerror)
146        rfErrorTrain = self.testRF(ytrain, Xtrain)
147        rfErrorTest = self.testRF(ytest, Xtest)
148        testError.append(rfErrorTest)
149        trainError.append(rfErrorTrain)
150    self.plot(ntrees, trainError, testError, olsErrorTrain,
           olsErrorTest)
151    self.logger.info("OLS RMS error on training dataset: %f, test
           dataset: %f", olsErrorTrain, olsErrorTest)
152
153
154 if __name__ == "__main__":
155     dirname = r"C:\prog\cygwin\home\samit_000\latex\book_stats\code\data"
156     pred = RandomForestPredictor(dirname, "SPY")
157     np.random.seed(32)
158     pred.findOptimalTrainingEstimators()
```

10.4.1 Code Explanation

Because the code is very similar to the one presented in Section 10.1, only the sections having random forest–specific code are explained in detail below. Remaining sections are identical to the code presented earlier.

1. Random forest regressor is created using the class **RandomForestPredictor** from library **sklearn.ensemble**. The constructor is provided with two arguments:

   - Number of estimators, **n_estimators**.
   - **random_state**: This value is used to initialize the random number generator inside sklearn. To promote reproducibility of results, it is recommended to provide a value for this argument.

2. For each setting on **n_estimators**, random forest is fitted to 90% of data. The remaining 10% of data comprises the test dataset.

Bibliography

1. White, H., 1980. A Heteroskedasticity-Consistent Covariance Matrix Estimator and a Direct Test for Heteroskedasticity. *Econometrica* 48 (4): 817–838.
2. Davidson, R., and J.G. MacKinnon. 1993. *Estimation and Inference in Econometrics*. Oxford: Oxford University Press.
3. Cribari-Neto, F. 2004. Asymptotic inference under heteroskedasticity of unknown form. *Computational Statistics and Data Analysis* 45: 215–233.
4. Newey, W.K., and K.D. West. 1987. A Simple, Positive Semi-definite, Heteroskedasticity and Autocorrelation Consistent Covariance Matrix. *Econometrica* 55 (3): 703–708.
5. Gallant, A.R. 1987. *Nonlinear Statistical Models* New York: Wiley.
6. Federal Reserve Bank of St. Louis. 2024. *DGS10*. Available at: https://fred.stlouisfed.org/series/DGS10. Accessed 1 Aug 2024.
7. Federal Reserve Bank of St. Louis. 2024. *DRCCLACBS*. Available at: https://fred.stlouisfed.org/series/DRCCLACBS. Accessed 1 Aug 2024.
8. Federal Reserve Bank of St. Louis. 2024. *TERMCBCCALLNS*. Available at: https://fred.stlouisfed.org/series/TERMCBCCALLNS. Accessed 1 Aug 2024.
9. Federal Reserve Bank of St. Louis. 2024. *CCLACBW027SBOG*. Available at: https://fred.stlouisfed.org/series/CCLACBW027SBOG. Accessed 1 Aug 2024.
10. Federal Reserve Bank of St. Louis. 2024. *PCEPI*. Available at: https://fred.stlouisfed.org/series/PCEPI. Accessed 1 Aug 2024.
11. Federal Reserve Bank of St. Louis. 2024. *GDP*. Available at: https://fred.stlouisfed.org/series/GDP. Accessed 1 Aug 2024.
12. Federal Reserve Bank of St. Louis. 2024. *DSPIC96*. Available at: https://fred.stlouisfed.org/series/DSPIC96. Accessed 1 Aug 2024.
13. Federal Reserve Bank of St. Louis. 2024. *MORTGAGE30US*. Available at: https://fred.stlouisfed.org/series/MORTGAGE30US. Accessed 1 Aug 2024.
14. Federal Reserve Bank of St. Louis. 2024. *SP500*. Available at: https://fred.stlouisfed.org/series/SP500. Accessed 1 Aug 2024.
15. Federal Reserve Bank of St. Louis. 2024. *PAYEMS*. Available at: https://fred.stlouisfed.org/series/PAYEMS. Accessed 1 Aug 2024.
16. Federal Reserve Bank of St. Louis. 2024. *GDPC1*. Available at: https://fred.stlouisfed.org/series/GDPC1. Accessed 1 Aug 2024.
17. Nadaraya, E.A. 1964. On Estimating Regression. *Theory of Probability and Its Applications* 9 (1): 141–2. https://doi.org/10.1137/1109020
18. Watson, G.S. 1964. Smooth Regression Analysis. *Sankhya: The Indian Journal of Statistics*, Series A. 26 (4): 359–372.
19. Aitchison, J., and C.G. Aitken. 1976. Multivariate Binary Discrimination by the Kernel Method. *Biometrika* 63 (3): 413–420.
20. Federal Reserve Bank of St. Louis. 2024. *OUPUTGAP*. Available at: https://fred.stlouisfed.org/graph/?g=f1cZ. Accessed 1 Aug 2024.

© Samit Ahlawat 2025
S. Ahlawat, *Statistical Quantitative Methods in Finance*,
https://doi.org/10.1007/979-8-8688-0962-0

21. Federal Reserve Bank of St. Louis. 2024. *NROU*. Available at: https://fred.stlouisfed.org/series/ NROU. Accessed 1 Aug 2024.
22. Fama, E., and K. French. 2015. A Five-Factor Asset Pricing Model. *Journal of Financial Economics* 116 (2015): 1–22.
23. Fama, E., and K. French. 1993. Common Risk Factors in the Returns on Stocks and Bonds. *Journal of Financial Economics* 33 (1993): 3–56.
24. French, K. 2024. *Current Research Returns*. Available at: https://mba.tuck.dartmouth.edu/ pages/faculty/ken.french/data_library.html. Accessed 1 Aug 2024.
25. Amit, Y., and D. Geman. 1997. Shape Quantization and Recognition with Randomized Trees. *Neural Computation* 9 (7): 1545–1588.
26. Deitterich, T. 2000. An Experimental Comparison of Three Methods for Constructing Ensembles of Decision Trees: Bagging, Boosting and Randomization. *Machine Learning* 40 (2): 139–157.
27. Brieman, L. 2001. Random Forests. *Machine Learning* 45 (1): 5–32.
28. UCI. 2022. *UC Irvine Machine Learning Repository*. Available at: https://archive.ics.uci.edu/ ml/index.php. Accessed 1 June 2022.
29. Moro, S., P. Cortez, and P. Rita. 2014. A Data-Driven Approach to Predict the Success of Bank Telemarketing. *Decision Support Systems* 62: 22–31. Elsevier.
30. Hansen, L.P. 1982. Large Sample Properties of Generalized Method of Moments Estimators. *Econometrica* 50 (4): 1029–1054.
31. Duffie, D., and K.J. Singleton. 1993. Simulated Moments Estimation of Markov Models of Asset Pricing. *Econometrica* 61 (4): 929–952.
32. Constantinides, G.M., and D. Duffie. 1996. Asset Pricing with Heterogeneous Consumers. *Journal of Political Economy* 104 (2): 219–240.
33. Jagannathan, R., G. Skoulakis, and Z. Wang. 2002. Generalized Method of Moments: Applications in Finance. *Journal of Business and Economic Statistics* 20 (4): 470–481.
34. Duffee, G.R. 2005. Time Variation in the Covariance Between Stock Returns and Consumption Growth. *Journal of Finance* 60 (4): 1673–1712.
35. Kingma, D.P., and J.L. Ba. 2015. ADAM: A Method for Stochastic Optimization. In *ICLR 2015*, arXiv:1412.6980v9.
36. Mikolov, T., K. Chen, G. Corrado, and J. Dean. 2013. Efficient estimation of word representations in vector space. arXiv:1301.3781v3.
37. Pennington, J., R. Socher, and C.D. Manning. 2014. GloVe: Global Vectors for Word Representation. In *Proceedings of the 2014 Conference on Empirical Methods in Natural Language Processing (EMNLP)*, 1532–1543. Doha: Association for Computational Linguistics.
38. Arora, S., Y. Li, Y. Liang, T. Ma, and A. Risteski. 2016. A Latent Variable Model Approach to PMI-based Word Embeddings. *Transactions of the Association for Computational Linguistics*, 4: 385–399.
39. Sargan, J.D. 1958. The Estimation of Economic Relationships Using Instrumental Variables. *Econometrica* 26 (3): 393–415. https://doi.org/10.2307/1907619.JSTOR1907619
40. Federal Reserve Bank of St. Louis. 2024. *DGS1MO*. Available at: https://fred.stlouisfed.org/ series/DGS1MO. Accessed 1 Aug 2024.
41. Federal Reserve Bank of St. Louis. 2024. *DGS30*. Available at: https://fred.stlouisfed.org/ series/DGS30. Accessed 1 Aug 2024.
42. Taylor, John B. 1993. Discretion versus policy rules in practice, 94905. Stanford: Stanford University.
43. Phillips, A.W. 1958. The Relation Between Unemployment and the Rate of Change of Money Wage Rates in the United Kingdom, 1861–1957. *Economica* 25 (100): 283–99.
44. Dupuis, David. 2004. The New Keynesian Hybrid Phillips Curve: An Assessment of Competing Specifications for the United States. *Bank of Canada Working Paper*, 2004–31, August 2004.
45. Federal Reserve Bank of St. Louis. 2024. *T10YIE*. Available at: https://fred.stlouisfed.org/ series/T10YIE. Accessed 1 Aug 2024.

46. Federal Reserve Bank of St. Louis. 2024. *A191RI1Q225SBEA*. Available at: https://fred.stlouisfed.org/series/A191RI1Q225SBEA. Accessed 1 Aug 2024.
47. Diebold, Francis X., Joon-Haeng Lee, and Gretchen C. Weinbach. 1993. Regime switching with time-varying transition probabilities. *Working Papers 93–12, Federal Reserve Bank of Philadelphia.*
48. UCLA Statistical Methods and Data Analysis. 2024. *Tobit Analysis | SAS Data Analysis Examples*. Available at: https://stats.oarc.ucla.edu/sas/dae/tobit-analysis/ Accessed 1 Aug 2024.
49. UCI. 2022. *Bank Marketing Data Set*. Available at: https://archive.ics.uci.edu/ml/datasets/Bank+Marketing. Accessed 1 June 2022.
50. Cont, R. 2001. Empirical Properties of Asset Returns: Stylized Facts and Statistical Issues. *Quantitative Finance* 1: 223–236.
51. Fama, F. 1965. The Behavior of Stock Market Prices. *Journal of Business* 38 (1965): 34–105.
52. Kon, S. 1984. Models of Stock Returns: A Comparison. *Journal of Finance* 39 (1984): 147–165.
53. Blattberg, R., and N. Gonedes 1974. A Comparison of the Stable and Student Distributions as Statistical Models for Stock Prices. *Journal of Business* 47 (1974): 244–280.
54. Bollerslev, T. 1986. Generalized Autoregressive Conditional Heteroscedasticity. *Journal of Econometrics* 31 (3): 307–327.
55. Bollerslev, T. 1987. A Conditionally Heteroskedastic Time Series Model for Speculative Prices and Rates of Return. *Review of Economics and Statistics* 69 (3): 542–547.
56. Engle, R.F. 1982. Autoregressive Conditional Heteroskedasticity with Estimates of the Variance of U.K. Inflation. *Econometrica* 50: 987–1008.
57. Nelson, D.B. 1991. Conditional Heteroskedasticity in Asset Returns: A New Approach. *Econometrica* 59: 347–370.

Index